Modern Spacecraft Dynamics & Control

Modern Spacecraft Dynamics & Control

MARSHALL H. KAPLAN

Department of Aerospace Engineering
The Pennsylvania State University

JOHN WILEY & SONS
New York   Santa Barbara   London   Sydney   Toronto

Library of Congress Cataloging in Publication Data:

Kaplan, Marshall H
    Modern spacecraft dynamics and control.

  1. Space flight.   2. Space vehicles—Attitude control
systems.   3. Astrodynamics.   I. Title.
TL790.K36   629.45   76–14859
ISBN 0-471-45703-5

Printed in the United States of America

10 9 8 7 6 5 4 3 2 1

TO ARLENE, STACEY, AND JASON

Preface

Spaceflight has been a reality for almost a full generation. Automated and manned vehicles that escape the earth's atmosphere and, sometimes, the pull of our terrestrial gravity are treated as casual, everyday occurrences by many of us. The technology of astronautics has, in fact, developed so rapidly since the first small satellite was sent into orbit that a void has developed in the availability of textbooks and references which explain relationships between the fundamentals of physics and the peculiar phenomena associated with spaceflight. This book provides a bridge that spans contemporary spacecraft maneuvering and control techniques with associated physical fundamentals. Mathematically oriented methods of *analytical dynamics* have been avoided because they usually relegate physical realities to secondary positions. Thus, concepts are developed from a Newtonian approach with all assumptions founded on sound logic. The material is structured and presented for students and practicing engineers who are interested in spacecraft dynamics and control. This book is primarily intended for use at the senior and graduate engineering levels. However, the treatment of contemporary problems should make it a sustaining reference source. In fact, mission planners and spacecraft designers may want to use it to expand their working knowledge. There are many *back-of-the-envelope* formulas for preliminary trajectory and attitude control calculations.

When used as a textbook, the material presented is appropriate for a course sequence of up to two semesters or three quarter terms. Exercises of varying difficulty are included to assist in developing insight and to test understanding of concepts. A basic working knowledge of vector algebra, matrices, Laplace transforms, and dynamics is assumed. Chapters 1 to 3 introduce fundamentals and divide study areas so that advanced concepts can be managed as the student progresses. Chapters 4 and 5 present spacecraft attitude control related technology. Both torque-producing actuators and sensors are discussed, and maneuvers about the vehicle center of mass are explained. Chapter 6 deals with automatic attitude control system concepts and includes a basic review of linear control theory. Chapters 7 and 8 represent relatively sophisticated approaches to handling two-body orbits, prediction of position and velocity, and orbital perturbations. Chapter 9 is included to illustrate three realistic problems of current concern. A logical distribution of material for a two semester sequence would begin with coverage of Chapters 1 to 4. The second term might include Chapters 5 and 6 and part of Chapter 9 if attitude dynamics and control are of major interest. If astrodynamics is to be emphasized, then Chapters 7 and 8 and Section 9.3 might be included in this second course. The first of three quarter term courses ought to include Chapters 1 to 3. Following terms could deal with Chapters 4 to 6, and then Chapters 7 to 9.

I thank the many people who offered guiding suggestions throughout the development of this book. In particular, I express my appreciation to Mrs. Charlotte Weldon for typing the manuscript and to Lt. Douglas Freesland for drafting the figures. Special thanks go to Dr. Gary Gordon and Thomas Patterson of COMSAT for their proofreading of the final manuscript.

MARSHALL H. KAPLAN

July 4, 1976

University Park, Pennsylvania

ACKNOWLEDGEMENTS

Section 1 of Chapter 9 is an adaptation of the article entitled ATTITUDE ACQUISITION MANEUVER FOR BIAS MOMENTUM SATELLITES by M. H. Kaplan and T. C. Patterson, *COMSAT Technical Review*, Vol. 6, No. 1, Spring 1976, pp. 1–23. © Communications Satellite Corporation 1976.

Section 3 of Chapter 9 is an adaptation of the article entitled INCLINATION CORRECTION STRATEGY WITH YAW SENSING VIA SUN ANGLE MEASUREMENTS by M. H. Kaplan, *COMSAT Technical Review*, Vol. 5, No. 1, Spring 1975, pp. 15–27. © Communications Satellite Corporation 1975.

Contents

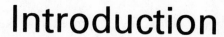

Introduction

It is appropriate to begin the subject of spacecraft dynamics and control with a review of the underlying basic principles from physics. Brief comments on the historical sequence of events which led to this modern technology are also in order. The general approach to solving problems treated here is outlined to permit anticipation of methods to be used. Coordinate systems and transformations which are basic to the entire text are defined and discussed.

1.1 HISTORICAL DEVELOPMENTS

The Copernican theory of *heliocentric* or sun-oriented motion opened the way for more exact and proper theories, which would have to be based on accurate observations of celestial bodies. Tycho Brahe, a Danish astronomer, provided this data by studying the motion of Mars in the late 1500's. One of Brahe's associates was Kepler, who developed the first general empirical laws of planetary motion. Newton came along many years later and created theoretical celestial mechanics. In doing this he proved that Kepler's laws were a necessary consequence of his own law of gravitation.

The study of spacecraft motion has been diversified into several subfields of investigation. Prior to the development and flight of man-made space vehicles, scientists pondered the natural motion of heavenly bodies, and this discipline is called *celestial mechanics.* The foundation of this field is simply the set of Newton's laws of motion, plus his universal law of gravitational attraction. Otherwise celestial mechanics is highly mathematical, partly because a great deal of work was carried out prior to the advent of high-speed digital computers. Application of these computers has made spaceflight practical. Other important developments include high thrust-to-weight ratio propulsion devices, space power and communications systems, and navigation techniques. Two general areas of contemporary study on the subject of spacecraft motion can be identified. These are *astrodynamics*, which considers particle motion in a gravity field, and *attitude control*, which is concerned with motion about the vehicle center of mass. Many sub-categories may be identified. For example, recent interest has been centered on optimal orbit maneuvers and modeling of structurally flexible spacecraft members.

1.2 PHYSICAL PRINCIPLES

The few physical principles underlying the study of spacecraft motion appear quite elementary in their basic form. It is the proper application of these concepts that permits development of a useful and important discipline. Therefore, these principles should be clearly set forth at this point.

1.2.1 Laws of Newton and Kepler

Newton may be identified as the one person responsible for formalizing the physical laws which completely determine the motion of a spacecraft. He established three laws of mechanics and one for gravitational attraction. Of the four, only three are of concern here:

(a) The rate of change of linear momentum of a body is just equal to the force applied,

$$\mathbf{F} = \frac{d}{dt}(m\mathbf{v})$$

where m is the mass and $\mathbf{v}$ is the velocity vector. If the mass is constant, then this is the familiar

$$\mathbf{F} = m\mathbf{a} \tag{1.1}$$

where $\mathbf{a} = d\mathbf{v}/dt$.

(b)   Accelerations can occur only in pairs. Thus, an object accelerating in the universe must be associated with an acceleration in the opposite direction. In other words, if there exists some force $\mathbf{F}_{12}$ of particle 2 imposed on 1, then there must also be present a second force $\mathbf{F}_{21}$ of 1 on 2, equal in magnitude and opposite in direction. If a two-particle system were looked upon as a complete system and no external force was applied, then the center of mass (or *barycenter*) would remain fixed or, if it were already moving, would continue at constant velocity.

(c)   Any two particles attract one another with a force of magnitude

$$F = G \frac{m_1 m_2}{r^2} \tag{1.2}$$

where m_1 and m_2 are the masses of the particles, r is the distance between them, and G is the universal constant of gravitation, $6.6695 \times 10^{-11} \, \mathrm{m^3/kg \cdot s^2}$.

Kepler provided the set of empirical laws which describe planetary motion, based on Brahe's observations:

(a)   The orbit of each planet is an ellipse with the sun at one focus (1609).
(b)   The radius vector joining sun to planet sweeps over equal areas in equal intervals of time (1609).
(c)   Planetary periods are proportional to the (mean distance to sun)$^{3/2}$ (1619).

1.2.2 Work and Energy

The concepts of work and energy are also important to the proper development of this subject. Work is a scalar quantity defined as the line integral along a path,

$$W_{12} = \int_{\mathbf{r}_1}^{\mathbf{r}_2} \mathbf{F} \cdot d\mathbf{r} \tag{1.3}$$

between position $\mathbf{r}_1$ and $\mathbf{r}_2$, as illustrated in Figure 1.1. Note that $\mathbf{F}$ is the force applied to particle m. Applying Newton's law (1.1) and noting that

$$d\mathbf{r} = \mathbf{v} \, dt$$

gives

$$W_{12} = \int_{t_1}^{t_2} m \frac{d\mathbf{v}}{dt} \cdot \mathbf{v} \, dt = \frac{1}{2} \int_{t_1}^{t_2} m \frac{d}{dt} (\mathbf{v} \cdot \mathbf{v}) \, dt$$

$$= \frac{1}{2} \int_{t_1}^{t_2} m \frac{d}{dt} v^2 \, dt = \tfrac{1}{2} m v^2 \Big|_{v_1}^{v_2}$$

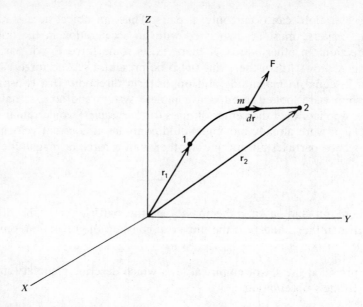

FIGURE 1.1 Line integral of force.

Thus, the work done on particle m is just the change in its kinetic energy, or

$$W_{12} = \tfrac{1}{2}mv_2{}^2 - \tfrac{1}{2}mv_1{}^2 \qquad (1.4)$$

If the integral of $\mathbf{F} \cdot d\mathbf{r}$ is zero when taken about any closed path, then $\mathbf{F}$ is said to be *conservative*. For such a force any closed path containing two given points, 1 and 2 shown in Figure 1.2, will result in

$$\oint \mathbf{F} \cdot d\mathbf{r} = 0$$

which can be written as

$$\int_{\mathbf{r}_1}^{\mathbf{r}_2} \mathbf{F} \cdot d\mathbf{r} + \int_{\mathbf{r}_2}^{\mathbf{r}_1} \mathbf{F} \cdot d\mathbf{r} = 0$$

$$\text{path } a \qquad \text{path } b$$

Since paths a and b are arbitrary, the integral from 1 to 2 (or energy change) is independent of path and a function of end points only. Thus, for a conservative force the work to go from one point to another is independent of the path taken.

The concept of *potential energy* $V(\mathbf{r}_1)$ can now be introduced as the work done by a conservative force in going from point $\mathbf{r}_1$ to some reference point $\mathbf{r}_0$,

$$V(\mathbf{r}_1) = \int_{\mathbf{r}_1}^{\mathbf{r}_0} \mathbf{F} \cdot d\mathbf{r} + V(\mathbf{r}_0) \qquad (1.5)$$

Since this reference point is arbitrary, select $V(\mathbf{r}_0) = 0$. A scalar potential can be associated with every point in space, if the forces acting are conservative. As a result the work done in going from $\mathbf{r}_1$ to $\mathbf{r}_2$ can be expressed in terms of the potentials:

$$W_{12} = \int_{\mathbf{r}_1}^{\mathbf{r}_2} \mathbf{F} \cdot d\mathbf{r} = \int_{\mathbf{r}_1}^{\mathbf{r}_0} \mathbf{F} \cdot d\mathbf{r} + \int_{\mathbf{r}_0}^{\mathbf{r}_2} \mathbf{F} \cdot d\mathbf{r}$$

which is

$$W_{12} = V(\mathbf{r}_1) - V(\mathbf{r}_2) \tag{1.6}$$

This implies that

$$\int_{\mathbf{r}_1}^{\mathbf{r}_2} \mathbf{F} \cdot d\mathbf{r} = -\int_{\mathbf{r}_1}^{\mathbf{r}_2} dV$$

or

$$\mathbf{F} \cdot d\mathbf{r} = -dV \tag{1.7}$$

which permits the force to be expressed as

$$\mathbf{F} = -\nabla V(\mathbf{r}) \tag{1.8}$$

Another property of a conservative force is that constant total energy is guaranteed. This can be demonstrated by combining equations (1.4) and (1.6),

$$\int_{\mathbf{r}_1}^{\mathbf{r}_2} \mathbf{F} \cdot d\mathbf{r} = T_2 - T_1 = -(V_2 - V_1)$$

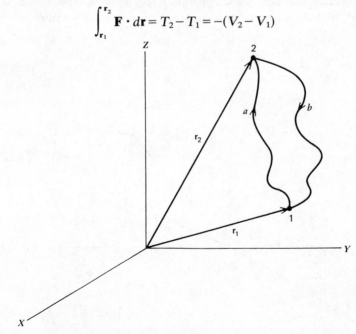

FIGURE 1.2 Potential between two points.

or

$$T_2 + V_2 = T_1 + V_1$$

where T_i, V_i are the kinetic and potential energies at point i, respectively. Thus, the right-hand side is the total energy at arbitrary point 1 and the left-hand side is the total energy at point 2. Since these are equal, energy is in fact conserved.

1.2.3 Angular Momentum

The concept of *moment of momentum* (or angular momentum) is another important idea in the development of both particle and rigid body dynamics. Consider a particle of mass m with linear momentum $\mathbf{p}$,

$$\mathbf{p} = m\dot{\mathbf{R}}$$

The moment of this momentum about an arbitrary point O, referring to Figure 1.3, is defined by

$$\mathbf{h}_o = \mathbf{r} \times m\dot{\mathbf{R}} \tag{1.9}$$

Since $\dot{\mathbf{R}} = \dot{\mathbf{R}}_o + \dot{\mathbf{r}}$, this becomes

$$\mathbf{h}_o = \mathbf{r} \times m\dot{\mathbf{r}} + \mathbf{r} \times m\dot{\mathbf{R}}_o \tag{1.10}$$

The first term on the right is the apparent angular momentum in the moving x, y, z frame, and the other term is a correction due to the motion of point O.

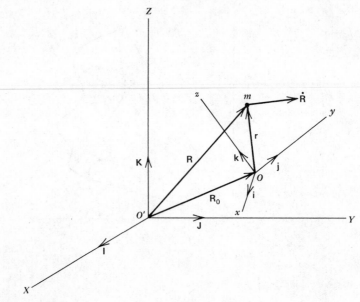

FIGURE 1.3 Angular momentum of m about O.

The rate of change of $\mathbf{h}_o$ is of particular importance in developing the equations of attitude motion. This takes the form

$$\dot{\mathbf{h}}_o = \frac{d}{dt}(\mathbf{r} \times m\dot{\mathbf{r}}) - \ddot{\mathbf{R}}_o \times m\mathbf{r} - \dot{\mathbf{R}}_o \times m\dot{\mathbf{r}} \qquad (1.11)$$

Each term on the right has a physical meaning. The first one is the rate of change of apparent angular momentum in the x, y, z frame. The second term represents the effect of acceleration of point O, and the last one represents a correction due to the velocity of point O. This rate of change of momentum can be related to an applied torque about O, defined as $\mathbf{M}_o$. The moment of a force acting on m about O is defined as

$$\mathbf{M}_o = \mathbf{r} \times \mathbf{F} \qquad (1.12)$$

where $\mathbf{F} = m\ddot{\mathbf{R}}$ for this case. Thus, $\mathbf{M}_o$ becomes

$$\mathbf{M}_o = \mathbf{r} \times m\ddot{\mathbf{R}} = \mathbf{r} \times m(\ddot{\mathbf{R}}_o + \ddot{\mathbf{r}}) \qquad (1.13)$$

Since $\dot{\mathbf{r}} \times \dot{\mathbf{r}} \equiv 0$, form (1.13) becomes

$$\mathbf{M}_o = \frac{d}{dt}(\mathbf{r} \times m\dot{\mathbf{r}}) - \ddot{\mathbf{R}}_o \times m\mathbf{r} \qquad (1.14)$$

Comparing this with expression (1.11) yields

$$\mathbf{M}_o = \dot{\mathbf{h}}_o + \dot{\mathbf{R}}_o \times m\dot{\mathbf{r}} \qquad (1.15)$$

One important observation can be made immediately from this result. If point O is fixed in space or $\mathbf{r}$ is constant, then

$$\mathbf{M}_o = \dot{\mathbf{h}}_o \qquad (1.16)$$

This will prove invaluable when later applied to attitude motion of spacecraft. If the applied torque is zero equation (1.16) indicates that $\mathbf{h}_o = $ constant, that is, angular momentum of a system is conserved under zero external torque conditions.

1.3 GENERAL APPROACH

The general approach to developing solutions to problems in spacecraft dynamics and control can be outlined in a manner which is useful as a guide throughout the text. Initially, appropriate and realistic engineering assumptions must be made. These are largely based on common sense, experience, and education. Most assumptions will involve a compromise (or trade-off) in which simplicity in handling a problem is purchased at a cost of some accuracy and realism. This is precisely where science and engineering differ.

Once assumptions are established, the physical principles discussed above may be applied to describe the problem as a set of differential equations representing a balance of accelerations and applied forces and torques. Typically, second order, ordinary differential equations will evolve from the fundamental situations studied here. Need for a highly accurate solution to many spacecraft problems may well lead to much more sophisticated equations requiring numerical treatment. If a closed form solution is not possible, special solutions may be obtainable by restricting the degrees of freedom or applying very restricted initial or boundary conditions. Very often these solutions are sufficient to draw engineering conclusions. Otherwise, numerical methods may provide a limited number of solutions for precisely specified conditions. Since the primary purpose of this book is to introduce the concepts of spacecraft dynamics and control in a manner which relates to physical interpretation, closed-form methods of solution will be emphasized. A great deal of insight may be gained by approaching such problems in this manner. Furthermore, a great many practical problems can be treated without the use of computers.

1.4 COORDINATE SYSTEMS AND TRANSFORMATIONS

1.4.1 Inertial Reference Frame

Several coordinate frames are used throughout this text. Each has a particular property which makes it appropriate to a limited number of applications. The fundamental coordinate system, to which all motion must be referred, is the *inertial frame*. In its most general sense it is a coordinate system fixed with respect to the stars. However, practical situations dictate only that the inertial frame be a reference coordinate set which guarantees required accuracy over the time interval of interest. For most problems considered here it is sufficient to select a coordinate frame which is not accelerating enough to disturb the problem solution beyond desired accuracy. For example, orbital motion about the earth can be referred to a coordinate system with origin at the center of the earth and one axis directed along a fixed celestial direction, such as the *first point of Aries*. One other axis would be normal to the ecliptic or the equatorial plane, and the third would complete the orthogonal set. When dealing with interplanetary flight within the solar system, the appropriate inertial frame is one with origin at the center of the sun and nonrotating axes relative to the stars.

1.4.2 Fundamental Transformations

Transformations between coordinate frames will be required for many problems. Both displacement and velocity components will be expressed in more than one coordinate system. Consider the absolute position of mass m in Figure 1.3. Assume the unit vectors to be $\mathbf{I}, \mathbf{J}, \mathbf{K}$ and $\mathbf{i}, \mathbf{j}, \mathbf{k}$ for the X, Y, Z and x, y, z frames, respectively. Then the position of m is $\mathbf{R} = \mathbf{R}_o + \mathbf{r}$, referred to the x, y, z system. These vectors can be written in component form as

$$\mathbf{R} = X\mathbf{I} + Y\mathbf{J} + Z\mathbf{K}$$
$$\mathbf{R}_o = X_o\mathbf{I} + Y_o\mathbf{J} + Z_o\mathbf{K}$$
$$\mathbf{r} = x\mathbf{i} + y\mathbf{j} + z\mathbf{k}$$

Each component of $\mathbf{R}$ can be expressed in terms of x, y, and z by taking the scalar product of $\mathbf{R}$ and each of the units vectors, $\mathbf{I}$, $\mathbf{J}$, and $\mathbf{K}$. Thus

$$\mathbf{R} \cdot \mathbf{I} = X = X_o + x\mathbf{I} \cdot \mathbf{i} + y\mathbf{I} \cdot \mathbf{j} + z\mathbf{I} \cdot \mathbf{k}$$
$$\mathbf{R} \cdot \mathbf{J} = Y = Y_o + x\mathbf{J} \cdot \mathbf{i} + y\mathbf{J} \cdot \mathbf{j} + z\mathbf{J} \cdot \mathbf{k} \qquad (1.17)$$
$$\mathbf{R} \cdot \mathbf{K} = Z = Z_o + x\mathbf{K} \cdot \mathbf{i} + y\mathbf{K} \cdot \mathbf{j} + z\mathbf{K} \cdot \mathbf{k}$$

The dot products, $\mathbf{I} \cdot \mathbf{i}$, $\mathbf{I} \cdot \mathbf{j}$, $\mathbf{I} \cdot \mathbf{k}$, $\mathbf{J} \cdot \mathbf{i}$, etc. are the 9 direction cosines representing the orientation of each axis of one frame with respect to each axis of the other. Consider a general two-dimensional situation, as shown in Figure 1.4. The direction cosines are

$$\mathbf{I} \cdot \mathbf{i} = \cos\theta = \mathbf{J} \cdot \mathbf{j}$$
$$\mathbf{I} \cdot \mathbf{j} = -\sin\theta = -\mathbf{J} \cdot \mathbf{i}$$

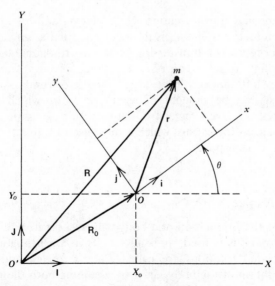

FIGURE 1.4 Two-dimensional coordinate transformation.

Since $\mathbf{R} = \mathbf{R}_o + \mathbf{r}$, $(\mathbf{R} - \mathbf{R}_o) \cdot \mathbf{I} = X - X_o = \mathbf{r} \cdot \mathbf{I}$ and $\mathbf{r} \cdot \mathbf{I} = x\mathbf{I} \cdot \mathbf{i} + y\mathbf{I} \cdot \mathbf{j} = x \cos \theta - y \sin \theta$. Thus

$$X - X_o = x \cos \theta - y \sin \theta$$

Similarly

$$Y - Y_o = x \sin \theta + y \cos \theta$$

In matrix notation this becomes

$$\begin{bmatrix} X - X_o \\ Y - Y_o \end{bmatrix} = \begin{bmatrix} \cos \theta & -\sin \theta \\ \sin \theta & \cos \theta \end{bmatrix} \begin{bmatrix} x \\ y \end{bmatrix} \tag{1.18}$$

The components x and y can now be expressed in terms of X and Y by inverting the transformation matrix,

$$\begin{bmatrix} x \\ y \end{bmatrix} = \begin{bmatrix} \cos \theta & \sin \theta \\ -\sin \theta & \cos \theta \end{bmatrix} \begin{bmatrix} X - X_o \\ Y - Y_o \end{bmatrix} \tag{1.19}$$

Notice that by displacing the origin of X, Y to X_o, Y_o a simple rotation through an angle θ completes the transformation. This led to the inversion of a simple matrix which represented that rotation. There is a very useful property of such matrices which will save much effort when working with three-dimensional cases. Any linear transformation, $\boldsymbol{\alpha}\mathbf{x} = \mathbf{x}'$ which has the property

$$\sum_i \alpha_{ij}\alpha_{ik} = \delta_{jk}(j, k = 1, 2, 3) \tag{1.20}$$

is called an *orthogonal* transformation. Note that α_{ij} and α_{ik} are the elements of transformation matrix $\boldsymbol{\alpha}$ and δ_{jk} is the Kronecker delta function. This type of transformation occurs when transferring from one orthogonal coordinate frame to another through a series of pure rotations, if both frames are either right-handed or left-handed. That is, elements α_{ij} and α_{ik} are direction cosines. Thus, $\boldsymbol{\alpha}$ can be thought of as an operator which transforms the component values of a given vector from one coordinate frame to another. The property of these transformations which is most useful is that the inverse of $\boldsymbol{\alpha}$ is simply equal to its transpose,

$$\boldsymbol{\alpha}^{-1} = \boldsymbol{\alpha}^T \tag{1.21}$$

1.4.3 Euler's Angles

When defining the orientation of a body with respect to a reference frame a series of pure rotations is used, and this results in an orthogonal transformation. The associated rotations are called *Euler's angles,* and they uniquely determine orientation of the body. Start by assuming both the reference X, Y, Z and body-fixed x, y, z frames coincide. One convenient sequence of rotations is illustrated in Figure 1.5 and can be listed as:

1. Rotation about Z axis through angle ψ to produce ξ', η', ζ' axes.
2. Rotation about ξ' axis through angle θ to produce ξ, η, ζ axes.
3. Rotation about ζ axis through angle ϕ to produce x, y, z axes.

Each rotation is characterized as an orthogonal transformation:

$$\begin{bmatrix} \xi' \\ \eta' \\ \zeta' \end{bmatrix} = \begin{bmatrix} \cos\psi & \sin\psi & 0 \\ -\sin\psi & \cos\psi & 0 \\ 0 & 0 & 1 \end{bmatrix} \begin{bmatrix} X \\ Y \\ Z \end{bmatrix} = \delta \begin{bmatrix} X \\ Y \\ Z \end{bmatrix} \qquad (1.22a)$$

$$\begin{bmatrix} \xi \\ \eta \\ \zeta \end{bmatrix} = \begin{bmatrix} 1 & 0 & 0 \\ 0 & \cos\theta & \sin\theta \\ 0 & -\sin\theta & \cos\theta \end{bmatrix} \begin{bmatrix} \xi' \\ \eta' \\ \zeta' \end{bmatrix} = \gamma \begin{bmatrix} \xi' \\ \eta' \\ \zeta' \end{bmatrix} \qquad (1.22b)$$

$$\begin{bmatrix} x \\ y \\ z \end{bmatrix} = \begin{bmatrix} \cos\phi & \sin\phi & 0 \\ -\sin\phi & \cos\phi & 0 \\ 0 & 0 & 1 \end{bmatrix} \begin{bmatrix} \xi \\ \eta \\ \zeta \end{bmatrix} = \beta \begin{bmatrix} \xi \\ \eta \\ \zeta \end{bmatrix} \qquad (1.22c)$$

The ξ axis, which is known as the *line of nodes* is the intersection of X, Y and x, y planes. Now it is possible to make the direct transformation from X, Y, Z to x, y, z by combining this sequence of rotations, that is,

$$\begin{bmatrix} x \\ y \\ z \end{bmatrix} = \beta \begin{bmatrix} \xi \\ \eta \\ \zeta \end{bmatrix} = \beta\gamma \begin{bmatrix} \xi' \\ \eta' \\ \zeta' \end{bmatrix} = \beta\gamma\delta \begin{bmatrix} X \\ Y \\ Z \end{bmatrix}$$

Note that $\beta\gamma\delta$ must remain in this order, because finite rotations cannot be represented as true vectors and are not commutative. This corresponds to the

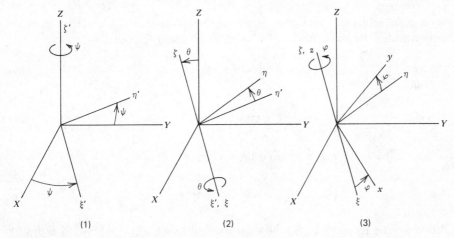

FIGURE 1.5 Construction of Euler's angles.

rotation sequence ψ, θ, and ϕ, which must also be taken in that sequence. Let $\alpha = \beta\gamma\delta$ or

$$\alpha = \begin{bmatrix} (\cos\phi\cos\psi - \sin\phi\cos\theta\sin\psi) & (\cos\phi\sin\psi + \sin\phi\cos\theta\cos\psi) & (\sin\phi\sin\theta) \\ (-\sin\phi\cos\psi - \cos\phi\cos\theta\sin\psi) & (-\sin\phi\sin\psi + \cos\phi\cos\theta\cos\psi) & (\cos\phi\sin\theta) \\ (\sin\theta\sin\psi) & (-\sin\theta\cos\psi) & (\cos\theta) \end{bmatrix} \quad (1.23)$$

Thus, α transforms the components of a vector expressed in the X, Y, Z frame to components expressed in the x, y, z frame:

$$\begin{bmatrix} x \\ y \\ z \end{bmatrix} = \alpha \begin{bmatrix} X \\ Y \\ Z \end{bmatrix} \quad (1.24)$$

If the x, y, z components are known and X, Y, Z components are to be determined, use property (1.21),

$$\begin{bmatrix} X \\ Y \\ Z \end{bmatrix} = \alpha^{\mathrm{T}} \begin{bmatrix} x \\ y \\ z \end{bmatrix} \quad (1.25)$$

The rates of change of Euler's angles are given by set $\dot{\psi}$, $\dot{\theta}$, and $\dot{\phi}$. A general angular velocity vector $\boldsymbol{\omega}$ can be written in component form

$$\boldsymbol{\omega} = \begin{bmatrix} \omega_x \\ \omega_y \\ \omega_z \end{bmatrix}$$

which is in terms of moving or body fixed coordinates x, y, z. These components can also be expressed as Euler rates by using Figure 1.6,

$$\omega_x = \dot{\theta}\cos\phi + \dot{\psi}\sin\theta\sin\phi \quad (1.26a)$$

$$\omega_y = -\dot{\theta}\sin\phi + \dot{\psi}\sin\theta\cos\phi \quad (1.26b)$$

$$\omega_z = \dot{\phi} + \dot{\psi}\cos\theta \quad (1.26c)$$

or

$$\begin{bmatrix} \omega_x \\ \omega_y \\ \omega_z \end{bmatrix} = \begin{bmatrix} \sin\theta\sin\phi & \cos\phi & 0 \\ \sin\theta\cos\phi & -\sin\phi & 0 \\ \cos\theta & 0 & 1 \end{bmatrix} \begin{bmatrix} \dot{\psi} \\ \dot{\theta} \\ \dot{\phi} \end{bmatrix} \quad (1.27)$$

Of course, the Euler rates can be expressed in terms of ω_x, ω_y, ω_z. In order to avoid coupling of the rates, only normal components of $\dot{\psi}$, $\dot{\theta}$, and $\dot{\phi}$ can be

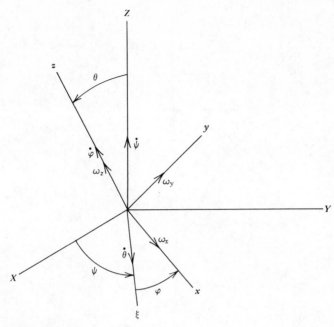

FIGURE 1.6 Euler rates.

used, because Euler rates are not orthogonal. With the aid of Figure 1.7 the three appropriate components are

$$\dot{\psi}\sin\theta \qquad \text{(which is normal to } \dot{\theta} \text{ and } \dot{\phi})$$

$$\dot{\phi}\sin\theta \qquad \text{(which is normal to } \dot{\psi} \text{ and } \dot{\theta})$$

$$\dot{\theta} \qquad \text{(which is already normal to } \dot{\psi} \text{ and } \dot{\phi})$$

The transformations are then easily obtained as,

$$\begin{bmatrix} \dot{\psi} \\ \dot{\theta} \\ \dot{\phi} \end{bmatrix} = \frac{1}{\sin\theta} \begin{bmatrix} \sin\phi & \cos\phi & 0 \\ \cos\phi\sin\theta & -\sin\phi\sin\theta & 0 \\ -\sin\phi\cos\theta & -\cos\phi\cos\theta & \sin\theta \end{bmatrix} \begin{bmatrix} \omega_x \\ \omega_y \\ \omega_z \end{bmatrix} \qquad (1.28)$$

Since $\dot{\psi}$, $\dot{\theta}$, and $\dot{\phi}$ are not along orthogonal coordinates, the transformation matrix of the rates is not orthogonal. Thus, expression (1.28) is not obtained from (1.27) by simply transposing the matrix.

1.5 RELATIVE MOTION

1.5.1 General Equation

The movement of a point when referred to an inertial frame is *absolute motion*, but when referred to a noninertial frame it is *relative motion*. There are

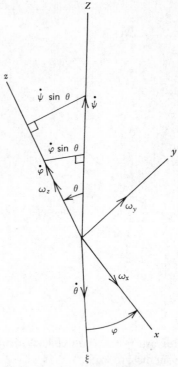

FIGURE 1.7 Euler rate components used for transformation to orthogonal set.

many situations in which it is much more convenient to express the motion of a vehicle in terms of a moving system. For example, satellite observations are made from the surface of the earth, which is moving relative to the orbital reference frame. Consider the general situation of Figure 1.8 in which the motion of point p is to be described with respect to the x, y, z frame. Start with the absolute velocity of p

$$\dot{\mathbf{R}} = \dot{\mathbf{R}}_o + \dot{\mathbf{r}} \tag{1.29}$$

where time derivatives are taken with respect to the inertial frame. Thus

$$\dot{\mathbf{R}} = \dot{X}\mathbf{I} + \dot{Y}\mathbf{J} + \dot{Z}\mathbf{K}$$

$$\dot{\mathbf{R}}_o = \dot{X}_o\mathbf{I} + \dot{Y}_o\mathbf{J} + \dot{Z}_o\mathbf{K}$$

$$\dot{\mathbf{r}} = \frac{d}{dt}(x\mathbf{i} + y\mathbf{j} + z\mathbf{k})$$

Since $\mathbf{i}$, $\mathbf{j}$, and $\mathbf{k}$ are attached to a rotating frame, the derivative of $\mathbf{r}$ must account for changes in direction of these unit vectors. This can be done by simply taking the derivative

$$\dot{\mathbf{r}} = \dot{x}\mathbf{i} + \dot{y}\mathbf{j} + \dot{z}\mathbf{k} + x\dot{\mathbf{i}} + y\dot{\mathbf{j}} + z\dot{\mathbf{k}} \qquad (1.30)$$

The rate of change of $\mathbf{i}$ can only be caused by $\boldsymbol{\omega}$ and must be normal to both $\boldsymbol{\omega}$ and $\mathbf{i}$, and be in the direction of rotation. Therefore,

$$\dot{\mathbf{i}} = \boldsymbol{\omega} \times \mathbf{i}$$

Similarly

$$\dot{\mathbf{j}} = \boldsymbol{\omega} \times \mathbf{j}$$

$$\dot{\mathbf{k}} = \boldsymbol{\omega} \times \mathbf{k}$$

Expression (1.30) now becomes

$$\dot{\mathbf{r}} = \left[\frac{d\mathbf{r}}{dt}\right]_b + \boldsymbol{\omega} \times \mathbf{r} \qquad (1.31)$$

where subscript b is used to denote differentiation in the moving frame. Thus, the first term on the right is the velocity of p with respect to point O. The general expression for absolute velocity of point p referred to the moving

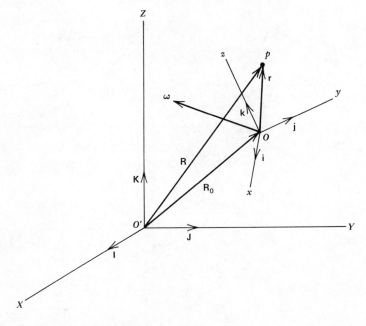

FIGURE 1.8 General relative motion.

frame is now obtained by combining expressions (1.29) and (1.31),

$$\dot{\mathbf{R}} = \dot{\mathbf{R}}_o + \left[\frac{d\mathbf{r}}{dt}\right]_b + \boldsymbol{\omega} \times \mathbf{r} \tag{1.32}$$

Expression (1.31) can alternatively be obtained by assuming a constant magnitude for $\mathbf{r}$ and considering only rotation of the x, y, z frame, as shown in Figure 1.9. The definition of $\dot{\mathbf{r}}$ is

$$\frac{d\mathbf{r}}{dt} = \lim_{\Delta t \to 0} \frac{(\mathbf{r} + \Delta\mathbf{r}) - \mathbf{r}}{\Delta t}$$

$$= \lim_{\Delta t \to 0} \frac{\Delta\mathbf{r}}{\Delta t}$$

Here $\Delta\mathbf{r} = \Delta\theta \, (r \sin \phi)\mathbf{1}$ with $\mathbf{1}$ parallel to $\Delta\mathbf{r}$. Take

$$\frac{\Delta\mathbf{r}}{\Delta t} = (r \sin \phi)\mathbf{1} \frac{\Delta\theta}{\Delta t}$$

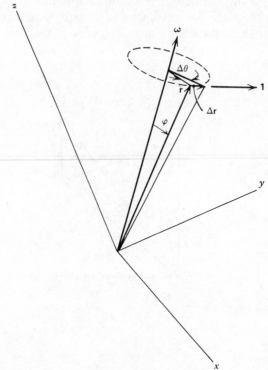

FIGURE 1.9 Rotation effects on derivatives.

since r is constant here. This expression can be taken to the limit,

$$\frac{d\mathbf{r}}{dt} = \lim_{\Delta t \to 0} (r \sin \phi)\mathbf{1} \frac{\Delta \theta}{\Delta t} = \omega(r \sin \phi)\mathbf{1}$$

which is also

$$\frac{d\mathbf{r}}{dt} = \boldsymbol{\omega} \times \boldsymbol{r}$$

This can be added to $[d\mathbf{r}/dt]_b$ when r is not constant.

To determine the absolute acceleration of point p referred to x, y, z, differentiate expression (1.29). Consider first the expression

$$\ddot{\mathbf{r}} = \frac{d}{dt}\left[\frac{d\mathbf{r}}{dt}\right]_b + \frac{d}{dt}(\boldsymbol{\omega} \times r) \tag{1.33}$$

whose operations are best carried out individually. Expand the first term on the right

$$\frac{d}{dt}\left[\frac{d\mathbf{r}}{dt}\right]_b = \frac{d}{dt}(\dot{x}\mathbf{i} + \dot{y}\mathbf{j} + \dot{z}\mathbf{k}) = \ddot{x}\mathbf{i} + \ddot{y}\mathbf{j} + \ddot{z}\mathbf{k} + \boldsymbol{\omega} \times \left[\frac{d\mathbf{r}}{dt}\right]_b$$

then define

$$\dot{\mathbf{r}}_b = \dot{x}\mathbf{i} + \dot{y}\mathbf{j} + \dot{z}\mathbf{k}$$

$$\ddot{\mathbf{r}}_b = \ddot{x}\mathbf{i} + \ddot{y}\mathbf{j} + \ddot{z}\mathbf{k}$$

The second term of expression (1.33) becomes

$$\frac{d}{dt}(\boldsymbol{\omega} \times r) = \dot{\boldsymbol{\omega}} \times \mathbf{r} + \boldsymbol{\omega} \times \dot{\mathbf{r}}$$

$$= \dot{\boldsymbol{\omega}} \times \mathbf{r} + \boldsymbol{\omega} \times \dot{\mathbf{r}}_b + \boldsymbol{\omega} \times (\boldsymbol{\omega} \times r)$$

Combining these expressions leads to

$$\ddot{\mathbf{R}} = \ddot{\mathbf{R}}_o + \ddot{\mathbf{r}}_b + 2\boldsymbol{\omega} \times \dot{\mathbf{r}}_b + \dot{\boldsymbol{\omega}} \times \mathbf{r} + \boldsymbol{\omega} \times (\boldsymbol{\omega} \times \mathbf{r}) \tag{1.34}$$

Each term on the right can be identified on a physical basis:

$\ddot{\mathbf{R}}_o$	is the acceleration of the origin of the moving frame.
$\ddot{\mathbf{r}}_b$	is the apparent acceleration of p in the moving frame.
$2\boldsymbol{\omega} \times \dot{\mathbf{r}}_b$	is the Coriolis acceleration due to the motion of p in x, y, z.
$\dot{\boldsymbol{\omega}} \times r$	is the acceleration of p due to $\boldsymbol{\omega}$ changing.
$\boldsymbol{\omega} \times (\boldsymbol{\omega} \times r)$	is the centrifugal acceleration due to the angle between $\boldsymbol{\omega}$ and $\mathbf{r}$.

1.5.2 Motion of the Earth's Surface

A special case of particular interest to this subject deals with motion observed from the earth's surface. Consider the situation shown in Figure 1.10.

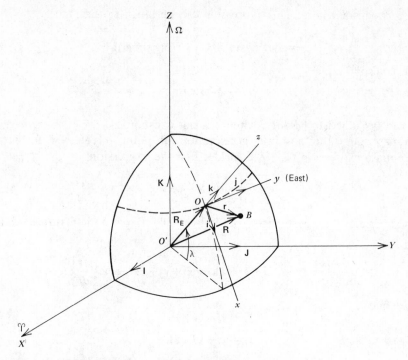

FIGURE 1.10 Motion over a ground station.

Body B is moving over the surface and is tracked by a station at point O. The inertial frame has its origin at O', X is parallel to the first point of Aries, Υ, and Z is along the polar axis. Latitude is measured by λ and $\mathbf{\Omega}$ is the earth's spin vector. The terms of equation (1.34) can be considered separately. By applying this equation to point O, get

$$\ddot{\mathbf{R}}_o = \mathbf{\Omega} \times (\mathbf{\Omega} \times \mathbf{R}_E)$$

Then other terms become

$$\ddot{\mathbf{r}}_b = \mathbf{a}_b$$

$$\mathbf{\omega} = \mathbf{\Omega}$$

$$\dot{\mathbf{r}}_b = \mathbf{v}_b$$

Finally, the absolute acceleration of B takes form

$$\ddot{\mathbf{R}}_B = \mathbf{\Omega} \times (\mathbf{\Omega} \times \mathbf{R}_E) + \mathbf{a}_b + \mathbf{\Omega} \times (\mathbf{\Omega} \times \mathbf{r}) + 2\mathbf{\Omega} \times \mathbf{v}_b \qquad (1.35)$$

In many situations of interest body B is small compared to earth and the only force acting is due to gravity. Then

$$\ddot{\mathbf{R}}_B = \frac{\mathbf{F}}{m}$$

where $\mathbf{F} = -mg\mathbf{R}/R$. The acceleration of B relative to O is $\mathbf{a}_b$, which is obtained from equation (1.35) as

$$\mathbf{a}_b = -g\frac{\mathbf{R}}{R} - \mathbf{\Omega}\times(\mathbf{\Omega}\times\mathbf{R}_E) - \mathbf{\Omega}\times(\mathbf{\Omega}\times\mathbf{r}) - 2\mathbf{\Omega}\times\mathbf{v}_b \qquad (1.36)$$

Since $\Omega = 0.728\times10^{-4}$ rad/s, the two centrifugal acceleration terms are negligible compared to the Coriolis term for low altitude orbits. Thus

$$\mathbf{a}_b \cong -g\frac{\mathbf{R}}{R} - 2\mathbf{\Omega}\times\mathbf{v}_b \qquad (1.37)$$

To illustrate the effect of this Coriolis acceleration, consider a satellite traveling due east over the ground station, that is,

$$\mathbf{v}_b = v\mathbf{j}, \qquad \mathbf{F} = -mg\mathbf{k}$$

Since

$$\mathbf{\Omega} = -(\Omega\cos\lambda)\mathbf{i} + (\Omega\sin\lambda)\mathbf{k} \qquad (1.38)$$

expression (1.37) is specialized to

$$\mathbf{a}_b \cong -(g - 2v\Omega\cos\lambda)\mathbf{k} + (2v\Omega\sin\lambda)\mathbf{i}$$

This indicates that an observer would notice the satellite turning to the south or to the right of its initial flight path. Furthermore, if the satellite were initially traveling due west over the station, it would again be observed to turn to its right. In general, any particle or vehicle traveling over the earth's surface has an apparent motion to its right in the Northern Hemisphere and to its left in the Southern Hemisphere due to the Coriolis acceleration. This phenomenon also causes motion about highs and lows in weather patterns.

EXERCISES

1.1 Derive a formula for the cosine of the angle between two position vectors represented by (x_1, y_1, z_1) and (x_2, y_2, z_2).

1.2 Show that the following identity is correct:

$$(\mathbf{a}\times\mathbf{b})\cdot(\mathbf{c}\times\mathbf{d}) = (\mathbf{b}\cdot\mathbf{d})(\mathbf{c}\cdot\mathbf{a}) - (\mathbf{b}\cdot\mathbf{c})(\mathbf{d}\cdot\mathbf{a})$$

1.3 What conditions are necessary to satisfy

$$\ddot{r} = |\ddot{\mathbf{r}}|$$

1.4 Demonstrate that the transformation given by expression (1.24) is orthogonal
 (a) by evaluating $\boldsymbol{\alpha}^{-1}$.
 (b) by applying condition (1.20) to $\boldsymbol{\alpha}$ given by (1.23).

1.5 Determine the inertial components of a point (2, 6, 8) in rotating system x, y, z of Figure 1.6, when $\psi = 45°$, $\theta = 45°$, $\phi = 30°$.

1.6 Determine the components of a point in the rotating frame of Figure 1.6 if its inertial components are $X = 3$ m, $Y = 2$ m, and $Z = 1$ m, when the Euler angles have values of $\psi = 30°$, $\theta = 30°$, $\phi = 60°$.

1.7 Consider the spinning top in the figure. Assume Euler rates of

$$\dot{\psi} = 2 \text{ rad/s}$$
$$\dot{\theta} = 0$$
$$\dot{\phi} = 180 \text{ rad/s}$$

when $\psi = 120°$, $\theta = 30°$, and $\phi = 90°$.
 (a) Determine the corresponding values of ω_x, ω_y, and ω_z.
 (b) For what value(s) of θ, would the inverse of the transformation in (a) be nonexistent?

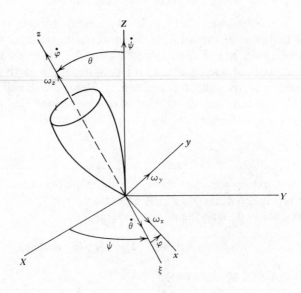

EXERCISES 1.7 and 1.8

1.8 Consider the spinning top shown. Assume angular rate components of

$$\omega_x = 0$$
$$\omega_y = 2 \text{ rad/s}$$
$$\omega_z = 20 \text{ rad/s}$$

when $\psi = 60°$, $\theta = 30°$, and $\phi = 120°$.
(a) Determine ω_X, ω_Y, and ω_Z.
(b) Determine $\dot{\psi}$, $\dot{\theta}$, and $\dot{\phi}$.

1.9 A sequence of two rotations is required to locate uniquely the axis of a vehicle or body. Consider the sequence ψ, θ as shown. The first rotation is through angle ψ about X, followed by a rotation through θ about η. The overall transformation is

$$\begin{bmatrix} x \\ y \\ z \end{bmatrix} = \boldsymbol{\alpha} \begin{bmatrix} X \\ Y \\ Z \end{bmatrix}$$

(a) Determine the elements of matrix $\boldsymbol{\alpha}$.
(b) Test for orthogonality by taking $\boldsymbol{\alpha}^{-1}$.
(c) If a vector has components $x = 0$, $y = 2$, $z = 0$, determine its components in the X, Y, Z frame.

1.10 At any given latitude λ, what combination of elevation angle β and azimuth angle α (measured from south) will result in no Coriolis acceleration on a ballistic projectile? Refer to the figure for nomenclature.

1.11 An ex-aerospace engineer and, of late, a Don Juan, Mr. Sat A. Lite, is challenged to a duel by an irate husband. He has the choice of weapons

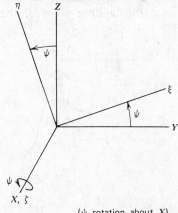
(ψ rotation about X)

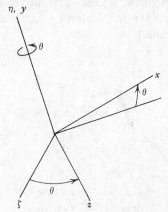
(θ rotation about η)

EXERCISE 1.9

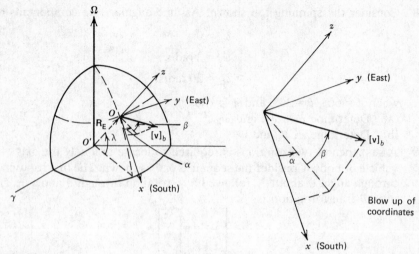

Blow up of coordinates

EXERCISE 1.10

and selects anti-aircraft artillery pieces, one mounted on each of two mountain tops in line-of-sight at the North Pole. The husband aims directly for Sat's heart by looking through the gunsights which correct for vertical gravity effects only. The figure depicts this situation. Neglecting air drag, will the 50 mm diameter projectile hit Sat if he does not move?

1.12 What is the angular deviation of a plumb line from the geocentric direction due to the earth's rotation? Assume a latitude $\lambda°\,N$. At what latitude is this deviation a maximum?

1.13 A bullet is fired vertically at a latitude of $50°\,N$ with a muzzle speed of 1000 m/s. Neglecting air friction, determine the landing point of the bullet.

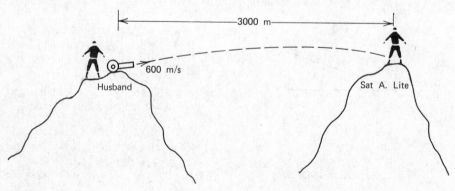

EXERCISE 1.11

REFERENCES

Baker, R. M. L., Jr., and M. W. Makemson, *An Introduction to Astrodynamics*, Second Edition, Academic Press, 1967, Chapters 1, 4.

Berman, A. I., *The Physical Principles of Astronautics*, Wiley, 1961, Chapters 3, 4.

Goldstein, H., *Classical Mechanics*, Addison-Wesley, 1950, Chapters 3, 4.

Thompson, W. T., *Introduction to Space Dynamics*, Wiley, 1961, Chapters 1, 2, 3.

2
Fundamental Spacecraft Dynamics

It is essential to develop basic concepts of spacecraft motion before considering specialized problems. This chapter introduces fundamental concepts and equations of motion for a satellite experiencing only simple gravitational forces. Other forces and all torques are assumed zero here. The concept of *rigid body* is defined, and all analytical results assume this type of body, with a single exception to illustrate energy dissipation effects on attitude stability.

2.1 GENERAL RIGID BODY MOTION

A *rigid body* can be defined as a system of particles whose relative distances are fixed with time. Thus, the internal potential energy of a rigid body is constant. Since a reference potential energy can be selected arbitrarily, rigid bodies may be regarded as having no internal potential energy. Although all structures have some flexibility, the rigid body concept has proven quite useful, because actual situations are usually treated by methods based on rigid body motion.

Consider the most general displacement of a free rigid body. It is apparent that this body can be moved from one orientation and position in space to another by first moving a selected point of the body from its initial location to its final position through a *translation*. Then a *rotation* about this point will bring the body to its final orientation. Thus, Chasles' theorem is evident: the most general displacement of a rigid body can be obtained by first translating the body, and then rotating it about a line. In other words, motion of a rigid body can be described as a combination of translation of the center of mass and rotation about the center of mass. Equations describing these two types of motion may be developed separately. With regard to spacecraft dynamics, translational motion under the influence of gravitational and other external forces becomes *orbital mechanics*. Rotation about the center of mass under the influence of applied torques becomes *attitude dynamics*. Proper application of Newton's linear momentum law, given by equation (1.1), provides the basis for orbital mechanics. Attitude motion of rigid bodies is based on the concept of angular momentum introduced for particles in Chapter 1 and discussed further in Section 2.3. Although it is possible that attitude and orbit dynamics may be coupled in a very limited number of situations, cases of interest here permit independent treatment of the two motions.

2.2 TWO-BODY AND CENTRAL FORCE MOTION

The problem of describing the motion of a number of mutually attracting bodies has been studied at length for many decades. Analytical results are quite limited, because the *n-body* problem has only 10 known integrals of motion: 3 velocity components, 3 position components, 3 angular momentum components, and kinetic energy. The relative motion of two bodies requires 6 of these, but the three-body problem requires 12 integrations for relative motion and 18 for the general solution. Therefore, only the two-body problem has an unrestricted solution. Special cases of the three-body problem have been treated in closed form, and these are known as *restricted three-body* problems, one of which is treated in Chapter 7.

2.2.1 General Solution to the Two-Body Problem

Consider the situation shown in Figure 2.1. Masses m_1 and m_2 could be two bodies whose minimum distance apart is large compared to their largest dimensions or could have spherically symmetric mass distributions and never touch each other. This restriction allows these masses to be treated as particles in the following analysis. The inertial coordinate frame is X, Y, Z and the two-body center of mass is at c, which is found by summing the moments of

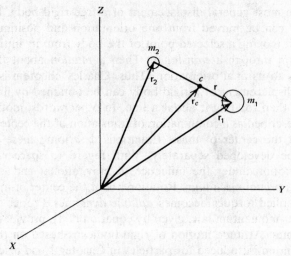

FIGURE 2.1 Two-body problem nomenclature.

mass about c. Thus, development of the appropriate differential equation begins by noting that

$$m_1(\mathbf{r}_1 - \mathbf{r}_c) + m_2(\mathbf{r}_2 - \mathbf{r}_c) = 0 \qquad (2.1)$$

Geometry dictates that

$$\mathbf{r}_2 = \mathbf{r}_1 - \mathbf{r}$$

which permits expression (2.1) to become

$$\mathbf{r}_1 - \mathbf{r}_c = \frac{m_2}{m_1 + m_2}\mathbf{r} \qquad (2.2)$$

By a similar treatment,

$$\mathbf{r}_2 - \mathbf{r}_c = \frac{-m_1}{m_1 + m_2}\mathbf{r} \qquad (2.3)$$

Expressions (2.2) and (2.3) give the distance of c from m_1 and m_2, respectively. If the force acting on m_1 is $\mathbf{F}_1$ and on m_2, $\mathbf{F}_2$, then

$$\mathbf{F}_1 = m_1\ddot{\mathbf{r}}_1 = m_1\ddot{\mathbf{r}}_c + \frac{m_1 m_2}{m_1 + m_2}\ddot{\mathbf{r}} \qquad (2.4a)$$

$$\mathbf{F}_2 = m_2\ddot{\mathbf{r}}_2 = m_2\ddot{\mathbf{r}}_c - \frac{m_1 m_2}{m_1 + m_2}\ddot{\mathbf{r}} \qquad (2.4b)$$

Mutual attraction requires that $\mathbf{F}_1 = -\mathbf{F}_2$. Satisfying this requirement yields

$$m_1\ddot{\mathbf{r}}_c = -m_2\ddot{\mathbf{r}}_c$$

which can only be valid if $\ddot{\mathbf{r}}_c = 0$. Therefore, the center of mass of the two-body system never accelerates. Applying this result to set (2.4) leads immediately to

$$\mathbf{F}_1 = -\mathbf{F}_2 = \frac{m_1 m_2}{m_1 + m_2}\ddot{\mathbf{r}}$$

Replacing $\mathbf{F}_1$ by the vector form of expression (1.2) for gravitational attraction yields the basic differential equation of motion for the two-body system

$$\frac{d^2\mathbf{r}}{dt^2} + \frac{\mu}{r^3}\mathbf{r} = 0 \tag{2.5}$$

where $\mu = G(m_1 + m_2)$.

The solution of equation (2.5) may be obtained by several methods. A straightforward vector approach is taken here. Crossing this expression with $\mathbf{r}$ yields

$$\mathbf{r} \times \frac{d^2\mathbf{r}}{dt^2} + \frac{\mu}{r^3}\mathbf{r} \times \mathbf{r} = 0$$

Remembering that $\mathbf{r} \times \mathbf{r} \equiv 0$ and

$$\frac{d}{dt}\left(\mathbf{r} \times \frac{d\mathbf{r}}{dt}\right) = \underbrace{\frac{d\mathbf{r}}{dt} \times \frac{d\mathbf{r}}{dt}}_{0} + \mathbf{r} \times \frac{d^2\mathbf{r}}{dt^2}$$

leads to

$$\frac{d}{dt}\left(\mathbf{r} \times \frac{d\mathbf{r}}{dt}\right) = 0 \tag{2.6}$$

Defining angular momentum per unit mass as

$$\mathbf{h} = \mathbf{r} \times \frac{d\mathbf{r}}{dt} \tag{2.7}$$

leads to the conclusion that $d\mathbf{h}/dt = 0$. Thus, angular momentum is conserved and three integrals of motion are $\mathbf{h} = $ constant vector. Since $\mathbf{h}$ is normal to both $\mathbf{r}$ and $d\mathbf{r}/dt$, it must be normal to the plane of motion. Furthermore, because $\mathbf{h}$ is constant, this plane must be inertially fixed. Next cross equation (2.5) with $\mathbf{h}$,

$$\frac{d^2\mathbf{r}}{dt^2} \times \mathbf{h} = -\frac{\mu}{r^3}\mathbf{r} \times \mathbf{h} = -\frac{\mu}{r^3}\mathbf{r} \times \left(\mathbf{r} \times \frac{d\mathbf{r}}{dt}\right)$$

Applying the standard identity for triple vector products and noting that

$$\mathbf{r} \cdot \frac{d\mathbf{r}}{dt} = r\frac{dr}{dt}$$

leads to

$$\frac{d^2\mathbf{r}}{dt^2} \times \mathbf{h} = \mu \frac{d}{dt}\left(\frac{\mathbf{r}}{r}\right) \tag{2.8}$$

Since $\mathbf{h}$ is constant, equation (2.8) may be integrated directly,

$$\frac{d\mathbf{r}}{dt} \times \mathbf{h} = \frac{\mu}{r}(\mathbf{r} + r\mathbf{e}) \tag{2.9}$$

where $\mathbf{e}$ is a constant of integration and is called the *eccentricity vector*. This provides three more integrals of motion. Taking the dot product of equation (2.9) and $\mathbf{h}$ leads to $\mathbf{e} \cdot \mathbf{h} = 0$, that is, the vector $\mathbf{e}$ lies in the orbit plane. The orientation of $\mathbf{e}$ in this plane is taken as a reference direction.

To complete the solution of equation (2.5), dot expression (2.9) with $\mathbf{r}$ and use the triple scalar product identity to obtain

$$r = \frac{h^2/\mu}{1 + e\cos\theta} \tag{2.10}$$

where

$$\cos\theta = \frac{\mathbf{r} \cdot \mathbf{e}}{re}$$

This is the general polar equation for a conic section with origin at a focus. Thus, Kepler's first law is proven. Notice that at the minimum value of r, $\theta = 0$. Thus, vector $\mathbf{e}$ is parallel to the direction of minimum r, and θ is measured from this point to the position in the orbit. This angle is known as the *true anomaly*.

In addition to the above solution, kinetic energy is easily expressed for the two-body system. Applying the usual definition of particle kinetic energy yields

$$T = \tfrac{1}{2}[m_1 v_1^2 + m_2 v_2^2]$$

where

$$v_1^2 = \dot{\mathbf{r}}_1 \cdot \dot{\mathbf{r}}_1, \qquad v_2^2 = \dot{\mathbf{r}}_2 \cdot \dot{\mathbf{r}}_2$$

Applying expressions (2.2) and (2.3) permits a more convenient form,

$$T = \frac{1}{2}\left[m_1 \dot{\mathbf{r}}_c \cdot \dot{\mathbf{r}}_c + \frac{m_1 m_2^2}{(m_1 + m_2)^2}\dot{\mathbf{r}} \cdot \dot{\mathbf{r}} + \frac{2m_1 m_2}{m_1 + m_2}\dot{\mathbf{r}} \cdot \dot{\mathbf{r}}_c \right]$$

$$+ \frac{1}{2}\left[m_2 \dot{\mathbf{r}}_c \cdot \dot{\mathbf{r}}_c + \frac{m_2 m_1^2}{(m_1 + m_2)^2}\dot{\mathbf{r}} \cdot \dot{\mathbf{r}} - \frac{2m_2 m_1}{m_1 + m_2}\dot{\mathbf{r}} \cdot \dot{\mathbf{r}}_c \right]$$

or

$$T = \left(\frac{m_1 + m_2}{2}\right)\dot{\mathbf{r}}_c \cdot \dot{\mathbf{r}}_c + \frac{1}{2}\left(\frac{m_1 m_2}{m_1 + m_2}\right)\dot{\mathbf{r}} \cdot \dot{\mathbf{r}} \tag{2.11}$$

The first term on the right side is just translational kinetic energy of the two-body system, which is of little interest here. The remaining term is *rotational kinetic energy* about c.

2.2.2 Central Force Motion

For all practical orbit situations involving man-made spacecraft, one of the masses in the two-body problem is much greater than the other. Thus, if $m_1 \gg m_2$, then the motion of m_2 about m_1 is essentially the motion of a particle in an inertially fixed field. This type of motion is known as *central force motion*. The force of attraction or repulsion is always directed to or from a fixed point in inertial space with magnitude solely a function of distance between the point of interest and the center of attraction. Solution of central force motion is easily obtained from two-body problem results for the inverse-square attraction law. Allowing m_1 to be much larger than m_2 leads to $\mu = Gm_1$ and permits expressions (2.2), (2.3), and (2.11) to become

$$\mathbf{r}_1 = \mathbf{r}_c$$

$$\mathbf{r}_2 = \mathbf{r}_1 - \mathbf{r}$$

$$T = \tfrac{1}{2} m_2 \dot{\mathbf{r}} \cdot \dot{\mathbf{r}}$$

where $\dot{\mathbf{r}}_c$ is assumed zero. Thus, central force motion results in a conic section orbit of form (2.10) with the focus at the center of attraction. Notice that the two-body problem may be treated as central force motion of a particle of mass $m_1 m_2 / (m_1 + m_2)$ at a distance r from the center of attraction.

A general central field need not specify the law of attraction or repulsion, but only that the force is a function of r. Such forces can be written as the gradient of gravitational potential U, which is the negative potential energy. Thus, U is defined by

$$\mathbf{F} = \nabla U \qquad (2.12)$$

where the form of U need be specified only as $U(r)$. The energy of a unit mass in a central field is given by

$$\mathscr{E} = T - U \qquad (2.13)$$

which is constant because $\mathbf{F}$ is conservative. Since angular momentum is conserved, motion is in a plane and kinetic energy per unit mass is

$$T = \frac{1}{2} \frac{d\mathbf{r}}{dt} \cdot \frac{d\mathbf{r}}{dt}$$

where $\mathbf{r} = r \mathbf{i}_r$. Referring to Figure 2.2,

$$\frac{d\mathbf{r}}{dt} = \dot{r} \mathbf{i}_r + r \frac{d}{dt} \mathbf{i}_r$$

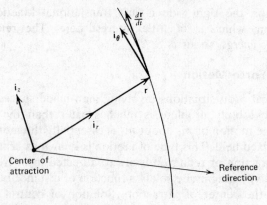

FIGURE 2.2 Radial and transverse components of plane motion.

Applying form (1.31) from the treatment of relative motion, this becomes

$$\frac{d\mathbf{r}}{dt} = \dot{r}\mathbf{i}_r + r\dot{\theta}\mathbf{i}_\theta$$

which permits writing kinetic energy in a more convenient form

$$T = \tfrac{1}{2}(\dot{r}^2 + r^2\dot{\theta}^2) \tag{2.14}$$

Total energy per unit mass now becomes

$$\mathscr{E} = \tfrac{1}{2}(\dot{r}^2 + r^2\dot{\theta}^2) - U(r) \tag{2.15}$$

Noting that $h = r^2\dot{\theta}$ and introducing a new variable as

$$\rho = \frac{1}{r}$$

permits the substitution

$$\frac{d\rho}{d\theta} = -\frac{1}{r^2}\frac{dr}{d\theta} = -\frac{\rho^2}{\dot{\theta}}\dot{r} = -\frac{\dot{r}}{h}$$

Thus, expression (2.15) becomes

$$\frac{2\mathscr{E}}{h^2} = \left(\frac{d\rho}{d\theta}\right)^2 + \rho^2 - \frac{2U^*(\rho)}{h^2}$$

where $U^*(\rho) = U(r)$. Solving for $d\theta$ gives

$$d\theta = \frac{\pm d\rho}{\sqrt{\dfrac{2}{h^2}[\mathscr{E} + U^*(\rho)] - \rho^2}} \tag{2.16}$$

which is not easily integrated for most forms of $U^*(\rho)$.

Now consider the inverse-square force law characterized by

$$U^*(\rho) = \mu\rho$$

Remember that $\mu = Gm$ for central force motion. Equation (2.16) takes on the form

$$\theta - \theta_0 = \pm \int_{\rho_0}^{\rho} \frac{d\rho}{\sqrt{\dfrac{2\mathscr{E}}{h^2} + \dfrac{2\rho\mu}{h^2} - \rho^2}}$$

which is directly integrated to

$$\theta - \theta_0 = \pm \sin^{-1} \left(\frac{\rho - \dfrac{\mu}{h^2}}{\sqrt{\dfrac{\mu^2}{h^4} + \dfrac{2\mathscr{E}}{h^2}}} \right) \Bigg|_{\rho_0}^{\rho} \qquad (2.17)$$

Define a parameter p as

$$p = \frac{h^2}{\mu} \qquad (2.18)$$

and let $\rho_0 = \mu/h^2 = 1/p$. Equation (2.17) becomes

$$\sin(\theta - \theta_0) = \pm \frac{\dfrac{1}{r} - \dfrac{1}{p}}{\sqrt{\dfrac{1}{p^2} + \dfrac{2\mathscr{E}}{p\mu}}}$$

Now replace θ_0 with $\omega \pm \pi/2$ and obtain

$$\cos(\theta - \omega) = \pm \frac{\dfrac{1}{r} - \dfrac{1}{p}}{\sqrt{\dfrac{1}{p^2} + \dfrac{2\mathscr{E}}{p\mu}}}$$

Rearranging gives

$$r = \frac{p}{1 \pm e \cos(\theta - \omega)} \qquad (2.19)$$

Where e is defined as

$$e = \sqrt{1 + \frac{2\mathscr{E}p}{\mu}} \qquad (2.20)$$

Quantity e is the *eccentricity* and is simply the magnitude of **e** defined in expression (2.9). Equation (2.19) again represents a conic section, although it was obtained by energy considerations. It is identical to form (2.10) if either $\omega = 0$ and the plus sign applies or $\omega = \pi$ and the minus sign applies. The selection of $\rho_o = 1/p$ requires that $\theta_o = \pm\pi/2$. Thus, $\omega = 0$ or π. The choice of the plus sign with $\omega = 0$ preserves the previous relationship between vector **e** and the direction of minimum r (i.e., they are parallel). A further result is that the value of r at $\theta = \pm\pi/2$ is just p, known as the *parameter* or the *semilatus rectum*.

Expression (2.20) relates eccentricity to energy and semilatus rectum for a given central body. Geometrically, e is primarily an indication of orbit shape. Figure 2.3 illustrates fundamental orbit configurations and Table 2.1 lists eccentricity and energy values for the different conic sections.

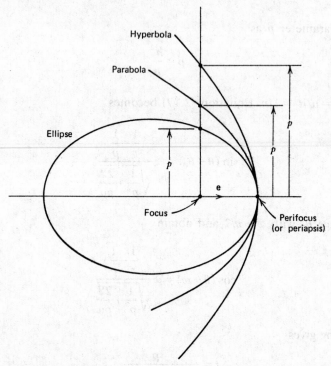

FIGURE 2.3 Conic sections.

TABLE 2.1 Orbit Classification

Range of e	Orbit Shape	Energy
$= 0$	Circle	< 0
$0 < e < 1$	Ellipse	< 0
$= 1$	Parabola	$= 0$
> 1	Hyperbola	> 0

2.2.3 Kepler's Time Equation

Vectors **h** and **e** together determine size, shape, and orientation of the conic with respect to the center of attraction. However, orbital motion is not completely defined even though **h** and **e** represent six integrals, because $\mathbf{h} \cdot \mathbf{e} = 0$. One of the six constants is a linear combination of the other five. So there are only five identified independent integrals of motion to this point. For completeness, the time dependency of the orbital position with respect to periapsis is required. These six independent parameters permit unique specification of orbit energy, shape, and orientation in inertial space, as well as spacecraft position at a given value of time. Only orbit shape and energy have been treated to this point. Next the relationship between time and position in orbit is considered. Orientation of the orbit plane in inertial space is left for Chapter 7.

Although the magnitude of **h** was already noted, it is derived by referring to Figure 2.4,

$$h = rv \sin\left(\mathbf{r}, \frac{d\mathbf{r}}{dt}\right)$$

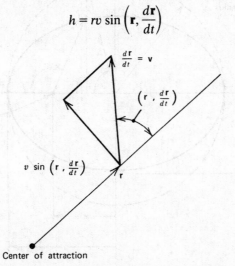

FIGURE 2.4 Nomenclature for angular momentum magnitude.

Noting that

$$v \sin\left(\mathbf{r}, \frac{d\mathbf{r}}{dt}\right) = r\frac{d\theta}{dt}$$

gives

$$h = r^2 \frac{d\theta}{dt} \tag{2.21}$$

Combining expressions (2.18), (2.19), and (2.21) to eliminate r and h yields

$$\sqrt{\frac{\mu}{p^3}}\, dt = \frac{d\theta}{(1 + e \cos\theta)^2} \tag{2.22}$$

which is usually interpreted as the differential equation of time since periapsis passage, t_p. Direct integration is difficult, thus, a new variable is introduced to replace θ.

Referring to Figure 2.5, the *eccentric anomaly* is expressed as

$$\cos\psi = \frac{ae + r\cos\theta}{a} \tag{2.23}$$

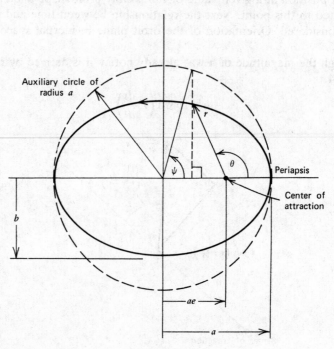

FIGURE 2.5 Nomenclature for eccentric anomaly.

where a is the *semimajor axis*. Since

$$p = a(1 - e^2) \tag{2.24}$$

from geometry, application of expression (2.19) leads to

$$\cos \psi = \frac{a - r}{ae} \tag{2.25}$$

Taking the time derivative of this gives radial speed in terms of $\dot{\psi}$,

$$\dot{r} = ae\dot{\psi} \sin \psi \tag{2.26}$$

Evaluation of energy at periapsis leads to a useful identity

$$\mathcal{E} = \frac{\dot{r}^2 + (r\dot{\theta})^2}{2} - \frac{\mu}{r} = -\frac{\mu}{2a} \tag{2.27}$$

Rearranging this leaves

$$\frac{ar^2\dot{r}^2}{\mu} = a^2 e^2 - (a - r)^2 \tag{2.28}$$

Now square expression (2.25) and combine the result with forms (2.26) and (2.28) to yield

$$r\dot{\psi} = \sqrt{\frac{\mu}{a}}$$

or

$$r \, d\psi = \sqrt{\frac{\mu}{a}} \, dt \tag{2.29}$$

Before integrating, r must be eliminated. This is easily accomplished by using equation (2.25). The resulting expression for time since periapsis passage is simply

$$\sqrt{\frac{\mu}{a^3}} \, t_p = \psi - e \sin \psi \tag{2.30}$$

remembering that $\psi = 0$ when $t_p = 0$. Define *mean motion* as

$$n = \sqrt{\frac{\mu}{a^3}} \tag{2.31}$$

to further simplify expression (2.30),

$$nt_p = \psi - e \sin \psi \tag{2.32}$$

This is known as *Kepler's equation* for relating time to position in orbit.

When eccentricity is less than 1.0, the orbit is closed and periodic. The associated period can be determined by considering the rate of area swept out by the radius vector which is the *areal velocity*. Referring to Figure 2.6, an element of swept area is

$$dA = \tfrac{1}{2}r(r\,d\theta)$$

Dividing by dt yields the areal velocity

$$\frac{dA}{dt} = \tfrac{1}{2}r^2\dot{\theta} \tag{2.33}$$

which is constant, because $h = r^2\dot{\theta} = $ constant. Thus, Kepler's second law is now proven. Integrating over the entire orbit gives the period τ as

$$\tau = \frac{2A}{h}$$

For an ellipse, $A = \pi ab$. To eliminate h, evaluate r at periapsis using form (2.10),

$$r_p = \frac{h^2/\mu}{1+e}$$

or

$$h = \sqrt{r_p\mu(1+e)} \tag{2.34}$$

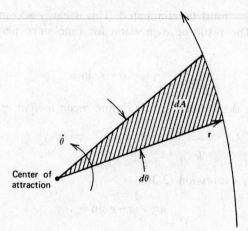

FIGURE 2.6 Areal velocity.

From geometry, $r_p = a(1-e)$ and $b = a\sqrt{1-e^2}$, leading to

$$\tau = 2\pi\sqrt{\frac{a^3}{\mu}} = \frac{2\pi}{n} \qquad (2.35)$$

which proves Kepler's third law. This appropriately completes the treatment of fundamental orbit relationships. Chapter 3 deals with orbital maneuvers and interplanetary transfers.

2.2.4 Commonly Used Constants

There are several astronomical constants used in examples and to illustrate results. Values assumed for most of the common ones are listed below.

Earth Related ($\oplus$)

$\mu_\oplus = 3.986 \times 10^5$ km^3/s^2 (gravitational constant)

$R_\oplus = 6,378$ km (mean radius)

$r_\oplus = 1.496 \times 10^8$ km ($= 1$ A.U.) $\Big\}$Heliocentric

$v_\oplus = 29.78$ km/s (mean velocity)

Moon Related ($\mathbb{C}$)

$\mu_\mathbb{C} = 4.887 \times 10^3$ km^3/s^2

$R_\mathbb{C} = 1,738$ km

$r_\mathbb{C} = 3.844 \times 10^5$ km $\Big\}$Geocentric

$v_\mathbb{C} = 1.02$ km/s

Sun Related ($\odot$)

$\mu_\odot = 1.327 \times 10^{11}$ km^3/s^2

$R_\odot = 6.98 \times 10^5$ km

2.3 ATTITUDE DYNAMICS AND EULER'S EQUATIONS

2.3.1 Angular Momentum of a Rigid Body

A basic treatment of particle motion in space should be balanced with an introduction to attitude dynamics and the fundamental equations of rigid body motion. This is begun by developing an expression for the moment of momentum (angular momentum) of a general rigid body. Consider the situation of Figure 2.7 in which body B contains fixed coordinate system x, y, z attached at

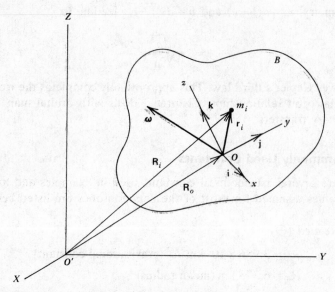

FIGURE 2.7 Rigid body moment of momentum.

its center of mass, O. A particle of mass m_i is located in the body by $\mathbf{r}_i$ and has absolute velocity of form (1.32),

$$\mathbf{v}_i = \mathbf{v}_o + \boldsymbol{\omega} \times \mathbf{r}_i + [\mathbf{v}_i]_B$$

where $\mathbf{v}_o = \dot{\mathbf{R}}_o$, $\mathbf{v}_i = \dot{\mathbf{R}}_i$ and $[\mathbf{v}_i]_B = 0$, because B is rigid. Applying the definition of moment of momentum, form (1.9), to m_i about O gives

$$\mathbf{h}_{o_i} = \mathbf{r}_i \times m_i \mathbf{v}_i = \mathbf{r}_i \times m_i (\mathbf{v}_o + \boldsymbol{\omega} \times \mathbf{r}_i)$$

If the body is thought of as a large number of small masses, the total moment of momentum about O of B is

$$\mathbf{h}_o = \sum_i \mathbf{h}_{o_i} = \sum_i \mathbf{r}_i \times m_i (\mathbf{v}_o + \boldsymbol{\omega} \times \mathbf{r}_i)$$

which is more conveniently written as

$$\mathbf{h}_o = \sum_i \mathbf{r}_i \times (\boldsymbol{\omega} \times \mathbf{r}_i) m_i - \mathbf{v}_o \times \sum_i m_i \mathbf{r}_i$$

However, the center of mass is defined as the point about which

$$\sum_i m_i \mathbf{r}_i = 0$$

This leads to

$$\mathbf{h}_o = \sum_i \mathbf{r}_i \times (\boldsymbol{\omega} \times \mathbf{r}_i) m_i \tag{2.36}$$

which is also true if $\mathbf{v}_o = 0$, that is, if point O is fixed in space. Now if m_i is allowed to decrease and the number of mass elements increases, expression (2.36) becomes an integral form,

$$\mathbf{h}_o = \int_B \mathbf{r} \times (\boldsymbol{\omega} \times \mathbf{r}) \, dm \tag{2.37}$$

This is evaluated by writing the vector products in x, y, z component form. First, consider the vector $\boldsymbol{\omega} \times \mathbf{r}$,

$$\boldsymbol{\omega} \times \mathbf{r} = (\omega_y z - \omega_z y)\mathbf{i} + (\omega_z x - \omega_x z)\mathbf{j} + (\omega_x y - \omega_y x)\mathbf{k}$$

and note that the three coefficients are the x, y, z components of linear momentum of a unit mass with respect to O, respectively. The complete integrand in component form becomes

$$\begin{aligned}
\mathbf{r} \times (\boldsymbol{\omega} \times \mathbf{r}) = & [\omega_x(y^2 + z^2) - \omega_y(xy) - \omega_z(xz)]\mathbf{i} \\
& + [-\omega_x(xy) + \omega_y(x^2 + z^2) - \omega_z(yz)]\mathbf{j} \\
& + [-\omega_x(xz) - \omega_y(yz) + \omega_z(x^2 + y^2)]\mathbf{k}
\end{aligned} \tag{2.38}$$

Each coefficient here is the x, y, or z component of angular momentum of a unit mass about point O, respectively.

Integration of form (2.38) over the body dimensions is strictly a function of mass distribution, while components of angular velocity are independent of body shape and internal location. Thus, any given rigid body can be characterized by a set of constants for the purpose of studying angular momentum and, ultimately, attitude motion. These constants are defined as

$$I_x = \int_B (y^2 + z^2) \, dm, \qquad I_y = \int_B (x^2 + z^2) \, dm, \qquad I_z = \int_B (x^2 + y^2) \, dm \tag{2.39}$$

$$I_{xy} = \int_B (xy) \, dm, \qquad I_{xz} = \int_B (xz) \, dm, \qquad I_{yz} = \int_B (yz) \, dm \tag{2.40}$$

where I_x, I_y, I_z are the *moments of inertia* of the body about the x, y, z axes, respectively, and I_{xy}, I_{xz}, I_{yz} are the *products of inertia* of B. Note that $I_{xy} = I_{yx}$, $I_{xz} = I_{zx}$, and $I_{yz} = I_{zy}$, and that the products of inertia may have positive or negative values while the moments of inertia can never be negative. Finally, the angular momentum of B in equation (2.37) becomes

$$\mathbf{h}_o = h_x \mathbf{i} + h_y \mathbf{j} + h_z \mathbf{k} \tag{2.41}$$

where the components of this vector have been defined as

$$h_x = I_x \omega_x - I_{xy} \omega_y - I_{xz} \omega_z \qquad (2.42a)$$

$$h_y = -I_{xy} \omega_x + I_y \omega_y - I_{yz} \omega_z \qquad (2.42b)$$

$$h_z = -I_{xz} \omega_x - I_{yz} \omega_y + I_z \omega_z \qquad (2.42c)$$

These appear to be a set of three simultaneous linear equations in $\boldsymbol{\omega}$, which may be written in matrix form as

$$\mathbf{h}_o = \begin{bmatrix} h_x \\ h_y \\ h_z \end{bmatrix} = \begin{bmatrix} I_x & -I_{xy} & -I_{xz} \\ -I_{xy} & I_y & -I_{yz} \\ -I_{xz} & -I_{yz} & I_z \end{bmatrix} \begin{bmatrix} \omega_x \\ \omega_y \\ \omega_z \end{bmatrix} \qquad (2.43)$$

The matrix containing moments and products of inertia is known as the *inertia tensor* and is identified here as **I**. Therefore, expression (2.43) takes the simple form

$$\mathbf{h} = \mathbf{I} \cdot \boldsymbol{\omega} \qquad (2.44)$$

where the subscript has been dropped and **h** is assumed to be taken about the center of mass hereafter, unless otherwise stated.

2.3.2 Rotational Kinetic Energy

The kinetic energy of a rigid body about its center of mass is referred to as *rotational kinetic energy*. This can be shown to be related to angular momentum **h** by considering the definition of kinetic energy of an element of mass dm,

$$dT = \tfrac{1}{2} v^2 \, dm$$

where v is the magnitude of absolute velocity of dm, and **v** is written as

$$\mathbf{v} = \mathbf{v}_o + \boldsymbol{\omega} \times \mathbf{r}$$

referring to Figure 2.7 for nomenclature. Taking the dot product

$$\mathbf{v} \cdot \mathbf{v} = v^2 = v_o{}^2 + (\boldsymbol{\omega} \times \mathbf{r}) \cdot (\boldsymbol{\omega} \times \mathbf{r}) + 2 \mathbf{v}_o \cdot (\boldsymbol{\omega} \times \mathbf{r})$$

gives

$$T = \frac{1}{2} \int_B v^2 \, dm = \tfrac{1}{2} m_B v_o{}^2 + \frac{1}{2} \int_B (\boldsymbol{\omega} \times \mathbf{r}) \cdot (\boldsymbol{\omega} \times \mathbf{r}) \, dm + \mathbf{v}_o \cdot \boldsymbol{\omega} \times \int_B \mathbf{r} \, dm$$

where the last integral must be zero because **r** is referred to the center of mass. The remaining two terms on the right side represent translational kinetic energy associated with a particle of mass m_B concentrated at the body center of mass and rotational kinetic energy about the center of mass. Thus, define

$$T_{\text{rot}} = \frac{1}{2} \int_B (\boldsymbol{\omega} \times \mathbf{r}) \cdot (\boldsymbol{\omega} \times \mathbf{r}) \, dm$$

$$= \frac{1}{2} \int_B [(\omega_y z - \omega_z y)^2 + (\omega_z x - \omega_x z)^2 + (\omega_x y - \omega_y x)^2] \, dm$$

Expanding this produces

$$T_{\text{rot}} = \frac{1}{2} \int_B [\omega_x^2 (y^2 + z^2) + \omega_y^2 (x^2 + z^2) + \omega_z^2 (x^2 + y^2) - 2\omega_x \omega_z xz$$

$$- 2\omega_y \omega_z yz - 2\omega_x \omega_y xy] \, dm$$

Applying definitions (2.39) and (2.40) leads to a form consisting of body constants and angular velocity components,

$$2T_{\text{rot}} = \omega_x^2 I_x + \omega_y^2 I_y + \omega_z^2 I_z - 2\omega_x \omega_z I_{xz} - 2\omega_y \omega_z I_{yz} - 2\omega_x \omega_y I_{xy} \qquad (2.45)$$

Notice that components of $\boldsymbol{\omega}$ can be extracted to get a form

$$2T_{\text{rot}} = \omega_x[\omega_x I_x - \omega_y I_{xy} - \omega_z I_{xz}]$$
$$+ \omega_y[-\omega_x I_{xy} + \omega_y I_y - \omega_z I_{yz}]$$
$$+ \omega_z[-\omega_x I_{xz} - \omega_y I_{yz} + \omega_z I_z]$$

which is simply

$$2T_{\text{rot}} = \boldsymbol{\omega} \cdot \mathbf{h} = \boldsymbol{\omega} \cdot \mathbf{I} \cdot \boldsymbol{\omega} \qquad (2.46)$$

If a body axis is selected such that $\boldsymbol{\omega}$ is parallel to it at the instant in which the rotational energy is calculated, then only one component of this vector is non-zero. Let ξ be an axis along $\boldsymbol{\omega}$ through the center of mass as shown in Figure 2.8. The rotational kinetic energy is then simply

$$T_{\text{rot}} = \tfrac{1}{2} I_\xi \omega^2 \qquad (2.47)$$

The moment of inertia about the axis of rotation ξ can be evaluated by comparing expressions (2.45) and (2.47)

$$I_\xi = \left(\frac{\omega_x}{\omega}\right)^2 I_x + \left(\frac{\omega_y}{\omega}\right)^2 I_y + \left(\frac{\omega_z}{\omega}\right)^2 I_z - 2\left(\frac{\omega_x}{\omega}\right)\left(\frac{\omega_z}{\omega}\right) I_{xz}$$

$$- 2\left(\frac{\omega_y}{\omega}\right)\left(\frac{\omega_z}{\omega}\right) I_{yz} - 2\left(\frac{\omega_x}{\omega}\right)\left(\frac{\omega_y}{\omega}\right) I_{xy} \qquad (2.48)$$

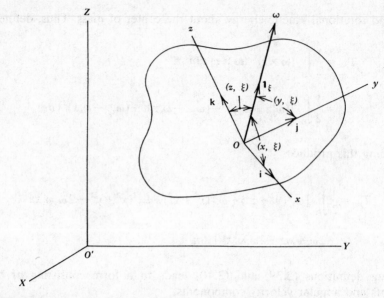

FIGURE 2.8 Moment of inertia about the axis of rotation.

Since $\boldsymbol{\omega} = \omega \mathbf{1}_\xi$, components of this vector can be determined by taking dot products

$$\left. \begin{array}{l} \omega_x = \boldsymbol{\omega} \cdot \mathbf{i} = \omega \mathbf{1}_\xi \cdot \mathbf{i} \\ \omega_y = \boldsymbol{\omega} \cdot \mathbf{j} = \omega \mathbf{1}_\xi \cdot \mathbf{j} \\ \omega_z = \boldsymbol{\omega} \cdot \mathbf{k} = \omega \mathbf{1}_\xi \cdot \mathbf{k} \end{array} \right\} \tag{2.49}$$

For convenience, let

$$l_{\xi x} = \mathbf{1}_\xi \cdot \mathbf{i}, \qquad l_{\xi y} = \mathbf{1}_\xi \cdot \mathbf{j}, \qquad l_{\xi z} = \mathbf{1}_\xi \cdot \mathbf{k}$$

which are simply the direction cosines locating ξ with respect to the x, y, z axes. Thus, comparing these definitions with forms (2.49) leads to

$$l_{\xi x} = \frac{\omega_x}{\omega}, \qquad l_{\xi y} = \frac{\omega_y}{\omega}, \qquad l_{\xi z} = \frac{\omega_z}{\omega}$$

Expression (2.48) can be rewritten as

$$I_\xi = l_{\xi x}{}^2 I_x + l_{\xi y}{}^2 I_y + l_{\xi z}{}^2 I_z - 2 l_{\xi x} l_{\xi z} I_{xz} - 2 l_{\xi y} l_{\xi z} I_{yz} - 2 l_{\xi x} l_{\xi y} I_{xy} \tag{2.50}$$

It should be noted that expression (2.47) is quite simple, but the use of I_ξ presents two serious disadvantages. Evaluation of I_ξ requires determination of the direction cosines and inertia properties with respect to a specified, body-fixed coordinate system. Secondly, the value of I_ξ is continuously changing, in general, because the axis of rotation is changing.

2.3.3 Principal Axes

The result given for rotational kinetic energy by expression (2.45) is known as a *quadratic form* in the components of $\boldsymbol{\omega}$. Notice that angular momentum components may be extracted by taking

$$h_i = \frac{1}{2} \frac{\partial}{\partial \omega_i} (2T_{\text{rot}}) \qquad (i = x, y, z) \tag{2.51}$$

and set (2.42) results directly. This set was also expressed in tensor form by equation (2.44). Since $\mathbf{I}$ is a symmetric matrix, certain theorems of linear algebra may be applied in order to express T_{rot} in terms of only the squares of $\boldsymbol{\omega}$ components, thus, eliminating products of inertia. The resulting expression for T_{rot} is known as a *canonical form*. First $\boldsymbol{\omega}$ must be expressed in terms of another coordinate system via the vector transformation

$$\boldsymbol{\omega} = \mathbf{a} \cdot \boldsymbol{\omega}' \tag{2.52}$$

where $\mathbf{a}$ is a square matrix of order 3. Combining this with form (2.46) gives

$$2T_{\text{rot}} = (\mathbf{a} \cdot \boldsymbol{\omega}') \cdot \mathbf{I} \cdot (\mathbf{a} \cdot \boldsymbol{\omega}')$$

or in matrix form

$$2T_{\text{rot}} = (\mathbf{a}\boldsymbol{\omega}')^T \mathbf{I}(\mathbf{a}\boldsymbol{\omega}') = \boldsymbol{\omega}'^T \mathbf{a}^T \mathbf{I} \, \mathbf{a} \, \boldsymbol{\omega}'$$

Defining $\mathbf{I}' = \mathbf{a}^T \mathbf{I} \, \mathbf{a}$, results in

$$2T_{\text{rot}} = \boldsymbol{\omega}'^T \mathbf{I}' \boldsymbol{\omega}' \tag{2.53}$$

This expression indicates that if the products of inertia are to be eliminated, the new coordinate system must be selected such that matrix $\mathbf{a}$ yields an $\mathbf{I}'$, which is diagonal,

$$\mathbf{I}' = \begin{bmatrix} I_1 & 0 & 0 \\ 0 & I_2 & 0 \\ 0 & 0 & I_3 \end{bmatrix} \tag{2.54}$$

These diagonal terms are known as the *principal moments of inertia* and corresponding axes are *principal axes*. Thus, $\mathbf{a}$ is the array of direction cosines between the original x, y, z coordinates and the principal axes, $1, 2, 3$. Since both are orthogonal and right (or left) handed, $\mathbf{a}$ is orthogonal and $\mathbf{a}^{-1} = \mathbf{a}^T$. The eigenvalues of $\mathbf{I}$ are the principal moments of inertia and may be found by evaluating

$$|\mathbf{I} - \lambda \mathbf{1}| = 0 \tag{2.55}$$

where **1** is the unit matrix. This is the characteristic equation for the inertia matrix. All that remains to completely evaluate principal inertias and axes is to determine the elements of **a**. If the unit vectors corresponding to λ_1, λ_2, λ_3 are $\mathbf{e}_1$, $\mathbf{e}_2$, $\mathbf{e}_3$, then, in tensor form,

$$\mathbf{I} \cdot \mathbf{e}_1 = \lambda_1 \mathbf{e}_1, \qquad \mathbf{I} \cdot \mathbf{e}_2 = \lambda_2 \mathbf{e}_2, \qquad \mathbf{I} \cdot \mathbf{e}_3 = \lambda_3 \mathbf{e}_3$$

This leads to an easy selection of elements of **a**. Construct this transformation matrix such that the elements of $\mathbf{e}_1$, $\mathbf{e}_2$, $\mathbf{e}_3$, are the elements of successive columns,

$$\mathbf{a} = \begin{bmatrix} e_{1x} & e_{2x} & e_{3x} \\ e_{1y} & e_{2y} & e_{3y} \\ e_{1z} & e_{2z} & e_{3z} \end{bmatrix} \tag{2.56}$$

where, for example, e_{1x}, e_{1y}, e_{1z} are the direction cosines of the three angles between $\mathbf{e}_1$ and the x, y, z axes, respectively. Furthermore, angular velocity components about the principal axes are obtained by inverting the matrix form of expression (2.52), $\boldsymbol{\omega}' = \mathbf{a}^{-1} \boldsymbol{\omega}$. Since **a** is orthogonal, this matrix expression becomes

$$\boldsymbol{\omega}' = \mathbf{a}^T \boldsymbol{\omega} \tag{2.57}$$

and the components of $\boldsymbol{\omega}'$ are

$$\omega_1 = e_{1x}\omega_x + e_{1y}\omega_y + e_{1z}\omega_z$$

$$\omega_2 = e_{2x}\omega_x + e_{2y}\omega_y + e_{2z}\omega_z$$

$$\omega_3 = e_{3x}\omega_x + e_{3y}\omega_y + e_{3z}\omega_z$$

where subscripts 1, 2, 3 refer to the principal axes.

As an example of the above process of principal axis determination, consider the case in which rotational kinetic energy has the form

$$2T_{\text{rot}} = 20\omega_x^2 + 30\omega_y^2 + 15\omega_z^2 - 20\omega_x\omega_y - 30\omega_x\omega_z \tag{2.58}$$

The associated inertia tensor is quickly constructed by comparing expression (2.58) to (2.45) and noting that products of inertia appear with negative signs, as defined in equation (2.43). Thus

$$\mathbf{I} = \begin{bmatrix} 20 & -10 & -15 \\ -10 & 30 & 0 \\ -15 & 0 & 15 \end{bmatrix}$$

The equations $(\mathbf{I} - \lambda \mathbf{1}) \cdot \boldsymbol{\omega} = 0$ take on the form

$$
\begin{aligned}
(20 - \lambda)\omega_x \quad &- 10\omega_y \quad &- 15\omega_z = 0 \\
-10\omega_x \quad &+ (30 - \lambda)\omega_y \quad &= 0 \\
-15\omega_x \quad & \quad &+ (15 - \lambda)\omega_z = 0
\end{aligned}
\tag{2.59}
$$

and characteristic equation (2.55) becomes

$$
\lambda^3 - 65\lambda^2 + 1025\lambda - 750 = 0
$$

The associated roots are 39.58, 24.65, and 0.77. Take one at a time to determine the eigenvectors. When $\lambda = \lambda_1 = 39.58$, set (2.59) becomes

$$
\begin{aligned}
-19.58\omega_x \quad &- 10\omega_y \quad &- 15\omega_z = 0 \\
-10\omega_x \quad &- 9.58\omega_y \quad &= 0 \\
-15\omega_x \quad & \quad &- 24.58\omega_z = 0
\end{aligned}
$$

The general solution of this set is

$$
\begin{aligned}
\omega_x &= c_1 \\
\omega_y &= -1.04c_1 \\
\omega_z &= -0.61c_1
\end{aligned}
$$

This can be written as

$$
\boldsymbol{\omega} = c_1 \mathbf{u}_1
$$

where

$$
\mathbf{u}_1 = \begin{bmatrix} 1 \\ -1.04 \\ -0.61 \end{bmatrix}
$$

The first eigenvector is

$$
\mathbf{e}_1 = \frac{\mathbf{u}_1}{u_1} = \begin{bmatrix} 0.64 \\ -0.67 \\ -0.39 \end{bmatrix}
$$

In a similar manner $\lambda = \lambda_2 = 24.65$ yields

$$
\mathbf{e}_2 = \begin{bmatrix} 0.38 \\ 0.71 \\ -0.59 \end{bmatrix}
$$

and $\lambda = \lambda_3 = 0.77$ leads to

$$
\mathbf{e}_3 = \begin{bmatrix} 0.67 \\ 0.23 \\ 0.71 \end{bmatrix}
$$

Matrix **a** of form (2.56) becomes

$$\mathbf{a} = \begin{bmatrix} 0.64 & 0.38 & 0.67 \\ -0.67 & 0.71 & 0.23 \\ -0.39 & -0.59 & 0.71 \end{bmatrix}$$

which gives the principal axis components of angular velocity as

$$\omega_1 = 0.64\omega_x - 0.67\omega_y - 0.39\omega_z$$
$$\omega_2 = 0.38\omega_x + 0.71\omega_y - 0.59\omega_z$$
$$\omega_3 = 0.67\omega_x + 0.23\omega_y + 0.71\omega_z$$

Of course, the principal moments of inertia are

$$I_1 = 39.58$$
$$I_2 = 24.65$$
$$I_3 = 0.77$$

and principal axes are located by the components of $\mathbf{e}_1$, $\mathbf{e}_2$, and $\mathbf{e}_3$. The rotational kinetic energy can be simply written as

$$2T_{rot} = 39.58\omega_1^2 + 24.65\omega_2^2 + 0.77\omega_3^2$$

This example illustrates the extent of calculations necessary to determine principal axes and inertias in a general situation. Fortunately, most bodies of interest have regular geometric shapes for which the principal axes are obtainable by inspection. A few useful rules regarding principal axes of homogeneous rigid bodies are now offered:

(a)   The axis of a body of revolution is a principal axis, and any transverse axis passing through the center of mass is also a principal axis.

(b)   The plane of symmetry contains two principal axes with the third being normal to this plane.

(c)   In general, the three principal axes passing through the center of mass include the axes of maximum and minimum inertia. These are referred to as the *major* and *minor* axes, respectively.

The use of principal axes and inertias simplifies many of the equations regarding rigid body motion. For example, the angular momentum components of set (2.42) become

$$h_1 = I_1\omega_1, \qquad h_2 = I_2\omega_2, \qquad h_3 = I_3\omega_3 \qquad\qquad (2.60)$$

and **h** is now

$$\mathbf{h} = I_1\omega_1\mathbf{e}_1 + I_2\omega_2\mathbf{e}_2 + I_3\omega_3\mathbf{e}_3 \qquad\qquad (2.61)$$

Finally, the moment of inertia about an axis of rotation is simply

$$I_\xi = l_{\xi 1}{}^2 I_1 + l_{\xi 2}{}^2 I_2 + l_{\xi 3}{}^2 I_3 \tag{2.62}$$

where $l_{\xi 1}$, $l_{\xi 2}$, $l_{\xi 3}$ are now direction cosines which orient angular velocity with respect to the principal axes.

Form (2.62) suggests a geometric surface may be used to describe variations in I_ξ with spin axis orientation relative to the principal axes. Define

$$\rho = \sqrt{\frac{1}{I_\xi}}$$

and rewrite equation (2.62) as

$$1 = I_1 (l_{\xi 1}\rho)^2 + I_2 (l_{\xi 2}\rho)^2 + I_3 (l_{\xi 3}\rho)^2 \tag{2.63}$$

Coefficients of principal inertias may be thought of as dimensions of a surface. Let $x = l_{\xi 1}\rho$, $y = l_{\xi 2}\rho$, $z = l_{\xi 3}\rho$, then

$$I_1 x^2 + I_2 y^2 + I_3 z^2 = 1 \tag{2.64}$$

which is the general equation of an ellipsoid with dimensions $\sqrt{1/I_1}$, $\sqrt{1/I_2}$, and $\sqrt{1/I_3}$. The surface associated with all possible values of inertia is called the *ellipsoid of inertia*, depicted in Figure 2.9. Each point on the surface represents a value of moment of inertia about the line joining that point to the body center of mass at point O. In a similar manner rotational energy

$$2 T_{\text{rot}} = I_1 \omega_1{}^2 + I_2 \omega_2{}^2 + I_3 \omega_3{}^2 \tag{2.65}$$

is associated with another ellipsoid. Let $\omega_1 = \xi'$, $\omega_2 = \eta'$, and $\omega_3 = \zeta'$, and expression (2.65) takes on the form

$$\frac{\xi'^2}{(\sqrt{2T/I_1})^2} + \frac{\eta'^2}{(\sqrt{2T/I_2})^2} + \frac{\zeta'^2}{(\sqrt{2T/I_3})^2} = 1 \tag{2.66}$$

which has dimensions $\sqrt{2T/I_1}$, $\sqrt{2T/I_2}$, and $\sqrt{2T/I_3}$, as shown in Figure 2.10. This ellipsoid is sometimes named after Poinsot and represents the locus of all possible angular velocity vectors which satisfy expression (2.65). Notice that all three dimensions of this ellipsoid are a factor of $\sqrt{2T}$ larger than those of the inertia ellipsoid. Thus, the two never touch, but are concentric about the body center of mass. Finally, one can associate angular momentum magnitude with an ellipsoid by noting that

$$h^2 = I_1{}^2 \omega_1{}^2 + I_2{}^2 \omega_2{}^2 + I_3{}^2 \omega_3{}^2 \tag{2.67}$$

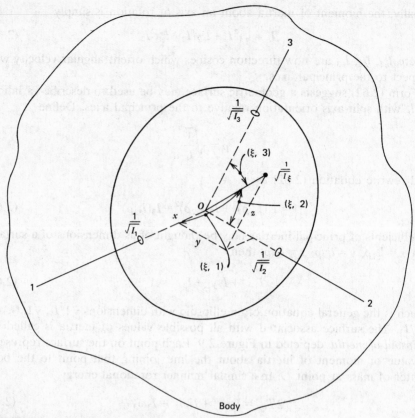

FIGURE 2.9 Ellipsoid of Inertia.

Again let $\omega_1 = \xi'$, $\omega_2 = \eta'$, $\omega_3 = \zeta'$ and get

$$\frac{\xi'^2}{(h/I_1)^2} + \frac{\eta'^2}{(h/I_2)^2} + \frac{\zeta'^2}{(h/I_3)^2} = 1 \tag{2.68}$$

which is an ellipsoid with dimensions h/I_1, h/I_2, and h/I_3 and represents the locus of all possible angular velocity vectors that satisfy expression (2.67). Thus, for any given free body with known energy and angular momentum values, three ellipsoids define all possible instantaneous values of moment of inertia and angular velocity. It is apparent that any $\boldsymbol{\omega}$ which satisfies energy must simultaneously satisfy angular momentum. Thus, the angular momentum ellipsoid of form (2.68) and the *Poinsot ellipsoid* of form (2.66) must intersect along a curve which is the locus of all $\boldsymbol{\omega}$'s that satisfy both requirements. The implications of this will be discussed later during a treatment of solutions to the free body equations of motion.

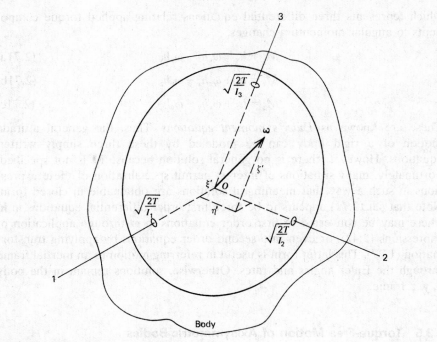

FIGURE 2.10 Poinsot ellipsoid.

2.3.4 Euler's Moment Equations

A discussion of rigid body momentum is naturally followed by development of the general equations of attitude motion. Begin by referring to equation (1.16) for the reaction of a torque about a point as the rate of angular momentum about that point. This result is valid for a rigid body when the point of interest is either fixed in space or is the center of mass. Torques may be thought of as applied about the center of mass without loss of generality. Thus

$$\mathbf{M} = \frac{d\mathbf{h}}{dt} \qquad (2.69)$$

where subscripts need not be displayed. Since this is the absolute rate of $\mathbf{h}$ about the center of mass,

$$\mathbf{M} = \left[\frac{d\mathbf{h}}{dt}\right]_b + \boldsymbol{\omega} \times \mathbf{h} \qquad (2.70)$$

Referring to Figure 2.7, this becomes

$$\mathbf{M} = (\dot{h}_x + \omega_y h_z - \omega_z h_y)\mathbf{i} + (\dot{h}_y + \omega_z h_x - \omega_x h_z)\mathbf{j} + (\dot{h}_z + \omega_x h_y - \omega_y h_x)\mathbf{k}$$

which represents three differential equations relating applied torque components to angular momentum changes,

$$M_x = \dot{h}_x + \omega_y h_z - \omega_z h_y \tag{2.71a}$$

$$M_y = \dot{h}_y + \omega_z h_x - \omega_x h_z \tag{2.71b}$$

$$M_z = \dot{h}_z + \omega_x h_y - \omega_y h_x \tag{2.71c}$$

These are known as *Euler's moment equations*. Thus, the general attitude motion of a rigid body can be modeled by these three simply written equations. However, there is no general solution because **M** is not specified. Fortunately, many situations of interest permit specialization of these expressions in such a way that meaningful solutions are obtainable in closed form. Note that set (2.71) appears to be three first order differential equations in **h**. These may be converted to first order equations in **ω** through application of expressions (2.42), or even into second order equations by applying transformation (1.27). This latter form is useful in referring motion to an inertial frame through the Euler angles and rates. Otherwise, solutions remain in the body x, y, z frame.

2.3.5 Torque-Free Motion of Axisymmetric Bodies

Since this chapter is concerned with torque-free attitude motion, only cases in which **M** = 0 should be treated here. First assume that the body frame coincides with the principal axes to simplify set (2.71):

$$M_1 = I_1 \dot{\omega}_1 + \omega_2 \omega_3 (I_3 - I_2) \tag{2.72a}$$

$$M_2 = I_2 \dot{\omega}_2 + \omega_1 \omega_3 (I_1 - I_3) \tag{2.72b}$$

$$M_3 = I_3 \dot{\omega}_3 + \omega_1 \omega_2 (I_2 - I_1) \tag{2.72c}$$

Then set all torque components equal to zero and consider the simple case of axisymmetry, which requires that $I_1 = I_2$ (a necessary but not sufficient condition). Equations (2.72) become

$$I_1 \dot{\omega}_1 + (I_3 - I_1) \omega_2 \omega_3 = 0 \tag{2.73}$$

$$I_1 \dot{\omega}_2 - (I_3 - I_1) \omega_1 \omega_3 = 0 \tag{2.74}$$

$$I_3 \dot{\omega}_3 = 0 \tag{2.75}$$

Although these are still nonlinear and coupled, a solution of motion can be obtained in the following manner. Component ω_3 is immediately available from equation (2.75) as

$$\omega_3 = n = \text{constant} \tag{2.76}$$

This permits linearization of the other two equations:

$$\dot{\omega}_1 + \lambda\omega_2 = 0$$
$$\dot{\omega}_2 - \lambda\omega_1 = 0$$

(2.77)

where λ is defined here as

$$\lambda = \left(\frac{I_3 - I_1}{I_1}\right)n$$

To solve equations (2.77) simultaneously, multiply the first by ω_1 and the second by ω_2, and add the resulting forms

$$\omega_1\dot{\omega}_1 + \omega_2\dot{\omega}_2 = 0$$

This leads immediately to

$$\omega_1\, d\omega_1 + \omega_2\, d\omega_2 = 0$$

which is easily integrated as

$$\omega_1{}^2 + \omega_2{}^2 = \text{constant}$$

(2.78)

This is recognized as the square of the $\boldsymbol{\omega}$ component in the 1, 2 plane, that is,

$$\omega_{12}{}^2 = \omega_1{}^2 + \omega_2{}^2$$

Referring to Figure 2.11 for nomenclature, combine this with result (2.76) to form

$$\omega^2 = \omega_{12}{}^2 + n^2 = \text{constant}$$

(2.79)

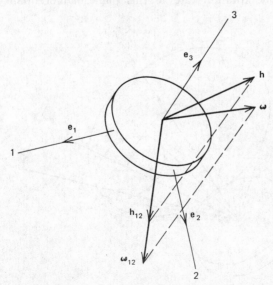

FIGURE 2.11 Axisymmetric body with principal axis coordinates.

Remembering that $\mathbf{h} = $ constant because $\mathbf{M} = 0$, and noting that $h_1 = I_1\omega_1$, $h_2 = I_1\omega_2$, $h_3 = I_3\omega_3 = I_3 n$, it is apparent that the component of $\mathbf{h}$ in the $1, 2$ plane is

$$\mathbf{h}_{12} = I_1(\omega_1\mathbf{e}_1 + \omega_2\mathbf{e}_2) = I_1\boldsymbol{\omega}_{12}$$

and $\mathbf{h}$ becomes

$$\mathbf{h} = I_1\boldsymbol{\omega}_{12} + I_3\omega_3\mathbf{e}_3 \qquad (2.80)$$

This implies that the three vectors, $\mathbf{h}$, $\boldsymbol{\omega}$, and $\mathbf{e}_3$ lie in one plane at all times, because $\boldsymbol{\omega}_{12}$ and $\mathbf{h}_{12}$ are parallel and $\boldsymbol{\omega}_3$, $\mathbf{h}_3$, and $\mathbf{e}_3$ are parallel. Furthermore, $\mathbf{h}$ is fixed in space, which requires that this plane rotates about $\mathbf{h}$, unless $\boldsymbol{\omega}$ is parallel to $\mathbf{h}$. This latter condition corresponds to *simple spin* about the axis of symmetry. In general, $\boldsymbol{\omega}$ and $\mathbf{h}$ are not colinear and two cones may be used to describe the motion. The path of $\boldsymbol{\omega}$ creates a *body cone* and a *space cone* as depicted for a body whose symmetry axis coincides with its major axis in Figure 2.12. Motion is simply described as the body cone rolling on the space cone with $\boldsymbol{\omega}$ corresponding to the line of tangency. This rolling motion takes place without slipping because of the relationship between $\boldsymbol{\omega}$ and $\mathbf{h}$. Note that the body cone is attached to the principal axis coordinate system of the body and aligned with the axis of symmetry, while the space cone is attached to $\mathbf{h}$ and is fixed in inertial space.

To complete the description of motion the rate of rotation of the $\mathbf{e}_3$, $\boldsymbol{\omega}$ plane, as well as θ and γ must be determined. Use of the linearized equations of set (2.77) leads to the rotation rate of this plane. Differentiate the first and

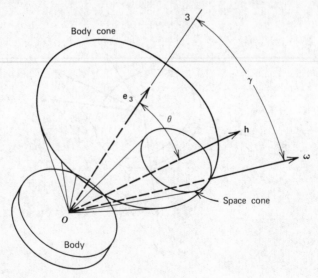

FIGURE 2.12 Cones used to describe motion.

substitute in the second to obtain the equation of a simple harmonic oscillator,

$$\ddot{\omega}_1 + \lambda^2 \omega_1 = 0$$

with the general solution

$$\omega_1 = \omega_1(0) \cos \lambda t + \frac{\dot{\omega}_1(0)}{\lambda} \sin \lambda t \tag{2.81}$$

Since $\omega_2 = -\dot{\omega}_1/\lambda$, the other component is

$$\omega_2 = \omega_1(0) \sin \lambda t - \frac{\dot{\omega}_1(0)}{\lambda} \cos \lambda t \tag{2.82}$$

Noting that $\omega_2(0) = -\dot{\omega}_1(0)/\lambda$ and introducing complex variable notation permits ω_{12} to become

$$\omega_{12} = \omega_1 + i\omega_2 = \left[\omega_1(0) - i\frac{\dot{\omega}_1(0)}{\lambda} \right](\cos \lambda t + i \sin \lambda t)$$

where $i = \sqrt{-1}$. This is also expressible in exponential notation as

$$\omega_{12} = [\omega_1(0) + i\omega_2(0)]e^{i\lambda t}$$

or simply

$$\omega_{12} = \omega_{12}(0)e^{i\lambda t}$$

This implies that ω_{12} rotates at rate λ with respect to the principal axes. Thus, an observer sitting on the body at the axis of symmetry 3 will see $\boldsymbol{\omega}$ rotate at rate λ in a positive sense about $\mathbf{e}_3$. Finally, angles θ and γ are determined from Figure 2.13,

$$\tan \theta = \frac{h_{12}}{h_3} = \frac{I_1 \omega_{12}}{I_3 n} \tag{2.83}$$

$$\tan \gamma = \frac{\omega_{12}}{\omega_3} = \frac{\omega_{12}}{n} \tag{2.84}$$

Both θ and γ are constant, and they are related by

$$\tan \theta = \frac{I_1}{I_3} \tan \gamma \tag{2.85}$$

Angle θ provides the body axis orientation with respect to an inertial direction and is sometimes called the *nutation angle*. These results demonstrate the utility of using cones to describe motion. For example, if $I_3 > I_1$ (for a disc), then $\gamma > \theta$ and the situation in Figure 2.12 applies. If $I_3 < I_1$ (for a rod), then $\gamma < \theta$ and the cones are arranged as shown in Figure 2.14. Here the space cone is outside the body cone. However, motion is still described as the body

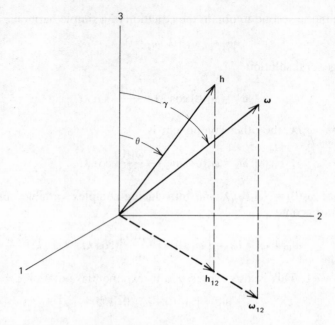

FIGURE 2.13 Geometry of angles θ and γ.

cone rolling without slip on the space cone. In summary, for a torque-free axisymmetric rigid body, the following properties of motion are maintained:

(1) $\mathbf{h}$, $\boldsymbol{\omega}$, ω_{12}, ω_3, θ, and γ are constant.
(2) $\boldsymbol{\omega}$, $\mathbf{h}$, and $\mathbf{e}_3$ are coplanar and this plane rotates about $\mathbf{h}$ at rate λ with respect to the body.

These results are significant, but body motion has not been referred to inertial coordinates as yet. This can be done with the use of Euler angles and rates. Consider the situation shown in Figure 2.15 for the case $I_3 > I_1$. Let the inertial Z axis be parallel to $\mathbf{h}$ for convenience. Using transformation (1.27) for Euler rates and noting that $\dot{\theta} = 0$, leads to

$$\omega_1 = \dot{\psi} \sin \theta \sin \phi$$
$$\omega_2 = \dot{\psi} \sin \theta \cos \phi$$
$$\omega_3 = \dot{\phi} + \dot{\psi} \cos \theta$$

Remembering result (2.76), these are differentiated to yield

$$\left. \begin{aligned} \dot{\omega}_1 &= \dot{\psi}\dot{\phi} \sin \theta \cos \phi + \ddot{\psi} \sin \theta \sin \phi \\ \dot{\omega}_2 &= -\dot{\psi}\dot{\phi} \sin \theta \sin \phi + \ddot{\psi} \sin \theta \cos \phi \\ \dot{\omega}_3 &= \ddot{\phi} + \ddot{\psi} \cos \theta = 0 \end{aligned} \right\} \qquad (2.86)$$

Before going on to rewrite Euler's equations notice that condition (2.79) leads to

$$\dot{\omega} = \frac{\omega_1\dot{\omega}_1 + \omega_2\dot{\omega}_2 + \omega_3\dot{\omega}_3}{\omega} = 0$$

which permits the equality

$$\omega\dot{\omega} = \dot{\psi}\sin\theta\sin\phi[\dot{\psi}\dot{\phi}\sin\theta\cos\phi + \ddot{\psi}\sin\theta\sin\phi]$$
$$+ \dot{\psi}\sin\theta\cos\phi[-\dot{\psi}\dot{\phi}\sin\theta\sin\phi + \ddot{\psi}\sin\theta\cos\phi]$$
$$= 0$$

This reduces to

$$\dot{\psi}\ddot{\psi}\sin^2\theta = 0$$

Since θ is not generally zero, it must be concluded that

$$\dot{\psi}\ddot{\psi} = 0$$

or

$$\dot{\psi} = \text{constant} \tag{2.87}$$

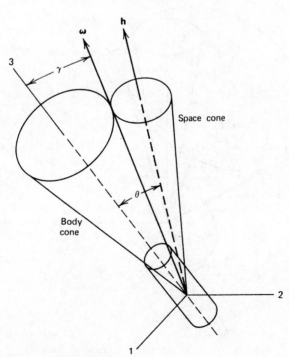

FIGURE 2.14 Motion of a thin rod.

where $\dot{\psi}$ is called the *precession speed*, that is, the rate of rotation of ξ in inertial space. Thus, set (2.86) becomes

$$\left.\begin{array}{l} \dot{\omega}_1 = \dot{\psi}\dot{\phi}\sin\theta\cos\phi \\ \dot{\omega}_2 = -\dot{\psi}\dot{\phi}\sin\theta\sin\phi \\ \dot{\omega}_3 = 0 \end{array}\right\} \qquad (2.88)$$

Since motion about axes 1 and 2 is coupled, either equation (2.73) or (2.74) should yield the desired inertial motion. Consider expression (2.73) with set (2.88) applied. The result reduces to

$$I_1\dot{\phi} + (I_3 - I_1)(\dot{\phi} + \dot{\psi}\cos\theta) = 0$$

which gives the constant precession rate as

$$\dot{\psi} = \frac{I_3\dot{\phi}}{(I_1 - I_3)\cos\theta} \qquad (2.89)$$

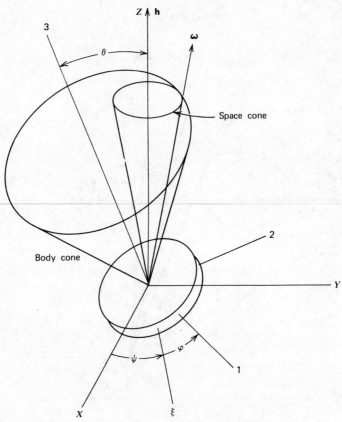

FIGURE 2.15 Motion referred to an inertial frame.

This leads to an interesting observation. If $I_3 > I_1$ (for a disc), then $\dot{\psi}$ and $\dot{\phi}$ are in opposite directions, and this is known as *retrograde precession*. Similarly, if $I_3 < I_1$ (for a rod), then $\dot{\psi}$ and $\dot{\phi}$ are in the same direction, thus, *direct precession* results. It would appear easy to demonstrate retrograde precession in the laboratory with the use of a large gyro. However, only $\dot{\psi}$ and n can be observed from an inertial frame, where $n = \dot{\phi} + \dot{\psi} \cos \theta$. Since $|\dot{\psi}| > |\dot{\phi}|$, for small values of θ, $\dot{\psi}$ and n have the same sign. Thus, retrograde precession cannot be easily observed.

2.3.6 General Torque-Free Motion

Torque-free motion of axisymmetric bodies has been easily described through the use of body and space cones. The general case of an unsymmetrical body in torque-free motion is somewhat more complicated. Nevertheless, it may be treated either analytically or geometrically. In order to gain physical insight into such motion, a geometric solution developed by Poinsot is presented here. The analytical solution requires the use of elliptic integrals and is of very limited interest in practical problems. Begin by assuming that all principal moments of inertia are distinct, $I_1 \neq I_2$, $I_2 \neq I_3$, and $I_3 \neq I_1$. Poinsot's geometric solution is developed from observations based on constant kinetic energy and angular momentum. Since

$$\boldsymbol{\omega} = \omega_1 \mathbf{e}_1 + \omega_2 \mathbf{e}_2 + \omega_3 \mathbf{e}_3$$
$$\mathbf{h} = I_1 \omega_1 \mathbf{e}_1 + I_2 \omega_2 \mathbf{e}_2 + I_3 \omega_3 \mathbf{e}_3$$

the dot product of these two vectors is simply

$$\boldsymbol{\omega} \cdot \mathbf{h} = I_1 \omega_1^2 + I_2 \omega_2^2 + I_3 \omega_3^2 = 2 T_{\text{rot}} \tag{2.90}$$

Dividing by h gives the component of $\boldsymbol{\omega}$ along h as

$$\boldsymbol{\omega} \cdot \frac{\mathbf{h}}{h} = \frac{2T}{h} = \text{constant} \tag{2.91}$$

This constant is interpreted as length d in Figure 2.16. Since d is constant, a plane s may be constructed normal to $\mathbf{h}$ through N. This plane is the *invariable plane* and the line between O and N is the *invariable line*, because they are both fixed in space. In fact, s is the locus of all possible $\boldsymbol{\omega}$s which satisfy condition (2.91) on energy and momentum. Remember that the Poinsot ellipsoid of Figure 2.10 and given by expression (2.66) represents the locus of all $\boldsymbol{\omega}$s which satisfy the energy requirement. This can be superimposed onto Figure 2.16 such that the ellipsoid intercepts the invariable plane at $\boldsymbol{\omega}$. The resulting situation is depicted in Figure 2.17. Motion is described as the Poinsot ellipsoid rolling, without slip, on the invariable plane. Body center of mass O remains a constant distance d away from s. To demonstrate that the ellipsoid

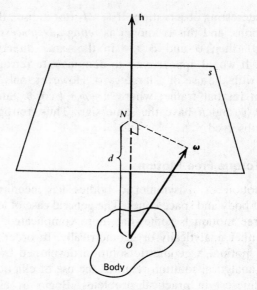

FIGURE 2.16 Definition of invariable plane and invariable line.

rolls on s, note the differential of equation (2.90),

$$d(2T) = d\boldsymbol{\omega} \cdot \mathbf{h} = 0$$

In other words, $d\boldsymbol{\omega}$ is normal to $\mathbf{h}$. Since $d\boldsymbol{\omega}$ is attached to $\boldsymbol{\omega}$ at the invariable plane, $d\boldsymbol{\omega}$ must lie in s. Also, changes in $\boldsymbol{\omega}$ must take place on the surface of the ellipsoid, requiring that $d\boldsymbol{\omega}$ be tangent to this surface. These two observations can only be satisfied simultaneously if the Poinsot ellipsoid is always

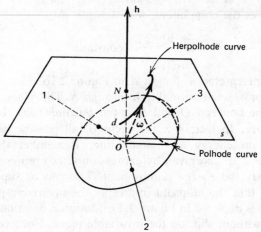

FIGURE 2.17 Poinsot ellipsoid rolls on the invariable plane.

tangent to s. Slip is not permitted because any movement of $\boldsymbol{\omega}$ on s corresponds to a change of spin vector orientation in body coordinates in order to maintain constant energy and momentum. The path of $\boldsymbol{\omega}$ on s is the inertial motion of this vector and is called the *herpolhode curve*. Since the Poinsot ellipsoid is attached to the body principal axes, the path of $\boldsymbol{\omega}$ on this surface is referred to the body coordinate frame and is called the *polhode curve*. Note that this situation degenerates for axisymmetric bodies to the space cone and body cone. Since the long axis of the Poinsot ellipsoid corresponds to the minor principal axis, Figure 2.17 indicates that if $I_1 = I_2$ and $I_3 < I_1$, this situation would be equivalent to Figure 2.14. For a case in which $I_3 > I_1 = I_2$, the 3-axis in Figure 2.17 would lie toward the short axis of the ellipsoid and a situation similar to that of Figure 2.18 would prevail. This is compatible with Figure 2.12 for a disc-shaped body.

In general herpolhode curves are not closed paths, because the Poinsot ellipsoid is not usually in circumferential phase with positions on s. However, polhode curves must be closed paths, because $\boldsymbol{\omega}$ can only return to its initial value by returning to its initial position after each circuit around the polhode. To further describe polhodes remember that these are the loci of $\boldsymbol{\omega}$s which satisfy both energy and momentum requirements, and must, therefore, be the intersection of the Poinsot and momentum ellipsoids. An equation for polhode curves may be derived by equating expressions (2.66) and (2.68) and rewriting as

$$I_1\left(I_1 - \frac{h^2}{2T}\right)\xi'^2 + I_2\left(I_2 - \frac{h^2}{2T}\right)\eta'^2 + I_3\left(I_3 - \frac{h^2}{2T}\right)\zeta'^2 = 0 \qquad (2.92)$$

In order to get real values of ξ', η', ζ', at least one coefficient in this expression must be negative. Therefore, $h^2/2T$ must have a value which lies between the

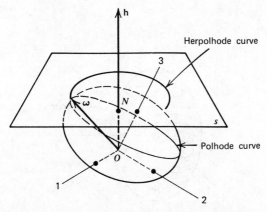

FIGURE 2.18 Poinsot ellipsoid orientation for a near disc-shaped body.

major and minor moments of inertia. Assume, without loss of generality, that $I_1 > I_2 > I_3$. Then the combination of momentum and energy, $h^2/2T$ must satisfy $I_1 > h^2/2T > I_3$ for realistic situations. Polhode paths are best visualized by mathematically eliminating one coordinate and studying the resulting two-dimensional shape. First eliminate ζ' by proper substitution of expression (2.66) into (2.92),

$$I_1(I_1 - I_3)\xi'^2 + I_2(I_2 - I_3)\eta'^2 = 2T\left(\frac{h^2}{2T} - I_3\right) \tag{2.93}$$

This is the polhode projection onto a plane normal to the 3-axis. The corresponding geometric shape, since the right hand side is positive and constant, is an ellipse. Similarly eliminating ξ' gives

$$I_2(I_1 - I_2)\eta'^2 + I_3(I_1 - I_3)\zeta'^2 = 2T\left(I_1 - \frac{h^2}{2T}\right) \tag{2.94}$$

which is also an ellipse. However, elimination of η' results in

$$I_1(I_1 - I_2)\xi'^2 - I_3(I_2 - I_3)\zeta'^2 = 2T\left(\frac{h^2}{2T} - I_2\right) \tag{2.95}$$

which is a hyperbola. Note that the right hand side can be either positive or negative, which will dictate the orientation of the branches, that is, whether they will open about ξ' or ζ'.

Figure 2.19 is a composite sketch of typical polhodes which satisfy these conditions. The value of $h^2/2T$ determines which polhode is actually followed. If $h^2/2T = I_3$, then $\xi' \equiv \eta' \equiv 0$ and the polhode is a point, which represents simple spin about the 3-axis. Similarly, if $h^2/2T = I_1$, $\eta' \equiv \zeta' \equiv 0$ and motion is

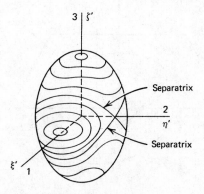

FIGURE 2.19 Polhode pattern on poinsot ellipsoid.

spin about the 1-axis. If $h^2/2T = I_2$, then expression (2.95) gives the asymptotes in a plane normal to the 2-axis, whose slopes are given by

$$\frac{\zeta'}{\xi'} = \pm \sqrt{\frac{I_1(I_1 - I_2)}{I_3(I_2 - I_3)}} \qquad (2.96)$$

These represent the two separatrices as projected onto the 1, 3 plane. This result implies motion may take place about the major or minor axis, but not about the intermediate axis. The *separatrices* are dividers which separate motion about the major axis from motion about the minor axis. They correspond to the case in which $h^2/2T = I_2$, the intermediate inertia.

2.4 STABILITY OF ROTATION ABOUT PRINCIPAL AXES

The foregoing discussion of polhode paths for an asymmetrical body brought about the concept of motion about a principal axis. It is the objective of this section to investigate the stability of such motion from an analytical approach. The term *stability* is used here to describe the motion of a torque-free, rigid body after being disturbed from steady spin. Thus, motion is stable if amplitudes of disturbed quantities are bounded by initial values. Existence of a stable motion is investigated by perturbing a known steady state. Thus, if $\boldsymbol{\omega}$ is initially parallel to the 1-, 2-, or 3-axis, then the torque-free form of set (2.72) indicates that the solution of motion is simply ω_1, ω_2, or ω_3, respectively, is a constant. Consider the case in which $\boldsymbol{\omega} = \omega_o \mathbf{e}_1$, with solution $\omega_1 = \omega_o$, $\omega_2 \equiv \omega_3 \equiv 0$. This represents a general case since the relative sizes of I_1, I_2, and I_3 have not been specified. If initial steady motion is perturbed then $\omega_1 = \omega_o + \varepsilon$ with ω_2 and ω_3 small like ε, and ω_o assumed constant. The equations of motion become

$$I_1 \dot{\varepsilon} = (I_2 - I_3)\omega_2 \omega_3 \qquad (2.97a)$$

$$I_2 \dot{\omega}_2 = (I_3 - I_1)(\omega_o + \varepsilon)\omega_3 \qquad (2.97b)$$

$$I_3 \dot{\omega}_3 = (I_1 - I_2)(\omega_o + \varepsilon)\omega_2 \qquad (2.97c)$$

Terms which have magnitude of the order ε^2 and higher may be neglected. Thus, the right-hand side of expression (2.97a) is set to zero, implying $\varepsilon \cong$ constant. The other two equations may also be rewritten as

$$\dot{\omega}_2 = \left(\frac{I_3 - I_1}{I_2}\right)\omega_0 \omega_3$$

$$\dot{\omega}_3 = \left(\frac{I_1 - I_2}{I_3}\right)\omega_0 \omega_2$$

Differentiating the first and eliminating $\dot{\omega}_3$ by the second gives

$$\ddot{\omega}_2 + \left[\frac{(I_1 - I_2)(I_1 - I_3)}{I_2 I_3} \omega_o^{\,2} \right] \omega_2 = 0$$

Similarly

$$\ddot{\omega}_3 + \left[\frac{(I_1 - I_2)(I_1 - I_3)}{I_2 I_3} \omega_o^{\,2} \right] \omega_3 = 0$$

Both of these equations represent simple harmonic oscillators with general solution

$$\omega_j = \Omega_{j1} e^{i\lambda t} + \Omega_{j2} e^{-i\lambda t} \qquad (j = 1, 2)$$

where λ is defined as

$$\lambda = \sqrt{\frac{(I_1 - I_2)(I_1 - I_3)}{I_2 I_3} \omega_o^{\,2}}$$

If λ is imaginary, ω_j will diverge and motion is unstable. Thus, λ must be real for stability, and ω_j will oscillate about the steady state. This is satisfied when the product $(I_1 - I_2)(I_1 - I_3) > 0$. Motion is stable about the 1-axis when $I_1 > I_2$ and $I_1 > I_3$ or when $I_1 < I_2$ and $I_1 < I_3$. In other words, motion about the major or minor principal axis is stable, but motion about the intermediate axis is unstable. These conclusions verify the indications offered by polhode curves in Figure 2.19.

2.5 INTERNAL ENERGY DISSIPATION EFFECTS

The strict definition of a rigid body does not permit any energy dissipation. However, all real spacecraft have, at least, some non-rigid properties. These include elastic structural deflection and liquid slosh due to accelerations about the center of mass. Many satellites have energy dissipators installed for the purpose of reducing energy, because damping of certain types of undesirable motion can be accomplished with such devices. Thus, an introduction to this topic seems appropriate here.

Consider a *semirigid* body, which has no moving parts but does dissipate energy. The ultimate motion of such a body can be easily determined by noting stability criteria and energy state. It is apparent that energy will dissipate until a minimum state is reached. From the preceding section the final state must be motion about the minor or major principal axis. However, checking values at both these possible states gives

$$2T = \frac{h^2}{I_{max}} \qquad \text{at the major axis}$$

$$2T = \frac{h^2}{I_{min}} \qquad \text{at the minor axis}$$

Since angular momentum must be conserved, motion about the major axis corresponds to a minimum energy state. Therefore, a semirigid body is stable only when spinning about its major axis.

This result is based on a *heuristic argument*, that is, one that satisfies physical constraints but is not rigorous. Many problems involving energy dissipation effects are treated by this approach, because rigorous arguments are at best very difficult to formulate and have not yet been developed for most practical situations. However, results of such treatments have been verified by actual flight experience. A classic example of the major axis stability phenomenon was encountered during the flight of the first U.S. satellite, Explorer I, whose configuration is shown in Figure 2.20. After only a few hours into the flight, radio signals indicated that a tumbling motion had developed and was increasing in amplitude in an unstable manner. It was concluded that the four turnstile wire antennae were dissipating energy; thus, causing a transfer of body spin axis from the axis of minimum inertia to a transverse axis of maximum inertia. This explanation is credited to R. N. Bracewell and O. K. Garriott of Stanford University. Garriott later became a Scientist-Astronaut in the Skylab mission, in which he carried out a simple experiment to demonstrate this phenomenon.

It is instructive to describe momentum transfer geometrically through the use of the inertia ellipsoid. Figure 2.21 depicts an ellipsoid for an asymmetrical body with distinct inertias. A single hypothetical polhode for constant angular momentum and decreasing energy is shown. Motion is initially about the minor

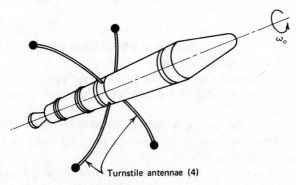

Turnstile antennae (4)

FIGURE 2.20 Explorer I configuration.

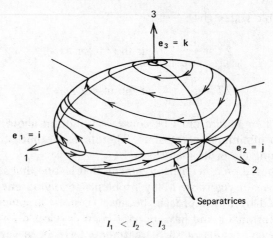

$$I_1 < I_2 < I_3$$

FIGURE 2.21 Hypothetical polhode of dissipation.

axis, but the instantaneous spin axis drifts away and eventually crosses one of the separatrices. Note that the choice of separatrix is purely arbitrary and unpredictable. The final motion may be either positive or negative spin about the major axis. Thus, if a semirigid body is initially spinning about its minor axis it will enter a *tumbling* mode and eventually reorient itself such that stable spin about its major axis is achieved. In this final state the positive major axis can be aligned with the angular momentum vector in either of two ways, distinguished by a 180° rotation about a transverse axis. Note that the angular momentum vector is constant in space so the body must reorient to align its major axis with the inertial direction of this vector. Application of energy dissipation to large-angle reorientation maneuvers is considered in Chapter 4.

EXERCISES

2.1 Find an equation for the speed v of a satellite in earth orbit as a function of energy $\mathscr{E}$ and radial distance r.

2.2 Consider a satellite in an elliptical orbit about earth.
  (a) Determine the value of r at the semiminor axis, as shown, in terms of a, e, p.
  (b) Determine the corresponding value of θ in terms of a, e, p.
  (c) Determine its angular momentum per unit mass at that point in terms of μ, a, e, p.
  (d) What fraction of the orbital period is required to travel from perigee to point b in terms of a, e, p, μ?

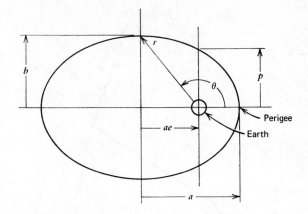

EXERCISE 2.2

2.3 Show that the magnitude of e in equation (2.9) is equal to the quantity in expression (2.20).

2.4 Carry out the steps required to derive the relationship between orbital energy and semimajor axis,

$$\mathscr{E} = -\frac{\mu}{2a}$$

2.5 Of two circular earth orbits, the higher one has greater energy $\mathscr{E}$. Explain why this orbit has lower velocity. Derive an equation relating circular orbit speed to radius.

2.6 Derive an equation for orbital period in terms of given initial speed v_0 and radius r_0.

2.7 Show that central force motion is confined to a plane fixed in space by using

$$\mathbf{F} = \frac{\mu}{r^3}\mathbf{r}, \qquad \dot{\mathbf{h}} = \mathbf{r} \times \mathbf{F}$$

where $\mathbf{F}$ is force per unit mass.

2.8 Derive an expression for eccentricity e in terms of initial speed v_0, radius r_0, and flight path angle β_0, as defined in the figure.

2.9 Show that the speed of a satellite in an elliptic orbit at either end of the minor axis is the same as circular speed at that point.

2.10 Prove that the flight path angle β is 45° when $\theta = 90°$ for all parabolic trajectories.

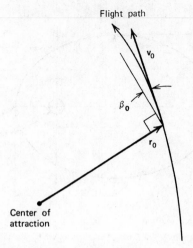

Flight path

v_0

β_0

r_0

Center of
attraction

EXERCISE 2.8

2.11 A satellite is tracked from ground stations and observed to have the following inertially referenced orbital properties at one point:

$$\text{altitude} = 662 \text{ km}$$

$$r\dot{\theta} = 7.0 \text{ km/s}$$

$$\dot{r} = 3.72 \text{ km/s}$$

(a) Determine orbital eccentricity e.
(b) Determine the value of true anomaly θ at this point.

2.12 An earth satellite is observed to have a perigee height of 122 km and apogee height of 622 km. Determine the orbital period and eccentricity.

2.13 Inertial position and velocity of a missile over the earth at a given point in time are observed to be

$$\mathbf{R} = (10^4 \mathbf{I} + 10^4 \mathbf{J}) \text{ km}$$

$$\dot{\mathbf{R}} = (2 \mathbf{I} + 5 \mathbf{J}) \text{ km/s}$$

where the inertial frame has origin at the earth's center and unit vectors are defined in Figure 1.10. Calculate
(a) angular momentum $\mathbf{h}$
(b) energy $\mathcal{E}$
(c) semilatus rectum p
(d) eccentricity e
(e) semimajor axis a

2.14 Show that the period of an orbit close to the surface of a homogeneous spherical planet is a function only of the planetary density.

2.15 A small satellite of a newly discovered planet moves along the spiral path described by

$$r\theta = C$$

where C = constant. What is the associated central force law for this type of motion? Hint: the radial component of acceleration in plane motion is $\ddot{r} - r\dot{\theta}^2$.

2.16 A manned spacecraft is sent to our nearest neighboring star, Alpha Centauri. After entering an orbit about its third planet a radio message is sent back to earth. Four years later the message is received here:

"Third planet is quite similar to earth in shape and our orbit is elliptical, but the planet is at the center rather than at a focus."

(a) Deduce the form of gravitational attraction which causes this type of orbit.

(b) Explain why angular momentum and energy are conserved.

2.17 Determine the length of a uniform cylinder of radius R such that its principal moments of inertia are equal.

2.18 Identify two typical body shapes which have equal principal inertias, that is, $I_1 = I_2 = I_3$.

2.19 Consider a rigid body with inertia tensor

$$\mathbf{I} = \begin{bmatrix} 30 & 0 & 0 \\ 0 & 40 & 0 \\ 0 & 0 & 20 \end{bmatrix} (\text{N} \cdot \text{m} \cdot \text{s}^2)$$

and angular velocity

$$\boldsymbol{\omega} = 10\mathbf{j} + 10\mathbf{k} \,(\text{rad/s})$$

Determine:

(a) The moment of inertia about an axis parallel to $\boldsymbol{\omega}$.

(b) The rotational kinetic energy T_{rot}.

2.20 Consider a rigid body experiencing angular motion associated with angular velocity $\boldsymbol{\omega}$. Inertia properties are given by

$$\mathbf{I} = \begin{bmatrix} 20 & -10 & 0 \\ -10 & 30 & 0 \\ 0 & 0 & 40 \end{bmatrix} (\text{N} \cdot \text{m} \cdot \text{s}^2)$$

and

$$\boldsymbol{\omega} = 10\mathbf{i} + 20\mathbf{j} + 30\mathbf{k} \,(\text{rad/s})$$

where nomenclature is that of Figure 2.7.

(a) What is the angular momentum of this body about point O, its center of mass?

(b) What is its rotational kinetic energy about O?

(c) Determine the principal moments of inertia.

2.21 Consider a rigid body with inertia tensor

$$\mathbf{I} = \begin{bmatrix} 30 & -I_{xy} & -I_{xz} \\ -10 & 20 & -I_{yz} \\ 0 & -I_{zy} & 30 \end{bmatrix} (\text{N} \cdot \text{m} \cdot \text{s}^2)$$

and angular velocity

$$\boldsymbol{\omega} = 10\mathbf{i} + 10\mathbf{j} + 10\mathbf{k}(\text{rad/s})$$

If

$$\mathbf{h} = 200\mathbf{i} + 200\mathbf{j} + 400\mathbf{k} \ (\text{N} \cdot \text{m} \cdot \text{s})$$

determine:

(a) Values of I_{zy}, I_{xy}, I_{xz} and I_{yz}.

(b) Moment of inertia about $\boldsymbol{\omega}$.

(c) Rotational kinetic energy.

(d) Principal moments of inertia.

2.22 Given that the rotational kinetic energy of a rigid body about its center of mass is

$$T_{\text{rot}} = \tfrac{1}{2}[25\omega_x^2 + 34\omega_y^2 + 41\omega_z^2 - 24\omega_y\omega_z]$$

where x, y, z is a specified body-fixed coordinate set.

(a) Determine the principal moments of inertia.

(b) Calculate the angles between x, y, z and principal axes 1, 2, 3.

(c) Determine the magnitude of angular momentum, h.

2.23 Consider a rigid body with axial symmetry and principal inertias $I_1 = I_2$, $I_3 > I_1$. If it is acted upon by a small body-fixed, transverse torque, $M_1 = M$, with $M_2 = M_3 = 0$, then:

(a) Write the differential equations of motion for this situation.

(b) Is ω_3 affected by M_1? Explain!

(c) How is the angular momentum vector affected by M_1?

2.24 For a rigid body in torque-free motion, derive the equations for ω_1, ω_2, and ω_3 when $h^2 > 2TI_2$ and $I_1 > I_2 > I_3$.

2.25 Consider the torque-free motion of a rigid body. For $h^2/2T < I_2$ when $I_1 > I_2 > I_3$, the angle θ_3 in the figure, made by the $\mathbf{h}$ vector and h_3, is represented by

$$\sin\theta_3 = \frac{1}{h}\sqrt{h_1^2 + h_2^2}$$

Show that the value of θ_3 varies between

$$\sin \theta_{3_{max}} = \sqrt{\frac{I_2 I_3}{I_2 - I_3}\left(\frac{1}{I_3} - \frac{1}{h^2/2T}\right)}$$

$$\sin \theta_{3_{min}} = \sqrt{\frac{I_1 I_3}{I_1 - I_3}\left(\frac{1}{I_3} - \frac{1}{h^2/2T}\right)}$$

Hint: Rewrite equations (2.93) to (2.95) in terms of h_1, h_2, and h_3.

2.26 Show that a thin disk thrown up spinning about its major axis with a small nutation angle will make two wobbles to every cycle of spin.

2.27 A rigid satellite in earth orbit is spinning about its axis of symmetry and has principal moments of inertia $I_1 = I_2$, I_3 with $I_3 < I_1$.
(a) If a micrometeorite hits the satellite and induces slight motion about the transverse axes, is the resulting motion stable?
(b) If $I_1/I_3 = 2$, and the nutation angle is $2°$, sketch the motion of this vehicle using space and body cones. Show values of θ and γ, and the body itself.

2.28 A satellite in earth orbit is spinning about its axis of symmetry and has principal moments of inertia $I_1 = I_2$, I_3 with $I_3 > I_1$. A micrometeorite hits this satellite and induces a wobble. If this vehicle experiences a dissipation of angular momentum while maintaining a constant rotational kinetic energy level, it will seek a state of minimum angular

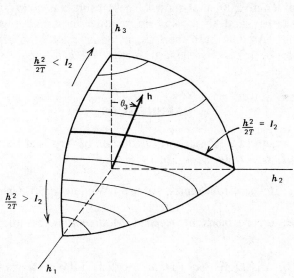

EXERCISE 2.25

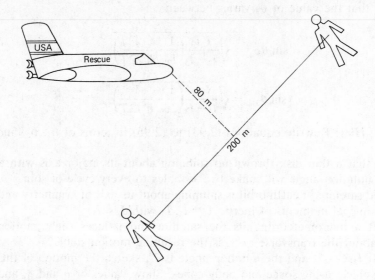

EXERCISE 2.29

momentum. Is the satellite still stable about the axis of symmetry?

2.29 During a space rescue mission an *astromedic* was trying to return an unconscious astronaut to his capsule when his maneuvering unit ran out of propellant. At that moment the two men were connected by a taut tether of length 200 m, as shown. The capsule was 80 m from the center of this tether and there was no relative motion between men and spacecraft. Assume both men are of equal mass, and discuss possible action to be taken by the astromedic. He can expect no help from anybody else.

REFERENCES

Bracewell, R. N., and O. K. Garriott, "Rotation of Artificial Earth Satellites," *Nature*, Vol. 82, *No. 4638*, Sept. 20, 1958, pp. 760–762.

Goldstein, H., *Classical Mechanics*, Addison-Wesley, 1950, Chapters 3, 4, 5.

Hildebrand, F. B., *Methods of Applied Mathematics*, Prentice-Hall, 1952, Chapter 1.

Kendrick, J. B., ed., *TRW Space Data*, 3rd Ed., TRW Systems Group, 1967, pp. 8–15.

Likins, P. W., "Effects of Energy Dissipation on the Free Body Motions of Spacecraft," Technical Report TR 32-860, Jet Propulsion Laboratory, Pasadena, California, July 1966.

Thomson, W. T., *Introduction to Space Dynamics*, Wiley, 1961, Chapters 3, 4.

Orbital Maneuvers

Fundamental principles of orbit mechanics were presented in Section 2.2. The treatment offered there may be interpreted as an introduction to celestial mechanics, because no maneuvers or changes in orbit were included. However, artificial satellites are now commonplace, and a new field of study has been created. In fact, the ability to maneuver through the use of propulsion devices has made spaceflight possible. This chapter is devoted to the determination of orbital state and estimation of velocity changes required to satisfy a mission objective. Only methods which do not require computer solution but permit sufficient accuracy for preliminary evaluations are presented here. Such techniques allow quick appraisal of mission designs by the project engineer and provide a great deal of insight for the student. Accurate methods of orbit establishment are reviewed in Chapter 7. As an initial step in developing maneuver estimation techniques, fundamental orbit parameter calculations are discussed. Then in-plane orbit adjustments and transfers are considered, assuming the use of impulsive velocity changes. Rotation of the orbit plane is included as the only out-of-plane maneuver to be exposed in this chapter. Interplanetary and lunar transfer estimation techniques are developed and example calculations presented.

3.1 ORBIT ESTABLISHMENT

Before orbital maneuvers can be implemented, properties of the initial orbit must be known. Typically, a launch vehicle will inject a payload into orbit at its upper-stage-burnout point. Through tracking from ground stations and telemetry from on-board sensors, the altitude, speed, and flight path angle may be determined accurately and quickly. Since the gravitational field of earth is known, orbital parameters can be calculated and compared with desired results. Errors in orbit insertion as well as mission objectives may require subsequent firing of thrusters on the payload to adjust or change the initial orbit completely.

3.1.1 Determination of Eccentricity and True Anomaly

Consider the situation in which the initial altitude, speed, and flight angle are known. This case is depicted in Figure 3.1 for an earth-based launch. Flight path angle β_0 is just the angle between $\mathbf{v}_0$ and the local horizon with positive values corresponding to $\mathbf{v}_0$ above the horizon. Thus, r_0, v_0, β_0 are assumed to be given or measured quantities. The first parameters to determine are eccentricity e and true anomaly at the burnout point θ_0, which locates the perigee. Then orbital period and time to perigee can be calculated. Referring to Section 2.2 and Figure 3.1, an expression for e can be derived as follows. Noting that

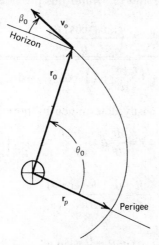

FIGURE 3.1 Launcher burnout conditions

angular momentum is constant around the orbit,

$$h = r^2 \dot{\theta} = r_0^2 \dot{\theta}_0$$

Here, $r_0 \dot{\theta}_0 = v_0 \cos \beta_0$, which leads to

$$h = r_0 v_0 \cos \beta_0 \tag{3.1}$$

Rewriting equation (2.27) as

$$\mathscr{E} = \frac{v_0^2}{2} - \frac{\mu}{r_0} \tag{3.2}$$

and using expressions (2.20) and (2.18) leads to

$$e^2 = 1 + \frac{2}{\mu^2} (r_0^2 v_0^2 \cos^2 \beta_0) \left(\frac{v_0^2}{2} - \frac{\mu}{r_0} \right)$$

By completing the square this becomes

$$e^2 = \left(\frac{r_0 v_0^2}{\mu} - 1 \right)^2 \cos^2 \beta_0 + \sin^2 \beta_0 \tag{3.3}$$

Next obtain θ_0 by evaluating equation (2.10) for a conic at the initial position,

$$r_0 = \frac{h^2/\mu}{1 + e \cos \theta_0}$$

and solve for $\cos \theta_0$

$$\cos \theta_0 = \frac{1}{e} \left(\frac{h^2}{\mu r_0} - 1 \right)$$

Substituting expressions (3.1) and (3.3) into this provides θ_0 in terms of given quantities,

$$\cos \theta_0 = \frac{\left(\dfrac{r_0 v_0^2 \cos^2 \beta_0}{\mu} - 1 \right)}{\sqrt{\left(\dfrac{r_0 v_0^2}{\mu} - 1 \right)^2 \cos^2 \beta_0 + \sin^2 \beta_0}} \tag{3.4}$$

Furthermore, $\sin \theta_0$ may be easily determined by noting that

$$v \sin \beta = \dot{r} = \frac{dr}{d\theta} \dot{\theta} = \frac{dr}{d\theta} \frac{h}{r^2} = -\frac{d}{d\theta} \left(\frac{h}{r} \right)$$

But

$$\frac{h}{r} = \frac{\mu(1 + e \cos \theta)}{h}$$

Thus,

$$v \sin \beta = \frac{\mu}{h} e \sin \theta \tag{3.5}$$

Evaluating this at the initial point gives

$$\sin \theta_0 = \frac{h v_0}{\mu e} \sin \beta_0 = \frac{r_0 v_0^2 \cos \beta_0 \sin \beta_0}{\mu \sqrt{\left(\frac{r_0 v_0^2}{\mu} - 1\right)^2 \cos^2 \beta_0 + \sin^2 \beta_0}} \tag{3.6}$$

Taking the ratio of expressions (3.6) and (3.4) yields

$$\tan \theta_0 = \frac{(r_0 v_0^2/\mu) \sin \beta_0 \cos \beta_0}{(r_0 v_0^2/\mu) \cos^2 \beta_0 - 1} \tag{3.7}$$

Expressions (3.3) and (3.7) indicate that the three initial conditions can be grouped into two parameters, $(r_0 v_0^2/\mu)$ and β_0. Use of these results in parametric form gives eccentricity and true anomaly values which are independent of the central body. Figure 3.2 displays all possible combinations of e, θ_0, β_0, and $r_0 v_0^2/\mu$ for elliptic orbits. This chart can be used for graphical determination of two parameters when the other two are known. Notice, however, that given values of e and rv^2/μ lead to two possible combinations of β and θ. This ambiguity can be resolved by deciding whether the spacecraft is approaching or leaving periapsis. Another interesting observation can be made by rearranging expression (3.2) as

$$\frac{r_0 v_0^2}{\mu} = \frac{2 \mathscr{E} r_0}{\mu} + 2$$

If $r_0 v_0^2/\mu = 2$, then $v_0 = \sqrt{2\mu/r_0}$, $\mathscr{E} = 0$, and $e = 1$. This corresponds to an escape trajectory which is independent of β_0 as long as the path does not intercept the central body of attraction. In contrast to this, consider the case when $r_0 v_0^2/\mu = 1$. Then $v_0 = \sqrt{\mu/r_0}$, which is the circular orbit speed. However, the orbit is not circular unless $\beta_0 = 0$.

The special case in which $\beta_0 = 0$ corresponds to the periapsis or apoapsis point, unless the orbit is circular. To analytically determine which situation exists, rewrite equations (3.3) and (3.7) with $\beta_0 = 0$,

$$e = \left| \frac{r_0 v_0^2}{\mu} - 1 \right| \tag{3.8}$$

$$\tan \theta_0 = \frac{0}{\left(\dfrac{r_0 v_0^2}{\mu} - 1 \right)}$$

Although the second expression appears to be zero, evaluation of θ_0 requires that the denominator be checked for sign. For example, if $r_0 v_0^2/\mu = 1$, then $e = 0$, and θ_0 is indeterminant. This corresponds to a circular orbit. If $r_0 v_0^2/\mu < 1$, then $\tan \theta_0$ approaches zero from the negative side. This corresponds to

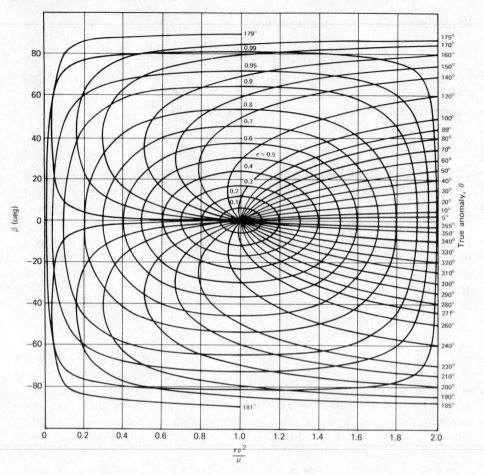

FIGURE 3.2 Graphic presentation of elliptic orbit parameters.

$\theta_0 = 180°$, and the initial position is at apoapsis. If $1 < (r_0 v_0^2/\mu) < 2$, then the orbit is still elliptic, but $\theta_0 = 0$ and the spacecraft is at periapsis initially. Obviously, if $r_0 v_0^2/\mu = 2$, then $e = 1$ and the orbit is parabolic. Finally, should it happen that $r_0 v_0^2/\mu > 2$, then $e > 1$ and this corresponds to a hyperbolic trajectory with initial position at the periapsis or *point of closest approach*.

3.1.2 Slightly Eccentric Orbits

There are many missions in which the target orbit is supposed to be circular, but the launch vehicle injects the payload with a small value of β_0 and a small error in v_0. No errors in r_0 are assumed for the moment. From equations (2.10) and (3.1) an expression for $e \cos \theta_0$ is obtained,

$$e \cos \theta_0 = \frac{r_0 v_0^2 \cos^2 \beta_0}{\mu} - 1$$

Equation (3.5) further permits

$$e \sin \theta_0 = \frac{r_0 v_0^2 \cos \beta_0 \sin \beta_0}{\mu}$$

Denote circular velocity as v_c and rewrite these forms

$$e \cos \theta_0 = \frac{v_0^2}{v_c^2} \cos^2 \beta_0 - 1$$

$$e \sin \theta_0 = \frac{v_0^2}{v_c^2} \cos \beta_0 \sin \beta_0$$

Since β_0 is small and $(v_0 - v_c)$ is small, only first-order terms need be retained. This leads to

$$e \cos \theta_0 \cong 2\left(\frac{v_0 - v_c}{v_c}\right)$$

$$e \sin \theta_0 \cong \beta_0$$

Solving for $\tan \theta_0$ and e gives

$$\tan \theta_0 = \frac{\beta_0 v_c}{2(v_0 - v_c)} \tag{3.9a}$$

$$e = \sqrt{4\left(\frac{v_0 - v_c}{v_c}\right)^2 + \beta_0^2} \tag{3.9b}$$

where e is also a small number here. The effect of $v_0 - v_c$ on semimajor axis can be quickly evaluated by solving energy equation (2.27) for a,

$$a = -\frac{\mu}{\left(v^2 - \dfrac{2\mu}{r_0}\right)}$$

and taking the difference equation holding r constant and e small,

$$\frac{\delta a}{a} \cong 2\frac{\delta v}{v}$$

where $\delta v = v_0 - v_c$. Thus, an excess of injection velocity over v_c causes an increase in a and orbital energy. The associated increase in period is obtained as

$$\frac{\delta \tau}{\tau} = \frac{3}{2}\frac{\delta a}{a} \tag{3.10}$$

which gives

$$\frac{\delta \tau}{\tau} = 3\frac{\delta v}{v}$$

as the increase due to δv. Finally, introduce an error in r_0. Its effect on semimajor axis can be handled in a similar manner, resulting in

$$\frac{\delta a}{a} = 2\frac{\delta r}{r}$$

The associated period increase is

$$\frac{\delta \tau}{\tau} = 3\frac{\delta r}{r}$$

3.2 ORBIT TRANSFER AND ADJUST

Orbits may be adjusted or changed by single or multiple thrust impulses. Such maneuvers are typically required to eliminate launch vehicle errors or to bring the spacecraft into a more desirable orbit. The analysis considered here is limited to a single plane; that is, no changes in direction of orbital angular momentum are permitted. Only impulsive thrust application is assumed, thus, velocity changes occur instantaneously. Single impulse changes are considered first. Then transfers between concentric circular orbits are treated, followed by a discussion of other coplanar transfers.

3.2.1 Single Impulse Adjustments

A single thrust impulse applied in the orbital plane can change both the eccentricity and energy simultaneously. However, transfer of a spacecraft to a new orbit is not possible with a single impulse unless the new orbit intercepts the original one. Nevertheless, single impulse maneuvers have found many useful applications. Begin the analysis by noting that eccentricity and true anomaly are given in terms of position, speed, and flight path angle by equations (3.3) and (3.7), respectively. Dropping subscripts and solving for rv^2/μ yields

$$\frac{rv^2}{\mu} = 1 \pm \sqrt{1 - \left(\frac{1-e^2}{\cos^2 \beta}\right)} \tag{3.11}$$

$$\frac{rv^2}{\mu} = \frac{1}{\cos^2 \beta - (\sin \beta \cos \beta / \tan \theta)} \tag{3.12}$$

Equation (3.11) can be thought of as presenting β as a function of rv^2/μ for constant e, while (3.12) gives β for constant θ. Thus, if a single impulse is to be executed such that e or θ is held constant, one of these two equations is appropriate for relating the new velocity to the new value of β. Nevertheless, expressions (3.11) and (3.12) can be bypassed completely by using Figure 3.2

and performing a graphical solution. Usually, a desired change is specified in terms of the semimajor axis. This is related to rv^2/μ through the energy equation. Since $\mathscr{E} = -\mu/2a$, expression (3.2) leads to

$$\frac{a}{r} = \frac{1}{2 - \left(\dfrac{rv^2}{\mu}\right)} \tag{3.13}$$

Once the point of impulsive change is selected r can be calculated. If a is specified, then expression (3.13) will yield the required speed v by

$$v = \sqrt{\frac{2\mu}{r} - \frac{\mu}{a}} \tag{3.14}$$

In order to uniquely determine the flight angle associated with an impulsive change some other characteristic of the maneuver must be specified.

As an example, consider a case in which an earth satellite is in an elliptical orbit with $e = 0.5$, $a/R_\oplus = 3.0$. It is required to change the semimajor axis to $a/R_\oplus = 4.0$ while maintaining e constant. Thrust impulse is applied just as apogee is reached. The value of $r_a/R_\oplus$ is obtained from the equation for a conic section,

$$\frac{r_a}{R_\oplus} = \frac{a}{R_\oplus}(1+e) = 4.5$$

before impulse. In order to increase a without changing e, the value of r_a must increase. Thus, the impulse must change the direction of velocity as well as its magnitude. This situation is illustrated in Figure 3.3. Applying expression (3.14) yields the new speed at point 1 (the original apogee) as

$$v_1 = 3.49 \text{ km/s}$$

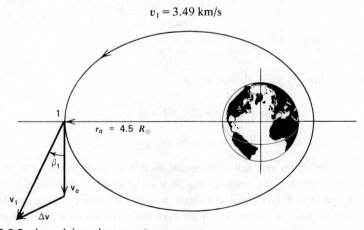

FIGURE 3.3 Impulsive change at apogee.

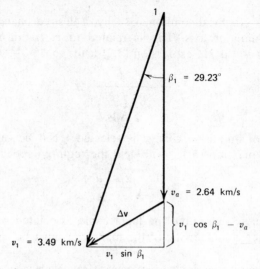

FIGURE 3.4 Velocity change calculation.

The original speed at apogee is

$$v_a = \sqrt{\frac{\mu}{a}\left(\frac{1-e}{1+e}\right)} = 2.64 \text{ km/s}$$

In order to calculate required Δv the value of β before and after firing must be determined. Here $\beta = 0$ initially and equation (3.11) should yield β_1,

$$\cos \beta_1 = \sqrt{\frac{1-e^2}{2\dfrac{r_a v_1^{\,2}}{\mu} - \left(\dfrac{r_a v_1^{\,2}}{\mu}\right)^2}}$$

which gives

$$\beta_1 = 29.23°$$

Referring to Figure 3.4, the value of Δv is

$$\Delta v = \sqrt{(v_1 \cos \beta_1 - v_a)^2 + v_1^{\,2} \sin^2 \beta_1}$$

or

$$\Delta v = 1.75 \text{ km/s}$$

Next, the value of θ_1 for this new orbit is obtained from expression (3.12) or (3.7) as $\theta_1 = 131.63°$. Notice that the perigee (line of apses) has been shifted by 48.37°. It is important to point out that there are other possible solutions to this problem. For example, v_1 could have been directed in the opposite direction as shown in Figure 3.5. This option requires that Δv be much greater and the required thrust impulse excessive.

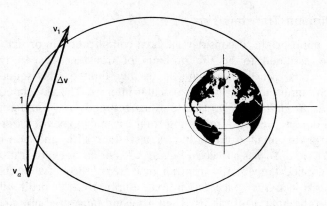

FIGURE 3.5 High impulse solution.

Therefore, the first solution represents a low impulse maneuver which satisfies the required objectives.

A graphical solution is also possible with the use of Figure 3.2. Begin by calculating the value of rv^2/μ. Before impulse this is 0.50 and after it is 0.88. Application of Δv has the affect of moving along the $e = 0.5$ curve from $\beta = 0$ and $rv^2/\mu = 0.5$ to a point where $\beta \cong 29°$ or $\beta \cong -29°$. Corresponding values of θ_1 are 132° and 228°, respectively. This points out still another possible solution. The associated Δv is, however, the same for either $\beta_1 = 29°$ or $\beta_1 = -29°$ because $\beta_0 = 0$. These two solutions are depicted in Figure 3.6.

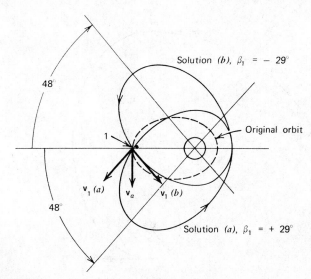

FIGURE 3.6 Two possible equal-impulse solutions.

3.2.2 Hohmann Transfers

It is now appropriate to consider the next complication in orbital maneuvering. This is assumed to be the problem of transfer between two circular concentric orbits. Since propellant is usually limited, the sequence which requires a minimum total Δv is of particular interest. The associated maneuver is known as a *Hohmann transfer* and corresponds to a minimum energy solution. It does represent a minimum total Δv maneuver for cases where the ratio of large to small orbit radius is less than 11.8 and consists of two impulses, Δv_1 and Δv_2, as shown in Figure 3.7. For an outward transfer the first impulse is applied tangentially in order to increase the initial circular velocity by Δv_1. This injects the spacecraft into an elliptic transfer orbit with apoapsis just equal to the final orbit radius. Then a second tangential impulse is applied to circularize the transfer orbit and complete the transition.

To calculate velocity change requirements, begin with equation (2.10) for conic sections and evaluate r at apoapsis and periapsis,

$$\frac{r_a}{r_p} = \frac{1+e}{1-e} \tag{3.15}$$

Solving for eccentricity of the transfer orbit yields

$$e = \frac{(r_a/r_p) - 1}{1 + (r_a/r_p)} \tag{3.16}$$

where r_a is the radius of the larger orbit and r_p, the radius of the smaller orbit.

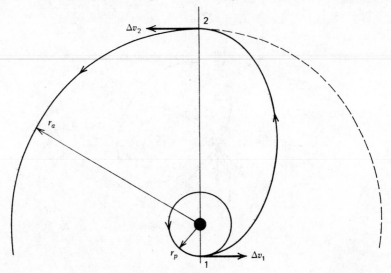

FIGURE 3.7 Hohmann transfer profile.

Note that Hohmann transfers are equally valid for transferring to an outer or inner orbit. No loss of generality is experienced if performance equations are based on an outward transfer. Since $\beta = 0$ at periapsis, equations (3.8) and (3.16) give

$$\frac{r_p v_p^2}{\mu} = 1 + \frac{(r_a/r_p) - 1}{1 + (r_a/r_p)} = \frac{2(r_a/r_p)}{1 + (r_a/r_p)} \tag{3.17}$$

This provides the required periapsis velocity to reach the outer orbit at apoapsis. Since the initial orbit is circular with radius r_p, the initial velocity is $\sqrt{\mu/r_p}$. Therefore, Δv_1 is simply

$$\Delta v_1 = v_p - \sqrt{\frac{\mu}{r_p}} = \sqrt{\frac{\mu}{r_p}} \sqrt{\frac{2(r_a/r_p)}{1 + (r_a/r_p)}} - \sqrt{\frac{\mu}{r_p}}$$

or

$$\Delta v_1 = \sqrt{\frac{\mu}{r_p}} \left[\sqrt{\frac{2(r_a/r_p)}{1 + (r_a/r_p)}} - 1 \right] \tag{3.18}$$

Apoapsis velocity is obtained by equating angular momentum values at the extremities of the transfer orbit, that is,

$$h = r_p v_p = r_a v_a$$

which gives

$$v_a = \frac{r_p}{r_a} \sqrt{\frac{\mu}{r_p}} \sqrt{\frac{2(r_a/r_p)}{1 + (r_a/r_p)}} \tag{3.19}$$

Since the final velocity should be $\sqrt{\mu/r_a}$, the value of Δv_2 is

$$\Delta v_2 = \sqrt{\frac{\mu}{r_a}} - v_a = \sqrt{\frac{\mu}{r_a}} \left[1 - \sqrt{\frac{2}{1 + (r_a/r_p)}} \right] \tag{3.20}$$

The total energy requirement for the Hohmann transfer is indicated by the sum of velocity increments,

$$\Delta v_T = \Delta v_1 + \Delta v_2$$

or

$$\Delta v_T = \sqrt{\frac{\mu}{r_p}} \left[\sqrt{\frac{2(r_a/r_p)}{1 + (r_a/r_p)}} \left(1 - \frac{r_p}{r_a} \right) + \sqrt{\frac{r_p}{r_a}} - 1 \right] \tag{3.21}$$

An inward transfer is handled by applying Δv_2 first to decrease initial circular velocity. Then Δv_1 is applied at periapsis to decrease velocity to its final value. Thus, energy is reduced twice.

As a point of interest, compare velocity increments for this type of transfer to that for escape from a low circular orbit. The energy of an escape parabola is zero, and the corresponding speed is $v_p = \sqrt{2\mu/r_p}$. Therefore, the required velocity increment for escape from circular orbit is

$$\Delta v_e = v_p - v_c = (\sqrt{2} - 1) \sqrt{\frac{\mu}{r_p}} \qquad (3.22)$$

If the same amount of energy is used to raise the satellite to another circular orbit, its final radius would be determined by equating expression (3.21) to (3.22) and solving for r_a/r_p. The result is that energy required to raise a circular orbit by a factor of

$$\frac{r_a}{r_p} = 3.4$$

is equal to that required to escape altogether.

To further illustrate the cost of raising a circular orbit compare the total energy required to transfer from an earth orbit with an altitude of 200 km to a *synchronous* orbit with that required to inject into an ellipse which will just reach the moon's orbit. The radius of a synchronous orbit is determined from its period, which is defined as one Sidereal day. Thus

$$\tau = 2\pi\sqrt{\frac{r_{syn}^3}{\mu}} = 23 \text{ h } 56 \text{ min } 4 \text{ s}$$

is used to obtain

$$r_{syn} = 42{,}164 \text{ km}$$

Since $r_p = R_\oplus + 200$ km, equations (3.18) and (3.20) give

$$\Delta v_1 = 2.45 \text{ km/s}$$
$$\Delta v_2 = 1.48 \text{ km/s}$$

which total

$$\Delta v_T = 3.93 \text{ km/s (synchronous transfer)}$$

For the lunar injection, $r_a = r_{\mathbb{C}} = 384{,}400$ km and equation (3.18) gives

$$\Delta v_1 = 3.13 \text{ km/s (translunar injection)}$$

Thus, on the basis of energy, a launch vehicle which is capable of placing a payload into synchronous orbit can alternatively send a more massive package to the moon. Furthermore, escape velocity increment from a 200 km circular orbit is only $\Delta v_e = 3.22$ km/s. Thus, the same type of launch vehicle can send a payload into a heliocentric orbit.

Optimality of the Hohmann Transfer is not an obvious property. Therefore, a proof of the minimum energy aspect of this two-impulse transfer is offered. Consider first, the minimum Δv required to reach a specified apoapsis distance r_a for an outward transfer case. This argument may be carried out in steps because the transfer orbit must reach the final orbit. It is obvious that this should occur at the apoapsis of transfer to minimize the total Δv, because any radial component of v as the transfer orbit passes the final orbit would have to be taken out. Assume a non-tangential impulse Δv_1 as depicted in Figure 3.8. The radial component of Δv_1 is v_{r_1} and transfer orbit energy is

$$\mathscr{E} = \frac{v_1^2}{2} - \frac{\mu}{r_p} = \frac{v_a^2}{2} - \frac{\mu}{r_a} \tag{3.23}$$

where

$$v_1^2 = v_{r_1}^2 + v_{\theta_1}^2, \ v_{\theta_1} = \frac{h}{r_p}$$

Thus, $\mathscr{E}$ becomes

$$\mathscr{E} = \frac{1}{2}\left(v_{r_1}^2 + \frac{h^2}{r_p^2}\right) - \frac{\mu}{r_p}$$

Note the inequality

$$\frac{1}{2}\left(v_{r_1}^2 + \frac{h^2}{r_p^2}\right) - \frac{\mu}{r_p} \geq \frac{1}{2}\left(\frac{h^2}{r_p^2}\right) - \frac{\mu}{r_p}$$

Referring to energy evaluated at apogee, this becomes

$$\frac{h^2}{2r_a^2} - \frac{\mu}{r_a} \geq \frac{1}{2}\left(\frac{h^2}{r_p^2}\right) - \frac{\mu}{r_p}$$

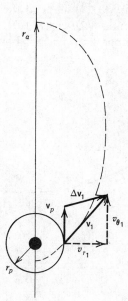

FIGURE 3.8 Nontangential impulse.

where equality holds when $v_{r_1} = 0$. Rearranging yields

$$\left(\frac{1}{r_p^2} - \frac{1}{r_a^2}\right)h^2 \le \frac{2\mu}{r_p} - \frac{2\mu}{r_a}$$

or

$$h \le \sqrt{\frac{2\mu r_p r_a}{r_a + r_p}} \qquad (3.24)$$

To relate Δv_1 to h use the geometry of Figure 3.8 and expression (3.23),

$$(\Delta v_1)^2 = \left(\frac{h}{r_p} - \sqrt{\frac{\mu}{r_p}}\right)^2 + v_{r_1}^2$$

$$= 2\left(\mathscr{E} + \frac{\mu}{r_p}\right) - 2h\sqrt{\frac{\mu}{r_p^3}} + \frac{\mu}{r_p}$$

This can be written in expanded form as

$$(\Delta v_1)^2 = \frac{1}{r_a^2}\left[h - r_a^2\sqrt{\frac{\mu}{r_p^3}}\right]^2 - r_a^2\left(\frac{\mu}{r_p^3}\right) + 2\mu\left(\frac{1}{r_p} - \frac{1}{r_a}\right) + \frac{\mu}{r_p} \qquad (3.25)$$

Furthermore, notice that

$$r_a^2\sqrt{\frac{\mu}{r_p^3}} = \sqrt{\frac{\mu r_a^4}{r_p^3}} > \sqrt{\mu r_a} > \sqrt{\frac{2\mu r_a r_p}{r_p + r_a}}$$

Comparing this with inequality (3.24) leads to the conclusion that h is always less than $r_a^2\sqrt{\mu/r_p^3}$ and is at most $\sqrt{2\mu r_p r_a/(r_a + r_p)}$. If $(\Delta v_1)^2$ versus h is plotted the optimum case becomes apparent. This is shown in Figure 3.9. It is clear

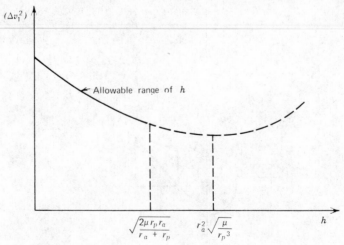

FIGURE 3.9 Range of h for transfer ellipse.

that minimum Δv_1 corresponds to

$$h = \sqrt{\frac{2\mu r_p r_a}{r_a + r_p}}$$

which is the value when $v_{r_1} = 0$. Thus, tangential impulse is indeed the optimum case for the first application of Δv. Similarly, it may be shown that the minimum impulse transfer from an ellipse into a circular orbit is a tangential impulse Δv_2 at r_a. This completes the proof that a Hohmann transfer requires minimum Δv for the 2-impulse sequence.

If, however, $r_a > 11.8\ r_p$, it is slightly more efficient to use a 3-impulse transfer as follows: (1) transfer by tangential impulse to a height well beyond r_a, (2) apply a small second forward tangential impulse at this point to bring periapsis up to r_a, (3) apply a reverse tangential impulse at r_a to circularize. The optimality of this sequence will not be rigorously proven, but a physical argument may indicate its logic. When r_a/r_p is excessive Δv_1 and Δv for escape become very close in value. Once at a point well beyond r_a the corresponding Δv to raise periapsis is very small. Thus, the final reverse Δv plus the extra energy applied at the first impulse may be less than that saved in raising the periapsis to r_a in the second impulse.

3.2.3 Other Coplanar Transfers

The next more complicated coplanar transfer is one in which the initial and final orbits are *coaxial* ellipses as shown in Figure 3.10. The minimum energy transfer between orbits 1 and 2 consists of Δv_{p_1} at the periapsis of the lower orbit and Δv_{a_2} at the apoapsis of the higher orbit. The reverse sequence should be followed when transferring from 2 to 1. Impulse requirements can be calculated in a straightforward manner. The periapsis velocity of orbit 1 is

$$v_{p_1} = \sqrt{\frac{\mu}{r_{p_1}}(1 + e_1)}$$

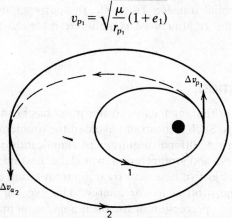

FIGURE 3.10 Transfer between coplanar coaxial elliptic orbits.

The transfer ellipse must satisfy equation (3.17)

$$\frac{r_{p_1} v_{p_t}^2}{\mu} = \frac{2(r_{a_2}/r_{p_1})}{1+(r_{a_2}/r_{p_1})}$$

which gives

$$v_{p_t} = \sqrt{\frac{\mu}{r_{p_1}} \left[\frac{2(r_{a_2}/r_{p_1})}{1+(r_{a_2}/r_{p_1})} \right]}$$

Therefore, the first impulse is

$$\Delta v_{p_1} = v_{p_t} - v_{p_1} \tag{3.26}$$

At apoapsis of the transfer ellipse velocity is obtained from conservation of angular momentum,

$$v_{a_t} = \frac{r_{p_1}}{r_{a_2}} v_{p_t} = \sqrt{\frac{\mu}{r_{a_2}} \left[\frac{2}{1+(r_{a_2}/r_{p_1})} \right]}$$

The required velocity of orbit 2 at this point is

$$v_{a_2} = \sqrt{\frac{\mu}{r_{a_2}} (1-e_2)}$$

Thus, the second impulse is

$$\Delta v_{a_2} = v_{a_2} - v_{a_t} \tag{3.27}$$

Summing forms (3.26) and (3.27) gives the total impulse of transfer.

Of course, this progression of increasingly more complicated transfer sequences could fill a whole chapter. The current literature contains many papers on the subject of orbital transfer. However, the purposes of this text are well served by limiting the treatment of coplanar transfers to the two presented above.

3.3 PLANE ROTATION

The treatment of orbital maneuvers to this point has been carefully confined to coplanar situations. Such a constraint avoided the problem of plane rotations during transfers. This additional requirement significantly increases the complexity of maneuvers, and optimization would be beyond the scope of this book. The primary objective here is to treat a pure rotation of the orbital plane without changing the orbit shape or energy. This type of maneuver can be viewed in two ways; as precession of the orbit angular momentum vector or as a direct rotation of the velocity vector. If the objective of this rotation is to

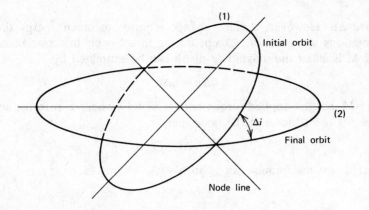

FIGURE 3.11 Typical plane rotation problem.

superimpose the initial orbit onto a specified plane (e.g., during a rendezvous mission between two spacecraft), then the point at which impulse is applied is critical to success. Consider the situation depicted in Figure 3.11. Here two circular orbits have a difference in inclination of Δi. An interceptor satellite in orbit (1) must maneuver into the target orbit (2). This can only be accomplished by rotating orbit (1) about the node line through an angle Δi. If impulse were applied at any other point, relative inclination may be reduced but cannot be brought to zero. A single impulsive rotation about the node line is also a minimum energy maneuver.

A plane rotation can be thought of as a precession of the angular momentum vector through angle Δi as shown in Figure 3.12. The required impulse is

FIGURE 3.12 Precession of the angular momentum vector.

related to $\Delta \mathbf{h}$. However, a plane change requires so much energy that only small rotations are practical, except during initial orbit injection maneuvers. Thus, if Δi is small the magnitude of $\Delta \mathbf{h}$ can be estimated by

$$\Delta h \cong h \Delta i$$

Since $\dot{\mathbf{h}} = \mathbf{M}$ and the applied torque here is $M = Fr$ where F is thrust acceleration and r is orbit radius at the node line,

$$\Delta h = h \, \Delta i = Fr \, \Delta t$$

But $F \, \Delta t$ is just the impulse Δv. Also $h = r v_\theta$, leading to

$$\Delta v = v_\theta \, \Delta i \qquad (3.28)$$

where v_θ is the transverse component of velocity evaluated at the node line. For a circular orbit $v_\theta = v$ and $\Delta v = v \, \Delta i$. For an elliptical orbit Δv is less where the orbit velocity is less. For example, if the node line is coincident with the line of apsides, thrust should be applied at the apoapsis. The other approach is to directly calculate the required velocity increment using a velocity vector diagram. To illustrate this method consider the situation in Figure 3.13 where Δi is small and circular orbits are assumed. The required velocity increment is quickly seen to be

$$\Delta v = v \, \Delta i$$

where v is the magnitude of either $\mathbf{v}_1$ or $\mathbf{v}_2$. Application of the technique to elliptical orbits represents a straightforward extension of this sequence.

As an example of Δv required for plane rotations, consider the transfer ellipse between a 200 km circular earth orbit and synchronous apogee. Assume the node line coincides with the line of apsides. A one-degree rotation is required. If this is executed at perigee where $v_p = 10.24 \, \text{km/s}$, then $\Delta v = 0.18 \, \text{km/s}$. However, if the impulse is applied at apogee where $v_a = 1.59 \, \text{km/s}$, then $\Delta v = 0.028 \, \text{km/s}$. This would represent a factor of 6.4 in propellant savings. In the actual launch of a synchronous communications satellite from Cape Canaveral a large plane change of about $28.5°$ is required. Since the node line does coincide with the major orbital axis this rotation may be done simultaneously with circularization at apogee. The total Δv required can be estimated

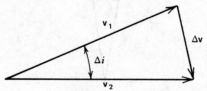

FIGURE 3.13 Rotation of the velocity vector at the node line.

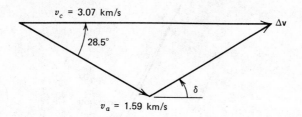

FIGURE 3.14 Synchronous orbit injection.

by using the vector diagram of Figure 3.14. Use of the cosing law leads to

$$\Delta v = 1.84 \text{ km/s}, \; \delta = 24.4°$$

To demonstrate the advantage of simultaneous rotation and circularization, assume these maneuvers are executed separately. First the inclination change would be performed. Application of the cosine law now yields a plane rotation velocity impulse of 0.78 km/s. Section 3.2 gives the circularization impulse as $\Delta v_2 = 1.48$ km/s. The sum of these is 2.26 km/s, or a 22.8% increase over the single impulse maneuver.

3.4 INTERPLANETARY TRANSFER AND HYPERBOLIC PASSAGE

Many missions of interest require a spacecraft to escape the earth, enter a partial heliocentric orbit, and encounter another planet. Such a sequence of orbital events is referred to as an *interplanetary transfer*. Typically, such flights will flyby, collide with, or enter an orbit about the target planet. An approximate method for dealing with the transfer, which is very useful in preliminary mission design, is presented here. This technique may be referred to as the method of *patched conics*. The process is one in which the gravity of each primary attracting body is considered in sequence while ignoring effects of other bodies. Individual conics are then matched to give the overall flight characteristics.

3.4.1 Hyperbolic Passage

A basic understanding of the *hyperbolic passage* is required before considering the whole interplanetary sequence. Geometry typical of hyperbolic trajectories is illustrated in Figure 3.15. Consider a small vehicle approaching a large body B with relative velocity $\mathbf{v}_\infty^-$ at great distance $(r \cong \infty)$. The motion of the large body and gravitational effects of other bodies are ignored. Several

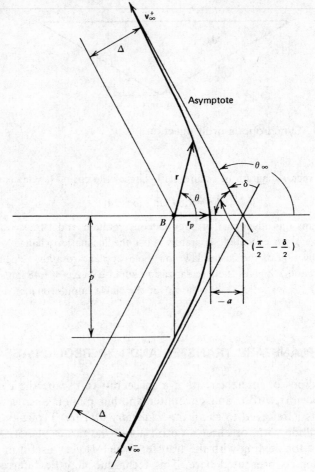

FIGURE 3.15 Geometry of hyperbolic passage.

parameters should now be defined:

    Δ = distance between B and the asymptotes
    r_p = radial distance of closest approach
    δ = deflection angle of $\mathbf{v}_\infty$
    θ_∞ = true anomaly of the asymptotes

For a hyperbola $e > 1$ and semimajor axis a is taken as negative to maintain the energy equation, $\mathscr{E} = -\mu/2a$. Thus, energy is positive and constant. It is useful to note that

$$\mathscr{E} = \frac{v^2}{2} - \frac{\mu}{r} = \frac{v_\infty^2}{2} = -\frac{\mu}{2a}$$

Therefore, the magnitude of $\mathbf{v}_\infty$ is the same for inbound and outbound legs,

$$|\mathbf{v}_\infty^-| = |\mathbf{v}_\infty^+| \tag{3.29}$$

with respect to B. Parameters θ_∞, δ, and e may be determined by using the conic section equation, written here as

$$r = \frac{a(1-e^2)}{1+e\cos\theta} \tag{3.30}$$

Now θ_∞ occurs when $r \to \infty$, which leads to

$$\cos\theta_\infty = \lim_{r\to\infty}\left\{\frac{1}{e}\left[\frac{a(1-e^2)}{r}-1\right]\right\} = -\frac{1}{e}$$

or

$$\theta_\infty = \cos^{-1}\left(-\frac{1}{e}\right) \tag{3.31}$$

From Figure 3.15

$$\frac{\pi}{2} - \frac{\delta}{2} = \pi - \theta_\infty$$

Thus

$$\theta_\infty = \frac{\pi}{2} + \frac{\delta}{2} \tag{3.32}$$

Equating expressions (3.31) and (3.32) yield

$$\frac{1}{e} = \sin\frac{\delta}{2} \tag{3.33}$$

From the energy relation the semimajor axis of passage is

$$a = -\frac{\mu}{v_\infty^2} \tag{3.34}$$

Angular momentum is quickly obtained as

$$v_\infty\Delta = \sqrt{\frac{\mu^2}{v_\infty^2}(e^2-1)}$$

Rearranging this gives another form of e,

$$e^2 = 1 + \frac{v_\infty^4\Delta^2}{\mu^2} \tag{3.35}$$

Still another form is obtained by solving equation (3.30) at periapsis and applying form (3.34),

$$e = 1 + \frac{r_p v_\infty^2}{\mu} \tag{3.36}$$

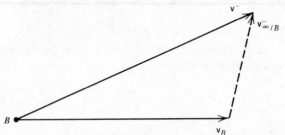

FIGURE 3.16 Relative velocity nomenclature.

If v_∞ and Δ are given or determinable from other information, equations (3.35) and (3.34) can be used to calculate e and a, respectively. Then r_p is obtainable from relation (3.36). The deflection angle δ is calculated from (3.33) and finally, θ_∞ is determined from (3.32) or (3.31). For scientific missions it is likely that the distance of closest approach would be specified. Furthermore, v_∞ is a function of interplanetary flight path and launch vehicle capability. This situation would require use of form (3.36) for e. Other parameters may be obtained by proper application of the relations presented.

Hyperbolic passage of a planetary body as in Figure 3.15 must be referred to some inertial frame in which the overall mission is taking place. In fact, the large body B must have some velocity itself, $\mathbf{v}_B$. Thus, the hyperbolic approach velocity, $\mathbf{v}_\infty^-$ is taken with respect to B and might well be written as $\mathbf{v}_{\infty/B}^-$. Similarly, the hyperbolic departure velocity may be written as $\mathbf{v}_{\infty/B}^+$. The net effect of body B during spacecraft passage through its gravity field is a deflection δ of $\mathbf{v}_{\infty/B}$. The relationship between $\mathbf{v}_{\infty/B}^-$, $\mathbf{v}_B$, and the inertial velocity of the vehicle before passing B, $\mathbf{v}^-$ is illustrated in Figure 3.16. Thus, the relative approach velocity is

$$\mathbf{v}_{\infty/B}^- = \mathbf{v}^- - \mathbf{v}_B \qquad (3.37)$$

Although passage of B does not affect the magnitude of relative velocity, there is a significant change in absolute velocity of a spacecraft. The complete vector diagram for the hyperbolic passage of Figure 3.15 is shown in Figure 3.17. This is a somewhat specialized case, because B is traveling along the direction of the

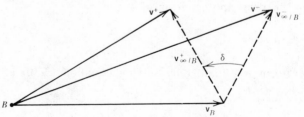

FIGURE 3.17 Inertial velocity change due to passage.

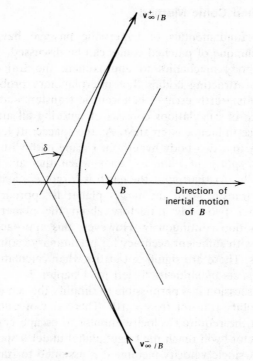

FIGURE 3.18 Passage behind body B.

hyperbola axis. Nevertheless, several important points can be observed from this situation. The absolute velocity of the spacecraft is reduced due to the direction of travel of B. Thus, passage *in front* of B resulted in a decrease of inertial orbital energy. If passage was behind B, as shown in Figure 3.18, then orbital energy would have increased. This is illustrated by the associated vector diagram in Figure 3.19. It is evident that a spacecraft can gain or lose energy during planetary passage. This is one reason for outer planet missions to include flybys of intermediate planets, and for Mercury probes to pass Venus first.

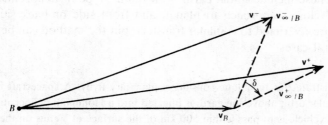

FIGURE 3.19 Passage with energy addition.

3.4.2 Patched Conic Method

Now that fundamentals of hyperbolic passage have been exposed, the complete technique of patched conics can be discussed. This approach permits use of two-body mechanics to approximate the trajectory of a spacecraft between two attracting bodies. For interplanetary probes, the trajectory has three segments: earth escape, heliocentric transfer, and planetary encounter. The sequence of calculations consists of ignoring all attracting bodies except the one whose influence is greatest. As the spacecraft leaves earth orbit it will be essentially in a two-body hyperbolic escape path with respect to earth. This gravitational *sphere of influence* ends when the sun becomes the primary attracting body. At that point the earth is forgotten, and the spacecraft is in a heliocentric transfer until the target planet is approached. The probe then enters another two-body trajectory about the planet. Obviously, the sun influences motion continuously. However, this approach permits quick hand calculations with sufficient accuracy for preliminary estimates of thrust impulse requirements. There are significant errors when calculating transfer times. The concept of *sphere of influence* is left for Chapter 7.

For this discussion it is permissible to simplify the sequence further. Consider an interplanetary transfer from earth. The order of calculations is as follows: (1) Establish the required velocity impulse to escape earth orbit such that the vehicle trajectory will reach the target planet under a specified time or impulse constraint. A single velocity increment is assumed to start the transfer mission from a low parking orbit. This is sufficient to carry out the entire flight. If only a flyby of the target planet is required, then no impulses will be needed during its passage. The trajectory from earth orbit to escape is assumed purely geocentric. (2) The earth is then *turned off* and the sun takes over. This is the heliocentric phase of flight. Initial velocity and direction conditions for this phase are transferred from the earth escape orbit, and the starting position is at earth. (3) When the target planet distance is reached, the sun is *turned off*, and the velocity vector at that point is used to generate initial conditions for the hyperbolic passage. Gravity of the planet takes over with these conditions at a large distance away. Several parameters of the overall flight profile must be specified. These include initial earth orbit radius, type of heliocentric transfer, distance of closest approach to planet, and front side or back side passage. Examples are restricted to coplanar transfers, but the method can be applied to more general cases.

To demonstrate this technique consider a planetary flyby. A spacecraft initially in a 200 km circular orbit above the earth is injected into a Hohmann heliocentric ellipse to Venus. This vehicle is to pass within 500 km of the surface of Venus on the sunlit side. There are several important parameters to be determined. Firstly, the required Δv to reach Venus is of primary concern. After passage the spacecraft will enter a new heliocentric orbit. Thus, its new perihelion, eccentricity, and period should be determined in case further use of its instruments is possible. Assume that the motion of Venus and the vehicle lie in the ecliptic plane. The initial Δv is determined by considering the heliocentric transfer requirements. Thus, absolute velocity of the spacecraft as it escapes earth $\mathbf{v}^+$ must be the apohelion velocity of the Hohmann transfer since Venus is in a lower orbit. This situation is depicted in Figure 3.20. The value of v^+ is easily determined from heliocentric transfer energy

$$\mathscr{E}_H = \frac{(v^+)^2}{2} - \frac{\mu_\odot}{r_\oplus} = -\frac{\mu_\odot}{2a_H}$$

where

$$2a_H = r_\oplus + r_\venus$$

This gives $a_H = 1.290 \times 10^8$ km. Since

$$v^+ = \sqrt{\mu_\odot \left(\frac{2}{r_\oplus} - \frac{1}{a_H} \right)}$$

for this case

$$v^+ = 27.30 \text{ km/s}$$

In order for an escaping spacecraft to reach v^+ with minimum Δv, the orbital velocity of earth must be used advantageously. This velocity is

$$v_\oplus = \sqrt{\frac{\mu_\odot}{r_\oplus}} = 29.78 \text{ km/s}$$

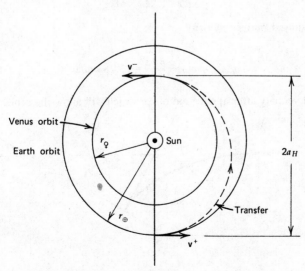

FIGURE 3.20 Heliocentric transfer to Venus.

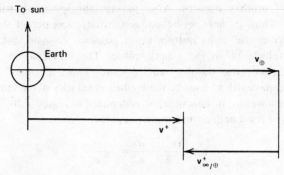

FIGURE 3.21 Velocity balance for earth escape.

Since $v_\oplus > v^+$ earth escape should be in the direction opposite to $\mathbf{v}_\oplus$, as shown vectorally in Figure 3.21. This illustrates how $\mathbf{v}^+$ is established by subtracting part of the earth's velocity. Thus, the hyperbolic excess velocity of earth escape, $v^+_{\infty/\oplus}$ is calculated simply from

$$|\mathbf{v}^+_{\infty/\oplus}| = |\mathbf{v}_\oplus - \mathbf{v}^+| = 2.48 \text{ km/s}$$

Next, the hyperbolic escape path from low circular orbit is considered. Only half of a normal passage need be studied, as depicted in Figure 3.22. Energy associated with this escape is

$$\mathscr{E} = \frac{v^2_{\infty/\oplus}}{2} = \frac{v^2_{p/\oplus}}{2} - \frac{\mu_\oplus}{r_{p/\oplus}}$$

which yields required perigee velocity

$$v_{p/\oplus} = \sqrt{v^2_{\infty/\oplus} + \frac{2\mu_\oplus}{r_{p/\oplus}}} = 11.28 \text{ km/s}$$

This is the total velocity after application of Δv which will allow the probe to just reach

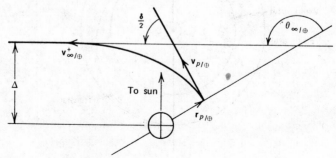

FIGURE 3.22 Earth escape hyperbola.

the orbit of Venus. Since the spacecraft was initially in a circular orbit,

$$\Delta v = v_{p/\oplus} - \sqrt{\frac{\mu_\oplus}{r_{p/\oplus}}} = 3.50 \text{ km/s}$$

In actual missions slight impulse errors and uncertainties usually require minor Δv adjustments during heliocentric transfer. These are known as *mid-course* corrections. It is important for ground controllers to determine the position of impulse application, $\theta_{\infty/\oplus}$. This is obtained via expression (3.36),

$$e = 1 + \frac{r_{p/\oplus} v_{\infty/\oplus}^2}{\mu_\oplus} = 1.10$$

and expression (3.31)

$$\theta_{\infty/\oplus} = \cos^{-1}\left(-\frac{1}{e}\right) = 155.2°$$

The deflection of velocity between $\mathbf{v}_{p/\oplus}$ and $\mathbf{v}_{\infty/\oplus}^+$ is $\delta/2$, and from equation (3.32) or (3.33),

$$\frac{\delta}{2} = 65.2°$$

This escape path is displaced toward the sun by an amount Δ, which is determined by using form (3.35),

$$\Delta = \frac{\mu_\oplus}{v_{\infty/\oplus}^2} \sqrt{e^2 - 1} = 2.97 \times 10^4 \text{ km}$$

The actual apohelion of transfer would then be slightly lower than $r_\oplus$, but this difference may be ignored due to the approximate nature of this technique. After earth escape heliocentric transfer commences. The velocity at Venus' orbit, ignoring the presence of Venus, is $\mathbf{v}^-$, whose magnitude is obtained from

$$h_H = r_\oplus v^+ = r_\varphi v^-$$

as

$$v^- = \frac{r_\oplus}{r_\varphi} v^+ = 37.71 \text{ km/s}$$

This is taken as the absolute velocity of approach to Venus, whose speed is

$$v_\varphi = \sqrt{\frac{\mu_\odot}{r_\varphi}} = 35.00 \text{ km/s}$$

Figure 3.23 illustrates the velocity vector balance as Venus is approached. It is apparent that $v_{\infty/\varphi} = 2.71$ km/s. Since the distance of closest approach is specified as 500 km, eccentricity of the passage is obtained directly from equation (3.36),

$$e = 1 + \frac{r_{p/\varphi} v_{\infty/\varphi}^2}{\mu_\varphi} = 1.14$$

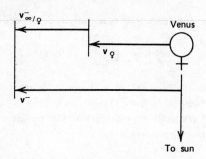

FIGURE 3.23 Velocity balance at Venus approach.

and equation (3.33) gives the deflection of $\mathbf{v}_{\infty/♀}$ as

$$\delta = 2\sin^{-1}\left(\frac{1}{e}\right) = 122.2°$$

Figure 3.24 illustrates the passage profile. Note that a sunlit side approach was specified and the effect is to deflect $\mathbf{v}_{\infty/♀}$ away from the sun. Now that $\mathbf{v}_{\infty/♀}^+$ has been determined, the resulting heliocentric orbit may be considered. To obtain the absolute velocity after encountering Venus, consider the complete velocity vector diagram of the passage in Figure 3.25. Heliocentric velocity components after encounter are given by

$$v_\theta^+ = v_♀ - v_{\infty/♀}^+ \cos(\pi - \delta) = 33.56 \text{ km/s}$$

$$v_r^+ = v_{\infty/♀}^+ \sin(\pi - \delta) = 2.29 \text{ km/s}$$

This gives the new flight path angle, β as

$$\beta = \tan^{-1}\left(\frac{v_r^+}{v_\theta^+}\right) = 3.9°$$

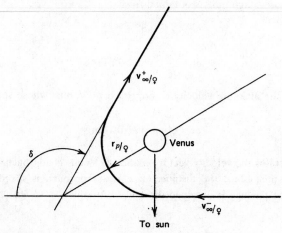

FIGURE 3.24 Hyperbolic passage at Venus.

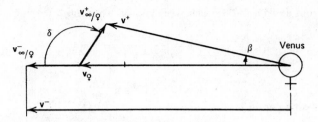

FIGURE 3.25 Complete velocity vector diagram of Venus passage.

and the new spacecraft speed as

$$v^+ = \sqrt{v_\theta^{+2} + v_r^{+2}} = 33.64 \text{ km/s}$$

The resulting heliocentric eccentricity is obtained from expression (3.3) or Figure 3.2 as

$$e = 0.10$$

True anomaly in the new heliocentric orbit can be calculated using equation (3.4) or graphically with e in Figure 3.2 as

$$\theta = 142.0°$$

and perihelion can be calculated from $r_p = h^2/[\mu(1+e)]$, where $h = r_\venus v^+ \cos\beta$,

$$r_{p/\odot} = 9.05 \times 10^7 \text{ km}$$

These results are depicted in Figure 3.26. Position of the spacecraft is 142° from perihelion immediately after Venus passage. Finally, the period of this new orbit can be

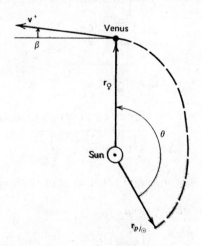

FIGURE 3.26 Perihelion position of new heliocentric orbit.

determined from equation (2.35) once a is known. Thus, use

$$a = \frac{h^2}{\mu_\odot(1-e^2)}$$

which gives

$$a = 1.01 \times 10^8 \text{ km}$$

The corresponding period is

$$\tau = 201.3 \text{ days}$$

Thus, the final orbit is nearly circular and has a period very close to that of Venus, which is 225 days.

3.4.3 Planetary Capture

Many scientific missions terminate with an orbit about a target planet B. This can be achieved by reducing energy during hyperbolic passage. One simple method of establishing a circular orbit with a single decelerating impulse is to adjust the approach parameter Δ such that the distance of closest approach is just equal to the final orbit radius. As this point is reached a Δv is applied to slow the vehicle. The value of Δv is

$$\Delta v = v_{p/B} - \sqrt{\frac{\mu_B}{r_{p/B}}}$$

If propellant is extremely limited, this maneuver may be optimized when the final orbit radius is not critical. The optimum value of $r_{p/B}$ to minimize required Δv is determined as follows. Write the energy expression as

$$\mathscr{E} = \frac{v_{\infty/B}^2}{2} = \frac{v_{p/B}^2}{2} - \frac{\mu_B}{r_{p/B}}$$

and extract

$$v_{p/B} = \sqrt{v_{\infty/B}^2 + \frac{2\mu_B}{r_{p/B}}}$$

The value of Δv may now be expressed in terms of $v_{\infty/B}$ and $r_{p/B}$ as

$$\Delta v = \sqrt{v_{\infty/B}^2 + \frac{2\mu_B}{r_{p/B}}} - \sqrt{\frac{\mu_B}{r_{p/B}}} \tag{3.38}$$

To minimize Δv take $\partial \Delta v / \partial r_{p/B}$ and set it to zero, noting that $v_{\infty/B}$ is fixed

$$\frac{\partial \Delta v}{\partial r_{p/B}} = \frac{-\mu_B/r_{p/B}^2}{\sqrt{v_{\infty/B}^2 + \frac{2\mu_B}{r_{p/B}}}} + \frac{\sqrt{\mu_B}/2}{r_{p/B}^{3/2}} = 0$$

This indicates that minimum Δv occurs when

$$r_{p/B} = \frac{2\mu_B}{v_{\infty/B}^2} \qquad (3.39)$$

and has the value

$$\Delta v_{\min} = \frac{v_{\infty/B}}{\sqrt{2}} \qquad (3.40)$$

3.5 LUNAR TRANSFER

As a first approximation in planning lunar transfer missions the patched conic method may be employed. Since the moon is relatively close to earth and the two bodies actually travel about each other, the results will be less accurate than those obtained for interplanetary transfer. In fact, this approach is not satisfactory for calculation of earth return trajectories due to the manner in which lunar gravity is handled. Similarly, perilune altitude cannot be accurately predicted. Nevertheless, this method is good for outbound Δv evaluations and permits insight into the problems of lunar transfer missions. The patched conic method used here is further simplified in that the sphere of influence of the moon is ignored until conditions at the lunar distance are established. This philosophy was also used in the interplanetary transfer analysis of the previous section.

Begin with the energy equation to determine the velocity increment required to reach the moon's orbit from circular parking orbit. Energy is expressed as

$$\mathscr{E} = \frac{v_p^2}{2} - \frac{\mu_\oplus}{r_p} = -\frac{\mu_\oplus}{2a}$$

Note that the apogee must be at least equal to the moon's radial distance $r_{\mathbb{C}}$ and its mass is ignored for this calculation. If this is a Hohmann transfer, then equation (3.18) can be used directly. In general, the required Δv is gived by

$$\Delta v = v_p - v_c = \sqrt{\mu_\oplus}\left[\sqrt{\frac{2}{r_p} - \frac{1}{a}} - \sqrt{\frac{1}{r_p}}\right]$$

This impulse was determined for a Hohmann transfer in Section 3.2.2 as 3.13 km/s, and the time of transfer is just half the period of the transfer ellipse,

$$t_{\text{transfer}} = \pi\sqrt{\frac{a^3}{\mu_\oplus}}$$

Here $a = (r_p + r_{\mathbb{C}})/2$, and since $r_p \cong 6600$ km, $a \cong 1.96 \times 10^5$ km. Then $t_{\text{transfer}} \cong$ 5.0 days. The associated angular momentum of the transfer is $h = r_p v_p = 7.2 \times 10^4$ km²/s. These numbers are based on the assumption that parking orbits have altitudes of only a few hundred kilometers. The Hohmann transfer requires the least impulse, but longest transfer time. Shorter times to the moon are possible by using transfer orbits with apogees beyond $r_{\mathfrak{c}}$. This, of course, requires extra Δv. However, the additional impulse applied at perigee above and beyond that for a Hohmann transfer is only a small amount. This implies that h is not increased significantly above $7.2 \times 10^4 \, \text{km}^2/\text{s}$ for higher energy transfers. Another way to look at this is to consider the velocity at $r_{\mathfrak{c}}$. Higher energy transfers lead to $v_r > 0$ at $r_{\mathfrak{c}}$, but v_θ at this point is just slightly greater than the v_a for the Hohmann transfer. Since $h = r_{\mathfrak{c}} v_\theta$, the assumption of constant angular momentum for all lunar transfers is valid within the approximate method used here. Thus, a constant value for v_θ of 0.19 km/s is assumed at the moon's distance.

Once v_r at $r_{\mathfrak{c}}$ is determined, then transfer ellipse parameters may be determined from equations of Section 3.1:

$$e = \sqrt{\left(\frac{r_{\mathfrak{c}} v_\theta^2}{\mu_\oplus} - 1\right)^2 + \left(\frac{r_{\mathfrak{c}} v_\theta v_r}{\mu_\oplus}\right)^2}$$

$$a = \frac{r_{\mathfrak{c}}^2 v_\theta^2}{\mu_\oplus(1 - e^2)}$$

The true anomaly upon reaching $r_{\mathfrak{c}}$ is

$$\tan \theta = \frac{r_{\mathfrak{c}} v_\theta v_r}{(r_{\mathfrak{c}} v_\theta^2 - \mu_\oplus)}$$

and the time to transfer, ignoring lunar gravity effects, is

$$t_{\text{transfer}} = \frac{\psi - e \sin \psi}{n}$$

where ψ is obtained from equation (2.23) and n is the mean motion of form (2.31). It is interesting to note that the eccentricity for the Hohmann transfer is $e \cong 0.97$ and any higher energy transfer would have a value

$$e \cong \sqrt{0.933 + 0.033 \, v_r^2}$$

Michielsen devised a graphical display for all lunar transfer information, including passage effects, on a single plot. This is given in Figure 3.27 with an example to illustrate its utility. Several items should be mentioned with respect to nomenclature. The v_r axis is calibrated in km/s, but this is directly related to transfer time. Thus, each value of v_r associated with the transfer orbit corresponds to a unique value of transfer time. The two vertical lines marking the constant values of v_θ at $r_{\mathfrak{c}}$ are calibrated in days to reach $r_{\mathfrak{c}}$. Transfers which have $v_\theta = +0.19$ km/s are *direct* and those with $v_\theta = -0.19$ km/s are *retrograde*.

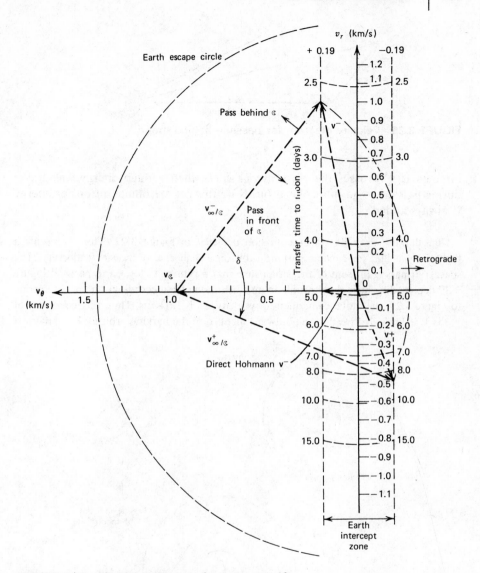

FIGURE 3.27 Michielsen's chart for lunar transfer.

Hyperbolic passage is handled in an analogous manner to that for planetary passage. As the spacecraft reaches $r_{\mathbb{C}}$ the earth is *turned off* and the moon *turned on*. A typical velocity vector diagram for lunar passage is shown in Figure 3.28. If the passage is in back of the moon δ is taken counterclockwise and, for the case shown, geocentric energy is increased. If v^+ is greater than $\sqrt{2}$ $v_{\mathbb{C}}$ earth escape occurs. The whole passage could be plotted on Figure 3.27

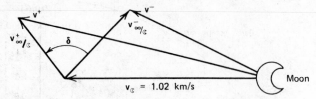

FIGURE 3.28 Velocity diagram for passage behind moon.

directly. If $\mathbf{v}^+$ crosses the *earth escape circle*, then enough energy was added during passage to go into solar orbit. Note that for a Hohmann transfer energy is always added.

Consider the example transfer mission depicted in Figure 3.27. This represents a flight from a 200 km circular parking orbit on a high energy transfer trajectory. The spacecraft passed in front of the moon such that it returns to low earth perigee. Figure 3.29 depicts the entire flight profile in an earth-moon fixed coordinate frame. Referring to Figure 3.27, the radial component of velocity, v_r^- is 1.0 km/s. Thus, the magnitude of $\mathbf{v}_{\infty/\mathfrak{C}}$ is 1.30 km/s. Notice that the outward flight from the parking orbit took 2.6 days. If

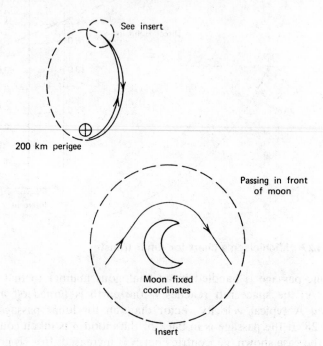

FIGURE 3.29 Lunar transfer and return.

passage is specified to be within 250 km of the moon's surface then deflection of $\mathbf{v}_{\infty/\mathbb{C}}$ may be calculated from equations (3.36) and (3.33), giving $e = 1.69$ and $\delta = 72.8°$. Now $\mathbf{v}^+_{\infty/\mathbb{C}}$ is plotted on the Michielsen chart as shown. Deflection δ is taken clockwise because of the front side passage, after which the vehicle velocity is $\mathbf{v}^+$ taken with respect to earth. Since $v^+_\theta = -0.19$ km/s, this new orbit is a retrograde return to low earth passage.

The type of transfer which results in a return to an earth reentry point has proven very useful on manned lunar flights. If for some reason spacecraft rockets cannot be fired upon reaching the moon, safe return to earth is guaranteed. Such transfers are referred to as *free-return* trajectories. The time for return to perigee is also obtainable from the chart. Since the return leg of such a mission can be thought of as a reverse of the outbound leg to the moon, the time to return is the same as the time to transfer to the moon if $\mathbf{v}^+$ were replaced by $-\mathbf{v}^+$. Thus, the time to return in the above example is about 3.4 days. Early Apollo flights were injected into free return trajectories of the type illustrated in Figure 3.30. Typically, the spacecraft would be sent on a 3-day outbound leg and make a front side passage such that it could enter a 3-day return leg if a failure occurred. The corresponding velocity vector diagram is shown in Figure 3.31. The required value of δ is 80.2° which corresponds to $r_{p/\mathbb{C}} = 2290$ km or a minimum altitude of 552 km above the moon. As confidence in the Apollo Command and Service Module (CSM) increased, these free-return trajectories were changed to accommodate lunar orbit entry

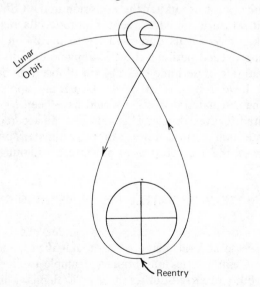

FIGURE 3.30 Apollo type free-return trajectory in earth-moon centered coordinates.

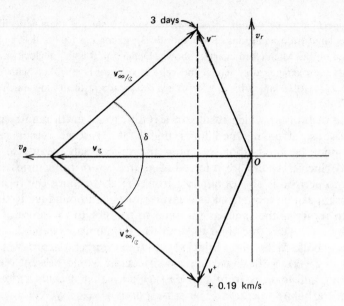

FIGURE 3.31 Apollo free-return vector diagram.

parameters. Apollo 13 had made a mid-course correction to leave its free-return path before experiencing the failure which aborted lunar landing. The lunar module engines were used after the explosion in the CSM to modify their course and permit return to an earth reentry altitude in a reasonable time.

Apollo 11 took 3 days on its outbound leg to the moon and 2.5 days to return. As they approached perilune the CSM engine fired to insert them into lunar orbit. The altitude of perilune was 114 km. If insertion had not occurred, then they would have returned to earth. Using the approximate method presented here, the calculated deflection would have been $\delta = 88°$, which is 8° higher than required for free-return. This indicates the accuracy available with this method. Note that return times can be estimated only when $|v_\theta^+| \leq$ 0.19 km/s. Otherwise the spacecraft does not return without thrusting.

3.6 RELATIVE MOTION OF SATELLITES IN NEIGHBORING ORBITS

The ability of two or more spacecraft to rendezvous is a technique of extreme importance to several missions. Although such maneuvers were demonstrated in the Gemini series, the primary example to illustrate the significance of this capability occurred during the Apollo flights. This was the return of the Lunar Excursion Module to lunar orbit and rendezvous with the Command and Service Module. Success of Apollo was directly linked to this maneuver. Other missions requiring this capability include assembly of orbital stations, personnel transfer, rescue, retrieval, inspection, and interception. In many of these cases the rendezvous phase involves a situation in which the two spacecraft are in neighboring near-circular orbits. Rather than describe the individual motion of these vehicles, it is much more convenient to consider their relative motions. This is the philosophy taken here. Rendezvous closure requirements in terms of relative motion and techniques for determining impulse components are discussed.

3.6.1 Equations of Relative Motion

Consider the positions of two close objects in neighboring near-circular orbits over an interval of time, as shown in Figure 3.32. If, for example, both orbits have the same period but slightly different eccentricities, then there is relative motion between the two vehicles. In an inertial frame motion of the two must be handled by r_1 and r_2, separately. If a coordinate system is attached to the object in the target orbit and allowed to rotate with orbit position, then motion of the chase satellite can be described with respect to the target. A typical example of this motion is illustrated in Figure 3.33. This shows inplane motion of a chase craft in an orbit with slightly higher eccentricity and period. The x-axis is along the radial direction from earth and y is along the target orbit path. A third axis, z is orbit normal. Both x and z components of relative motion are assumed small. However, the resulting equations of motion do not limit the value of y. These equations are developed by assuming the target

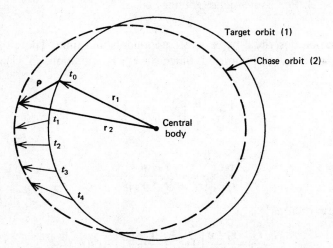

FIGURE 3.32 Relative position of two satellites in an inertial frame.

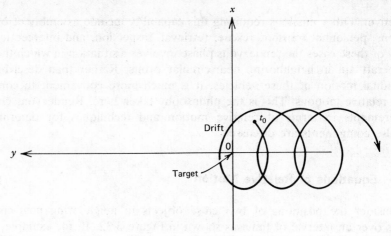

FIGURE 3.33 Drift of the chase craft with respect to the target.

body is in an unperturbed orbit,

$$\ddot{\mathbf{r}}_1 = -\frac{\mu \mathbf{r}_1}{r_1^3} \qquad (3.41)$$

The chase craft may be influenced by a perturbing force (e.g., thrusters, drag, etc.). Thus

$$\ddot{\mathbf{r}}_2 = -\frac{\mu \mathbf{r}_2}{r_2^3} + \mathbf{f} \qquad (3.42)$$

where $\mathbf{f}$ is force per unit mass, and $|\mathbf{f}|$, $|\mathbf{r}_1 - \mathbf{r}_2|$ are small. Relative position is simply $(\mathbf{r}_2 - \mathbf{r}_1)$. For convenience, define

$$\boldsymbol{\rho} = \mathbf{r}_2 - \mathbf{r}_1$$

where $\boldsymbol{\rho}$ can be described by x, y, z components and time. Therefore, slightly noncoplanar orbits are permitted. Since $\ddot{\boldsymbol{\rho}} = \ddot{\mathbf{r}}_2 - \ddot{\mathbf{r}}_1$, equations (3.41) and (3.42) give

$$\ddot{\boldsymbol{\rho}} = \frac{\mu}{r_1^3}\left[\mathbf{r}_1 - \frac{r_1^3}{r_2^3}\mathbf{r}_2\right] + \mathbf{f} \qquad (3.43)$$

Since $\mathbf{r}_2 = \mathbf{r}_1 + \boldsymbol{\rho}$,

$$\frac{\mathbf{r}_2}{r_2^3} = \frac{\mathbf{r}_1 + \boldsymbol{\rho}}{(r_1^2 + 2\mathbf{r}_1 \cdot \boldsymbol{\rho} + \rho^2)^{3/2}}$$

which is rewritten as

$$\frac{\mathbf{r}_2}{r_2^3} = \frac{\mathbf{r}_1 + \boldsymbol{\rho}}{r_1^3}\left[1 - \frac{3}{2}\left(\frac{2\mathbf{r}_1 \cdot \boldsymbol{\rho}}{r_1^2}\right)\right] + O(\rho^2)$$

Substituting into equation (3.43) gives

$$\ddot{\boldsymbol{\rho}} = \frac{\mu}{r_1^{\,3}} \left[-\boldsymbol{\rho} + 3\left(\frac{\mathbf{r}_1}{r_1} \cdot \boldsymbol{\rho} \right) \frac{\mathbf{r}_1}{r_1} \right] + \mathbf{f} + O(\rho^2) \qquad (3.44)$$

Since $\ddot{\boldsymbol{\rho}}$ is the absolute acceleration of $\boldsymbol{\rho}$ it can be written in terms of the x, y, z coordinates through application of equation (1.34),

$$\ddot{\boldsymbol{\rho}} = \ddot{\boldsymbol{\rho}}_b + 2\boldsymbol{\omega} \times \dot{\boldsymbol{\rho}}_b + \boldsymbol{\omega} \times (\boldsymbol{\omega} \times \boldsymbol{\rho}) + \dot{\boldsymbol{\omega}} \times \boldsymbol{\rho} \qquad (3.45)$$

where $\boldsymbol{\omega}$ is the orbit rate with magnitude $\dot{\theta}$. Here $\dot{\theta} \cong n = \text{constant}$, thus, $\dot{\boldsymbol{\omega}} \times \boldsymbol{\rho}$ is ignored. Also note that the conic equation can be written as

$$r_1 = a(1 - e^2)[1 - e \cos \theta + (e \cos \theta)^2 - \cdots]$$

or

$$r_1 = a(1 - e \cos \theta) + O(e^2)$$

where e is small, and a is the semimajor axis of the target orbit. Then

$$\frac{\mu}{r_1^{\,3}} \cong \frac{\mu}{a^3} (1 - e \cos \theta)^{-3} \cong \frac{\mu}{a^3} (1 + 3e \cos \theta)$$

Neglecting products of ρ and e, as well as terms of order e^2 and ρ^2, equations (3.44) and (3.45) yield three component equations of relative motion,

$$\ddot{x} - 2n\dot{y} - 3n^2 x = f_x \qquad (3.46)$$

$$\ddot{y} + 2n\dot{x} = f_y \qquad (3.47)$$

$$\ddot{z} + n^2 z = f_z \qquad (3.48)$$

These are sometimes called *Hill's equations*. Since y does not appear explicitly it need not be a small quantity provided it is measured circumferentially, as illustrated in Figure 3.34. Thus, for a given difference in true anomaly $\Delta\theta$, the

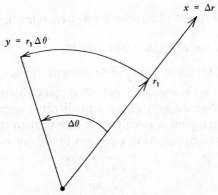

FIGURE 3.34 Definition of x and y components of relative motion.

value of y is

$$y = r_1 \, \Delta\theta$$

3.6.2 Special Solutions

Hill's equations [expressions (3.46) to (3.48)] are not solvable in general. However, several special cases can be handled without difficulty. Consider the force-free situation, $\mathbf{f} \equiv 0$. The third equation represents simple harmonic motion in the z direction. This corresponds to a slight inclination difference between target and chase orbits. Equation (3.47) can be integrated immediately to give

$$\dot{y} + 2nx = \text{constant} = \dot{y}_0 + 2nx_0$$

Substituting this into expression (3.46) yields

$$\ddot{x} + n^2 x = 2n(\dot{y}_0 + 2nx_0)$$

which can now be integrated to

$$x(t) = \frac{\dot{x}_0}{n} \sin nt - \left(\frac{2\dot{y}_0}{n} + 3x_0\right) \cos nt + \left(\frac{2\dot{y}_0}{n} + 4x_0\right) \tag{3.49}$$

This leads to

$$y(t) = \frac{2\dot{x}_0}{n} \cos nt + \left(\frac{4\dot{y}_0}{n} + 6x_0\right) \sin nt + \left(y_0 - \frac{2\dot{x}_0}{n}\right)$$
$$- (3\dot{y}_0 + 6nx_0)t \tag{3.50}.$$

and to complete the solution,

$$z(t) = z_0 \cos nt + \frac{\dot{z}_0}{n} \sin nt \tag{3.51}$$

From equations (3.49) and (3.50) the force-free rates in-plane are

$$\dot{x} = \dot{x}_0 \cos nt + (2\dot{y}_0 + 3nx_0) \sin nt \tag{3.52}$$

$$\dot{y} = -2\dot{x}_0 \sin nt + (4\dot{y}_0 + 6nx_0) \cos nt - (3\dot{y}_0 + 6nx_0) \tag{3.53}$$

Notice that y-oscillation is a quarter period ahead of x-oscillation with double the amplitude, and that y has a linear drift directed opposite to the constant term in x. If the target orbit is circular and $x_0 = 0$, then the eccentricity of the chase orbit is obtained by applying equation (3.9b) for near-circular orbits,

$$e = \frac{\sqrt{4\dot{y}_0^2 + \dot{x}_0^2}}{na}$$

where a can be replaced by r_1.

To illustrate the utility of Hill's equations consider the nominal launch of a satellite into a circular orbit at 200 km with corresponding speed of 7.784 km/s. Perigee altitude and latitude of perigee are to be found for the case in which injection is due south at an altitude of 222 km and latitude of 30°N with velocity 7.869 km/s and flight path angle $-0.5°$. The target orbit has $a = r_1 = 6578$ km and $n = 1.18 \times 10^{-3}$ rad/s. Since orbits are symmetric about major axes, let $\beta = +0.5°$ to simplify the sign situation. Initial conditions for Hill's equations are $y_0 = 0$, $x_0 = 22$ km, $\dot{x}_0 = v_0 \sin \beta_0$, and $\dot{y}_0 = v_0 \cos \beta_0 - v_c - nx_0$. Notice that $\dot{y}_0$ must account for rotation of the x, y coordinate system by including $-nx_0$. The initial rates become $\dot{y}_0 = 0.0587$ km/s, $\dot{x}_0 = 0.0687$ km/s. At perigee, $\dot{x} = 0$, which implies that equation (3.52) should give the time of perigee passage:

$$\tan nt_p = -\frac{x_0}{n\left(\dfrac{2\dot{y}_0}{n} + 3x_0\right)} = -0.35$$

or

$$t_p = -287 \text{ sec}$$

where the negative indicates time to perigee. Therefore, true anomaly of this initial position is

$$-\theta \cong nt_p = -19.4°$$

which corresponds to a latitude of 10.6° N, because the sign of β_0 is negative. Now perigee height is obtained by setting $nt = -19.4°$ in equation (3.49),

$$x(t_p) = 12 \text{ km}$$

Therefore, perigee altitude is 212 km. In general, Hill's equations give excellent results for this type of problem, because motion is referred to a close target orbit.

Hill's equations can also be applied to the rendezvous problem. For a given initial position of the chase craft a proper combination $\dot{x}_0$, $\dot{y}_0$ will allow it to drift to the target if both orbits are coplanar. A slight difference in inclination will result in the chase vehicle crossing the target plane twice per cycle of the orbit. Therefore, a *drift-in* type rendezvous with inclination difference would require contact to occur when the two orbits crossed. Thus, it would seem advisable to eliminate the z component of motion before closure of the two craft. Once the planes coincide the problem is to induce a drift which will result in x and y components simultaneously reaching zero after some reasonable time interval. If the chase vehicle is allowed to drift after inducing initial values of $\dot{x}_0$ and $\dot{y}_0$, then the rendezvous problem reduces to manipulating the force-free solutions (3.49) and (3.50). To extract correct values of $\dot{x}_0$ and $\dot{y}_0$ for given x_0, y_0, and time to rendezvous, set these two expressions to zero and solve simultaneously:

$$\dot{y}_0 = \frac{[6x_0(nt - \sin nt) - y_0]n \sin nt - 2nx_0(4 - 3 \cos nt)(1 - \cos nt)}{(4 \sin nt - 3nt) \sin nt + 4(1 - \cos nt)^2} \quad (3.54)$$

$$\dot{x}_0 = -\frac{nx_0(4 - 3 \cos nt) + 2(1 - \cos nt)\dot{y}_0}{\sin nt} \quad (3.55)$$

As an example of a rendezvous calculation consider the following situation. A chase/interceptor satellite is at a position $x_0 = 50$ km and $y_0 = 100$ km from a target spacecraft in a circular earth orbit with a period of 88.8 min, which corresponds to an altitude of 222 km and $n = 1.178 \times 10^{-3}$ rad/s. Inserting the known values into equations (3.54) and (3.55) permits calculation of $\dot{y}_0$ and $\dot{x}_0$ once a value of t is selected. Figure 3.35 presents all permissible combinations of initial velocity components for rendezvous

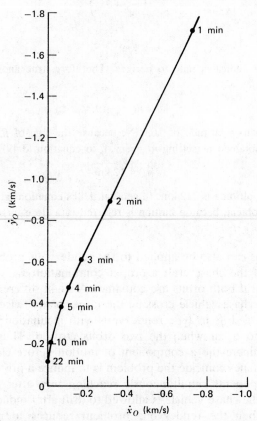

FIGURE 3.35 Relation between time-to-rendezvous and initial velocity components for example case.

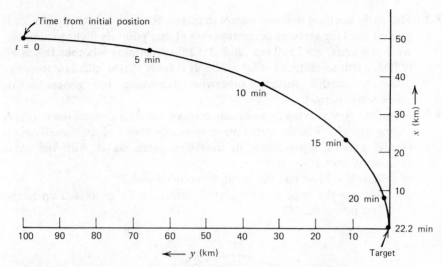

FIGURE 3.36 Example of rendezvous profile for 22.2 min transfer.

times between 1 and 22 minutes. A profile of the relative approach trajectory is shown for a 22.2 min (quarter period) transfer in Figure 3.36.

During missions involving satellite inspection and crew rescue, a *standoff position* is sometimes desirable. This is a location in the target vicinity from which operations required during the mission can be carried out. An ideal relative position with respect to fuel usage is one which requires no thrusting to maintain the position. Such a situation exists when the chase craft is located in the target orbit, but with nonzero value of y.

EXERCISES

3.1 The Skylab space station was launched with a Saturn V booster such that at burn-out of the last stage it had

$$v_0 = 9.25 \text{ km/s}$$

$$r_0 = 7000 \text{ km}$$

$$\beta_0 = 0°$$

(a) A circular orbit was desired. Did this achieve such an orbit? What is the eccentricity of the attained orbit?

(b) What is the value of r at perigee?

(c) Calculate the following parameters of this orbit; $\mathcal{E}$, a, p, r_a, h

3.2 The early warning defense system detects a Russian launched vehicle. If ground tracking stations determine that at one point its flight parameters are $v = 9$ km/s, $r = 7500$ km, and $\beta = 25°$, determine whether this is an ICBM, earth satellite, or solar probe. If it is an ICBM calculate its range over the earth's surface, otherwise determine the geocentric or heliocentric period.

3.3 A vehicle A is moving in a circular orbit of radius a around the earth. A second vehicle B is instantaneously vertically above A at a small height H and moving horizontally in the same plane as A with the same velocity as shown.
(a) Which vehicle has the greater orbital period?
(b) What is the approximate period difference $\delta\tau$ expressed up to the first power of H?

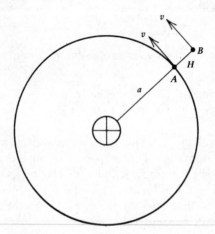

EXERCISE 3.3

3.4 A satellite is launched from earth such that the booster burn-out occurs at an altitude of 422 km with a velocity $v_0 = 9.5$ km/s and flight path angle $\beta_0 = 10°$.
(a) Determine θ_0 and e.
(b) Determine a and r_p.
(c) As the satellite passes perigee its rocket is fired in order to circularize the orbit. However, the resulting velocity was 4% off, and $\beta = 0.04$ radian at the end of thrusting. What is the resulting value of e? (*Hint:* Notice that e should be very small.)

3.5 The Apollo-Soyuz flight plan included booster burn-out parameters of $r_0 = 7000$ km, $v_0 = 7.55$ km/s, and $\beta_0 = 20°$.
(a) Approximate θ_0, e, ψ, and a for these conditions.

(b) After one orbit an adjustment of eccentricity to 0.5 was planned without changing the line of apsides. Approximate β and v just after applying the impulse.

(c) Sketch the original and new orbits and show the point at which the orbit changed.

3.6 A lunar flight was initially launched into an orbit with $\beta_0 = 20°$, $v_0 = 8.28$ km/s, and $r_0 = 7000$ km.

(a) Calculate θ_0 and e for this orbit.

(b) After one orbit the space-craft was injected into the translunar trajectory without changing the line of apsides. Approximate β and v just after applying the impulse if the new apogee is at the lunar distance. What is the Δv required for this injection?

3.7 What is the minimum Δv required to bring a synchronous satellite back to earth reentry altitude (about 75 km)?

3.8 A satellite transfers from a low circular orbit of radius 7000 km to a circular 12-hour orbit.

(a) What is the radius of the final orbit?

(b) If a Hohmann transfer is used, calculate eccentricity of the transfer ellipse and total required velocity increment.

3.9 Suppose that the moon becomes a target of the anti-conservationists and they think we ought to get rid of it.

(a) How much Δv is required to permit the moon to escape the earth?

(b) The conservationists think we ought to bring the moon closer to earth. What is the minimum total Δv required to lower its present circular orbit to 10^5 km?

3.10 An earth satellite crosses the equatorial plane when its true anomaly is 45°. This orbit has a period of 318 min and an eccentricity of 0.5. Its initial inclination is 10° with respect to the equator. What minimum velocity increment is required to make this an equatorial orbit?

3.11 A comet makes a close approach to earth on a hyperbolic passage. Its velocity vector with respect to earth is deflected by 60°. If its velocity at infinity with respect to earth is $v_{\infty/\oplus} = 2$ km/s, then determine:

(a) Eccentricity of the passage, e.

(b) True anomaly of the asymptotes, $\theta_{\infty/\oplus}$.

(c) Semimajor axis of passage, a.

(d) Altitude of closest approach.

(e) Velocity with respect to earth at the point of closest approach, $v_{p/\oplus}$.

3.12 The earth's mean heliocentric velocity is 29.78 km/s.

(a) Assuming that meteors travel in parabolic heliocentric orbits, show that the speed of approach $v_{\infty/\oplus}$ of meteors towards earth lies between 12.3 km/s and 71.9 km/s.

(b) Assume one of the 12.3 km/s meteors passes the earth with closest distance equal to 1.5 earth radii (from the earth's center). How far is its approach asymptote from the earth's center?

(c) How much is its path deflected as seen from the earth?

(d) How much is its path deflected as seen in a heliocentric inertial frame?

3.13 A spaceship after escaping the earth has a heliocentric velocity equal to that of the earth but in a direction, in the plane of the earth's orbit, inclined away from the sun at some acute angle with the earth's motion. What is the period of the spacecraft's heliocentric orbit?

3.14 Assuming a one-impulse transfer from low earth orbit, which flight requires more impulsive Δv; a Mars flyby or a Venus flyby? Explain your conclusion.

3.15 What is the minimum Δv required to send a vehicle out of a circular orbit around earth at altitude 200 km in the ecliptic plane into a Hohmann transfer to Jupiter? Assume Earth and Jupiter have circular orbits in the ecliptic plane around the Sun. What is the angular distance (around the earth) from perigee to escape from Earth, $\theta_{\infty/\oplus}$?

3.16 The Pioneer X spacecraft flew by Jupiter and on to solar system escape. During the planning phase of this mission the question of using a Hohmann transfer to Jupiter came up. No Δv was to be added after leaving low earth orbit except for minor mid-course corrections. Determine whether or not the spacecraft could pick up enough energy via a Hohmann transfer and Jupiter flyby to escape the solar system.

3.17 A Mariner spacecraft is sent on a Mars flyby mission via a direct Hohmann heliocentric transfer orbit from Earth.

(a) Assume a Mars passage within 590 km of the surface and approximate parameters e, δ, and $\theta_{\infty/\delta}$ relative to the passage through Mars' gravity.

(b) What are the new components of heliocentric orbital velocity of Mariner if the spacecraft passes on the sunny side of Mars?

(c) Repeat (b) for the case of passage on the dark side.

3.18 A nuclear waste disposal spacecraft is to be sent from low circular earth orbit in order to carry radioactive waste either out of the solar system or crashing into the sun. It is your responsibility to decide minimum Δv requirements for such a mission. Assume impulsive thrusts only and answer the following questions:

(a) If no planetary swingby assists are permitted, which mission requires less propellant, that is, solar impact or solar escape?

(b) Compare propellant requirements for Mercury and Jupiter swingbys to see if these change your decision. Which requires less Δv? *Hint*: Assume Hohmann heliocentric transfers. Mercury will aid in the solar crash and Jupiter in solar system escape.

(c) We would like to use the spacecraft again by separating the waste before planetary swingby. Describe the sequence of events required to return this ship to earth orbit. Assume a Mercury swingby and the waste requires guidance from this ship until halfway to Mercury.

3.19 An unmanned spacecraft is sent to the moon on a 10-day transfer from a low circular earth orbit with altitude of 222 km.
(a) What is the required Δv to accomplish this transfer?
(b) What is the value of $r_{p/\mathfrak{c}}$ for a free-return to low earth orbit? Does this require a front-side or back-side passage?
(c) How long does the return take?
(d) Sketch the entire trajectory in earth-moon fixed coordinates.

3.20 A sight-seeing space shuttle is sent on a round trip to the moon from a low earth orbit. Use the Michielsen chart to obtain:
(a) The deflection by the moon, δ for a direct outward leg of 3 days and a direct return (to a low earth orbit) leg of 3 days.
(b) Should the craft pass in front of or behind the moon?
(c) What is the radial distance of closest approach to the moon?

3.21 Since the Russian version of the Saturn V booster has not been developed, they may decide to try a manned lunar mission with a smaller launch vehicle. In order to do this, they would probably use a minimum energy transfer from low earth orbit (200 km altitude) to the moon's distance. In addition, if they cannot afford to carry enough fuel to perform a lunar orbit insertion burn and land a capsule, they may make a lunar approach that will allow them to just graze the lunar surface ($r_{p/\mathfrak{c}} \cong R_\mathfrak{c}$). At $r_{p/\mathfrak{c}}$ they might drop an anchor to slow down and land.
(a) If they pass behind the moon and the anchor and all rockets fail, what is their final orbit?
(b) If they go around the front of the moon, would their fate be different? Explain.

3.22 Apollo 17 altered its initial free-return translunar trajectory in order to provide an opportunity for a more precise landing on the lunar surface. The point of closest approach (LOI-1 point) was $r_{p/\mathfrak{c}} = 1849$ km. Apollo 17 passed in front of the moon.
(a) Use the Michielsen chart to determine distance of the hyperbolic asymptote to the moon and eccentricity of the hyperbolic approach trajectory for a direct outward transfer of 3.4 days.
(b) Assuming the initial circular earth parking orbit had an altitude of 200 km, determine Δv to reach the moon.
(c) If the spacecraft had entered a lunar orbit of perilune altitude 111 km and apolune altitude of 314 km, determine Δv at the LOI-1 point if this point becomes the perilune position.

(d) If the spacecraft had left lunar orbit for a 3-day direct return to earth reentry from a circular lunar orbit of altitude 111 km, determine the Δv required to leave this orbit, and approximate the earth reentry velocity.

(e) Apollo 13 had the same translunar trajectory, but a major failure was experienced enroute to the moon after leaving the free-return trajectory and before arriving at the moon. If all of Apollo 13's rockets were then inoperable, determine its final orbit and fate of Lovell, Haise, and Swigert.

(f) As it turned out the Lunar Module descent engine on Apollo 13 did work and was fired 2.5 days into the lunar trajectory. The spacecraft approached the moon with $v_\theta = 0$, $v_r = +0.244$ km/s. If no more firing was done the ship would have returned to low earth orbit on a direct 3-day return after passing in front of the moon. Determine δ and $r_{p/\mathbb{C}}$ by using the Michielsen chart.

3.23 Vehicles A and B are moving in circular orbits, as shown, about earth.

(a) If at one instant A and B are colinear with the center of earth, as in the figure, must their orbits be in the same plane? Explain.

(b) Which has the greater orbital velocity? Explain.

(c) Which has the greater orbital period? Explain.

(d) Does B ever get ahead of A?

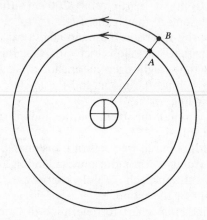

EXERCISE 3.23

3.24 Plot the path of satellite B with respect to A if A is in a circular earth orbit of radius 9000 km and initially both A and B have identical and parallel velocities.

3.25 A synchronous satellite over the Pacific Ocean at 200° East longitude is to be moved to the Indian Ocean at 80° East in two months.

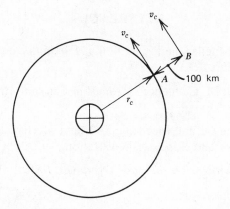

EXERCISE 3.24

(a) What are the total velocity requirements, including both initial and final velocity changes?

(b) Should the initial acceleration be east or west? Why?

3.26 A synchronous satellite is placed into its orbit with slight errors in inclination and eccentricity. Sketch a typical ground track for this situation. The *ground track* is the projection of the orbit position onto the earth's surface.

3.27 A satellite is initially in a circular orbit about the earth. It experiences a constant small drag deceleration along the orbital path, $f_y = -D$ while $f_x = f_z = 0$.

(a) Write the three equations which describe the perturbed motion of this satellite with respect to its motion without drag.

(b) Solve these differential equations so that x, y, and z components are expressed as functions of time.

(c) Sketch the in-plane part of this motion in x, y coordinates.

3.28 On a typical satellite intercept and destroy mission the final rendezvous phase begins with initial values of position components relative to the target of

$$x_0 = 25 \text{ km}$$

$$y_0 = -75 \text{ km}$$

If the target satellite is in a circular, 100 minute orbit, determine appropriate values of $\dot{x}_0$ and $\dot{y}_0$ for rendezvous, and plot an approach profile similar to Figure 3.36 for a closure time of:

(a) 10 minutes

(b) 25 minutes

(c) 40 minutes

(d) 60 minutes

REFERENCES

Bate, R. R., D. D. Mueller, and J. E. White, *Fundamentals of Astrodynamics*, Dover, 1971, Chapters 7,8.

Brouwer, D., and G. M. Clemence, *Methods of Celestial Mechanics*, Academic Press, 1961, Chapter 7.

Dunning, R. S., "The Orbital Mechanics of Flight Mechanics," NASA SP-325, National Aeronautics and Space Administration, 1973, Chapter 3.

Thomson, W. T., *Introduction to Space Dynamics*, Wiley, 1961, Chapter 4.

Attitude Maneuvers

Discussions of rigid body motion in Chapter 2 were limited to torque-free situations. The only reference to changing the attitude of a spinning body was made in connection with energy dissipation. Most spacecraft will require the execution of attitude maneuvers which adjust the angular momentum vector during at least one phase of their mission. Many of these vehicles will be spinning during part or all of their lifetime in space. Methods of carrying out maneuvers involving motion about the center of mass are the primary concern of this chapter. Both applied torque and dissipative techniques are presented. In addition to implementation of such maneuvers, attitude determination methods and associated sensors are discussed, because it is important to measure the spacecraft attitude before and after momentum vector adjustments. As an example of a set of maneuvers related to a realistic mission, a typical attitude acquisition sequence is described.

4.1 MOMENTUM PRECESSION AND ADJUSTMENT FOR A RIGID SPACECRAFT

Early satellites which actively employed momentum control included the Syncom series, which were also the first synchronous communications satellites. They were spin-stabilized and the configuration is illustrated in Figure 4.1. A discussion of advantages related to spin stabilization appears in Chapter 5. For the moment, a brief historical sketch of the satellite development will suffice. The concept of a spin-stabilized, 24-hour satellite was first proposed by the Hughes Aircraft Company in the fall of 1959. The idea is credited to Dr. Harold Rosen and its development was guided by D. D. Williams, both of Hughes. The use of spin stabilization in this orbit, as compared to *body-stabilized* attitude control, was seen as a means of achieving the difficult but extremely desirable mission at an early date, with existing boosters at relatively low cost. After a great deal of preliminary study, the concept was accepted by NASA and sponsored in the form of Project Syncom. The first launch, Syncom 1, was attempted on February 14, 1963. Two separate propulsion systems were incorporated. The initial design philosophy was to use the cold gas (nitrogen) system for maneuvers requiring pulsed operation, because its pulse rise and decay times were less than 10 milliseconds. The hydrogen peroxide system had the same configuration but was to be used for larger impulse corrections and as backup to the nitrogen system for pulsed operations. Unfortunately, Syncom 1 did not make it into its final orbit. Just prior to apogee motor burnout all signals were lost abruptly. The investigation following this failure suggested that one of the nitrogen tanks had exploded. To avoid a repeat of this, pressure in these tanks was reduced on Syncom 2, which was launched successfully on July 26, 1963 into a synchronous, but inclined, orbit. In the case of Syncom 3, launched on August 19, 1964, orbit plane changes at transfer orbit perigee and apogee were executed to achieve an equatorial synchronous (geostationary) orbit. Since the hydrogen peroxide system had proven highly successful in pulsed mode operation on Syncom 2, Syncom 3 incorporated two such systems and cold gas was not used at all.

The thruster location for momentum precession is typically near the rim and parallel to the spin axis as illustrated in Figure 4.2. To reorient the satellite this thruster is fired for a short interval through angle $\Delta\phi$. Since thrust is parallel to the spin axis with a moment arm of the satellite radius, a net precession torque impulse results. The magnitude of $\Delta\phi$ is important because the effectiveness of the thrust in producing the desired torque decreases as the $\cos(\Delta\phi/2)$. In other words, if $\Delta\phi = 2\pi$ there would be no net precession of **h**. To illustrate this further, compare torque impulse taken about a reference direction assuming a constant thrust level throughout the impulse,

$$I_{\text{torque}} = 2\int_0^{\Delta\phi/2} FR \cos\phi \frac{d\phi}{\omega} = \frac{2FR}{\omega}\sin\left(\frac{\Delta\phi}{2}\right) \tag{4.1}$$

FIGURE 4.1 Syncom, the world's first synchronous communications satellite (Hughes Aircraft Company photo).

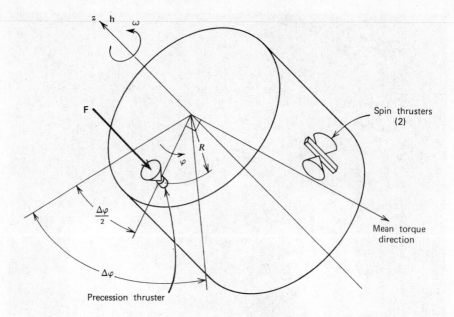

FIGURE 4.2 Momentum precession and spin thruster locations.

to the thrust impulse,

$$I_{\text{thrust}} = 2\int_0^{\Delta\phi/2} F\frac{d\phi}{\omega} = \frac{F\Delta\phi}{\omega} \qquad (4.2)$$

by taking the ratio

$$\frac{I_{\text{torque}}}{I_{\text{thrust}}} = R\frac{\sin\left(\dfrac{\Delta\phi}{2}\right)}{\left(\dfrac{\Delta\phi}{2}\right)} \qquad (4.3)$$

Notice that the ideal torque impulse is RI_{thrust}. Degradation of thrust utilization efficiency is vividly depicted in Figure 4.3. It is apparent that a high level of thrust is desirable for a specified impulse. If this interval of thrust is repeated in synchronism with the spin rate, the momentum vector can be precessed through any desired angle. Each torque impulse corresponds to a rotation of **h** through an angle approximated by I_{torque}/h.

There are many missions which require adjustment of the momentum magnitude, that is, spin adjustment. Several methods have been developed for accomplishing this. The use of thrusters mounted normal to the spin axis, as

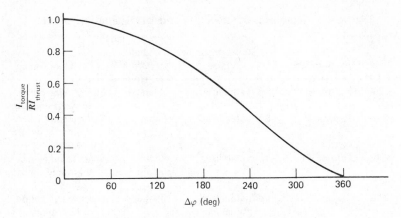

FIGURE 4.3 Thrust effectiveness in momentum precession.

shown in Figure 4.2, is typical. Two such thrusters are required if upward and downward adjustments are to be made. Expendable mass devices, called *yo-yos*, have been used extensively for completely despinning payloads. These are discussed in Section 5.3.

4.2 REORIENTATION WITH CONSTANT MOMENTUM

4.2.1 Energy Dissipation Effects

Many spacecraft have experienced reorientations without the use of thrusters. Unfortunately, a large portion of these were unwanted and unexpected maneuvers caused by unanticipated energy dissipation. This section surveys the adverse effects of energy dissipation and is followed by a potential application of reorientation via dissipation. Early spacecraft designs were small and mechanically simple, and were therefore treated as rigid bodies when predicting attitude motion. Even Explorer I, the first U.S. satellite, defied this simplified approach and quickly tumbled end over end, as described in Section 2.5. Since then spacecraft builders have been aware of the dangers associated with the rigid body assumption. Nevertheless, there have since been several cases of anomalous behavior due to structural flexibility and/or dissipation. Table 4.1 lists just four of the many cases. Notice that in the second and third examples a reduction of angular momentum was experienced as a result of periodic structural bending. Solar energy produced this flexing and solar pressure torques caused a net despin over a large number of spacecraft revolutions. Of particular interest is the sequence of events associated with Applications

TABLE 4.1 Examples of Flexibility and/or Dissipation Effects

Year	Satellite	Control System Type	Adverse Effect	Probable Cause
1958	Explorer 1	Spin stabilized	Unstable	Energy dissipation in whip antennas
1962	Alouette 1	Spin stabilized	Rapid spin decay	Solar torque on thermally deformed vehicle
1964	Explorer XX	Spin stabilized	Rapid spin decay	Solar torque on thermally deformed vehicle
1969	ATS 5	Spin stabilized with active nutation damper[a]	Unstable	Energy dissipation in heat pipe

[a] For orbit injection only; satellite was designed for gravity stabilization after gaining orbit.

Technology Satellite 5 (ATS 5). The spacecraft was launched from Kennedy Space Center on August 11, 1969. In the transfer orbit to synchronous altitude it was spinning about the minor principal axis. Stabilization was accomplished through active nutation control which uses thrusters to limit excursions of the spin axis from the angular momentum direction. Figure 4.4 illustrates the satellite configuration. As it reached apogee the solid propellant kick motor was fired to inject it into final orbit. Loss of this propellant caused the inertia

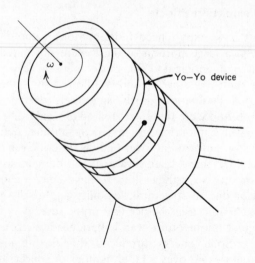

FIGURE 4.4 ATS 5 configuration.

ratio to change, but the spin axis remained the minor axis. However, this shift caused the rate of energy dissipation to unexpectedly increase significantly. The active stabilization system could no longer maintain a small nutation angle and the spacecraft quickly began to tumble. This was caused by unanticipated dissipation in a heat pipe which became a very efficient damper under these special conditions.

The original launch sequence had called for apogee motor case ejection just after expending its propellant. This would shift inertias such that the spin axis would become the major axis, insuring stability until later despin and deployment of long booms to permit gravity gradient stabilization. Unfortunately, ATS 5 entered a *flat-spin* state before the apogee motor case was ejected. It was apparent that ejection of this casing would still shift inertias such that the satellite would again reorient itself. However, it might then be upside down with spin in the opposite direction. There was a 50–50 chance of this and no way to predict the outcome. Since the spacecraft was eventually to be completely despun, a yo-yo device was mounted on the drum. This will despin the satellite only if it is spinning in a prescribed direction. Thus, if it reoriented after ejecting the case such that spin was reversed, the yo-yo could not be used. No despin thrusters were mounted on the spacecraft. Needless to say, when the casing was ejected, the satellite reoriented upside down. It has never been despun, although studies considering sending another spacecraft up to despin it have been made.

Methods for modeling dissipation effects can be grouped into three types. The first is based on a rigid spacecraft model with an *energy sink*. The second involves modeling of the dampers and other dissipators analytically and is referred to as the *discrete parameter* method. In the third method a modal model is used, that is, motions are described in terms of the normal modes of deformation of a slightly flexible, lightly damped structure. This last method comes from vibration analysis techniques and is especially useful when treating structural damping. Since analysis of flexibility effects is not addressed here, only the first two methods are discussed in this chapter. The *energy sink* approach requires that the spacecraft is assumed to have no moving parts which dissipate energy. Of course, in reality there must be some motion to cause dissipation, but this is ignored in the analysis. With such contradictions one would expect this method to give, at most, good engineering approximations if assumptions are based on physically allowable processes. Euler's equations, set (2.71), for rigid body motion must be assumed appropriate here. However, the first integral of these equations is a statement of conservation of rotational energy. This cannot be satisfied with dissipation present. One can argue that motion over any precession cycle is nearly the same as that of a rigid body with the same momentum and energy. This argument can be applied repeatedly with incremental reductions in rotational energy after each cycle.

This reasoning forms the basis of the energy sink approximation. If the use of this method is acceptable, then the major difficulty is in the selection of a dissipation rate. This usually requires some physical insight, and possibly some empirical knowledge, to make an appropriate choice. Nevertheless, the method has been applied in both large angle reorientation analyses and stability determinations.

A simple example situation will illustrate the fundamentals of the energy sink approach. Consider a symmetrical satellite with a configuration like that of Figure 2.11 and inertias $I_1 = I_2 = 10$ N·m·s^2 and $I_3 = 18$ N·m·s^2. It is initially spinning about the 1-axis at a rate $\omega_1(0) = 20$ rad/s. An energy damper is assumed, and the satellite will eventually be spinning about the 3-axis. It is highly desirable to determine the time required for this transformation of spin axes. Taking the energy sink approach requires that the dissipation rate be modeled. Assume that this can be done and the kinetic energy dissipation rate is found to be proportional to $\cos\theta$, where θ is the nutation angle of Figure 2.12. Take the relation

$$\dot{T} = -10\cos\theta \text{ N·m·s}$$

as the specific dissipation model. In order to relate this to satellite kinetics remember that

$$h^2 = I_1^2(\omega_1^2 + \omega_2^2) + I_3^2\omega_3^2$$

and

$$2T = I_1(\omega_1^2 + \omega_2^2) + I_3\omega_3^2$$

which may be combined to give

$$h^2 - 2TI_1 = I_3(I_3 - I_1)\omega_3^2$$

To express this in terms of θ, note that $I_3\omega_3 = h\cos\theta$, thus,

$$h^2 - 2TI_1 = \frac{h^2}{I_3}(I_3 - I_1)\cos^2\theta \tag{4.4}$$

The effect of energy dissipation on θ is simply obtained by differentiating expression (4.4) and remembering that h is constant,

$$\dot{T} = \frac{h^2}{I_3}\left(\frac{I_3}{I_1} - 1\right)(\sin\theta\cos\theta)\,\dot{\theta} \tag{4.5}$$

When $\dot{T}$ is negative and $I_3/I_1 > 1$, the nutation angle decreases. This is consistent with the result of Section 2.5. Equating the specified dissipation rate to equation (4.5) leads to the desired time integral,

$$-10\int_0^t dt = \frac{h^2}{I_3}\left(\frac{I_3}{I_1} - 1\right)\int_{90°}^0 \sin\theta\,d\theta$$

Since $h = I_1\omega_1(0) = 200$ N·m·s, the transformation time is $t = 178$ seconds.

If the spacecraft is modeled with full rigor, including dissipator motion and other moving parts, then the equations of motion may be solved on a digital computer and accurate simulations will result. This is essentially the *discrete parameter* approach. If large external torques or attitude thrusters are applied, internal dissipation is not a significant factor. However, if the satellite is freely rotating with specified dissipation mechanisms, an accurate result can be expected. Bodies nominally spinning about their major axes would incorporate a specific damper to attenuate small nutations. For such cases the damper can usually be modeled without difficulty. Thus, the discrete parameter approach can be used to study stability of nominal motion. In fact, this method is used effectively in Chapter 5 to study stability of multispin bodies. Severe limitations of the discrete parameter method are encountered when trying to accurately model complex spacecraft structural elements which act as unspecified dampers. Dissipation typically occurs as a consequence of friction in joints, hysteresis losses due to stress oscillations, and viscous fluid flow in tanks, for example. Attempts at comprehensive treatment of structural damping have been made in the literature on vibrational analysis. One approach is to use a model composed of rigid bodies or mass points connected by linear springs and dashpots. Although modeling is very difficult, the failure of this approach is due to the virtual impossibility of obtaining valid empirical data on the properties of equivalent dashpots connecting discrete masses of the system.

4.2.2 Large Angle Reorientation with Passive Damping

As a potential application of energy dissipation for reorientation and an illustration of the discrete parameter method, an example attitude maneuver associated with orbit insertion is presented. The technique described here may be thought of as the Apogee Motor Assembly with Paired Satellites (AMAPS). It makes use of the major axis stability phenomenon to passively transform spin axis orientation during deployment of a pair of satellites into high circular orbits from either a low orbiting space shuttle or a launch vehicle which performs the apogee transfer injection from a low parking orbit. Active attitude control beyond the apogee transfer injection is limited only to guaranteeing the payload reorients in the positive direction. Elimination of this pointing ambiguity, experienced in the ATS 5 case, will be discussed separately in Section 5.3. Although there are significant design advantages associated with this technique, the discussion is limited to development of a discrete parameter model in order to estimate spin axis transfer times for at least one example payload. Energy dissipation is assumed to be provided by a viscous ring damper mounted about and centered on the major axis. There are two basic configurations with which this technique may be associated. When used with a launch vehicle no separate apogee transfer injection motor is required. Such a

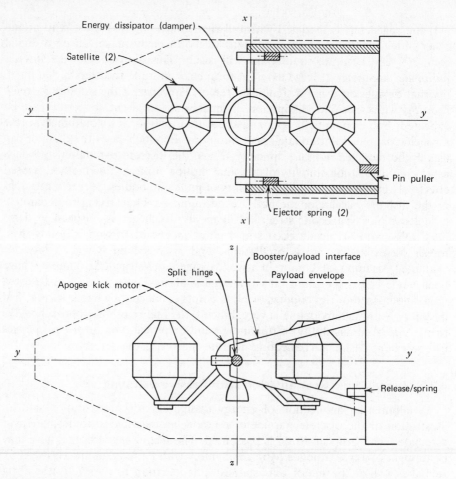

FIGURE 4.5 AMAPS configuration for launch vehicle applications.

scheme is illustrated in Figure 4.5. Either the upper stage of the booster is spin stabilized, or a spin table is used to spin-up AMAPS about its yy-axis. Immediately after injection into a high-altitude transfer orbit the payload attitude is changed so that the momentum vector is reoriented for later apogee firing, and then the release sequence is executed. Since the assembly is stable about its major axis, energy dissipation would result in the transfer of spin to that axis while keeping the direction of momentum fixed in inertial space. It is assumed that the zz-axis is the body axis of maximum inertia and yy is the axis of minimum inertia. The zz-axis will tend to align itself with the angular momentum direction if the yy-axis is perturbed when separated from the launch vehicle. Figure 4.5 shows how initial conditions can be achieved in order to start the transfer of the rotation axis as the assembly leaves the launch vehicle. A pin puller device is activated which allows a spring to start rotational motion about the xx-axis. This induces a perturbation, starting the transfer of rotation axes. When the split hinges line up with the axle flats, springs will push the assembly away from the upper stage, and the payload is then freely precessing. The axially mounted damper will then absorb energy in such a way as to transfer spin axes. The rate of axis transfer for a given payload is a function of initial spin rate and the energy dissipator configuration. When the assembly has stabilized and is rotating only about the zz-axis, this axis will be parallel to the original yy-direction, and the assembly will be in the proper orientation for apogee firing. After the apogee burn, the two satellites can be separated and, if desired, spun up simultaneously by transferring angular momentum of the assembly to the individual satellites. Further momentum adjustments and positioning around the orbit can be done with onboard satellite propulsion systems.

The other basic configuration useful with the AMAPS technique is associated with deployment from a low orbiting space shuttle. This scheme is illustrated in Figure 4.6. The payload package consists of two satellites, an apogee kick motor, and an apogee transfer injection motor. This assembly is deployed from the cargo bay and spun up about the y-axis by a deploy/retrieve device. The payload package is then separated from this device, and the apogee transfer motor is fired shortly thereafter. This is followed by reorientation with the limited attitude control capability of the injection motor subsystem, and then payload separation and perturbation occur. The apogee transfer and axis transformation sequences are shown schematically in Figure 4.7.

The time of spin transfer from the axis of minimum inertia to that of maximum inertia is a function of energy dissipation rate. This, in turn, depends

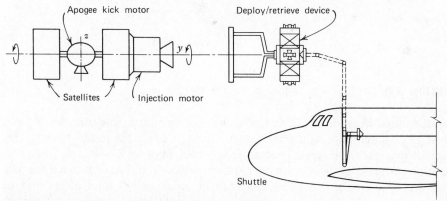

FIGURE 4.6 AMAPS package leaving deploy/retrieve device.

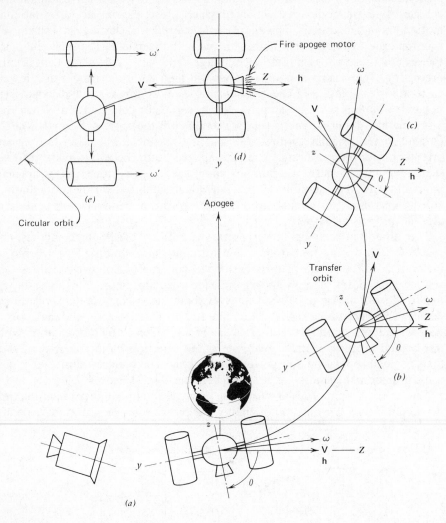

FIGURE 4.7 Deployment sequence for shuttle scheme.

on the initial nutation angle and magnitude of angular momentum for a given AMAPS configuration and absorber. An accurate approach to determining this time is to apply the discrete parameter approach in order to simulate free body motion with energy dissipation over the time interval of interest. Since the axis transformation time should be less than or equal to the time of transfer to apogee (typically about 5 hr), this situation could be represented by a two-point boundary value problem with fixed terminal time. However, the end conditions on orientation are satisfied naturally by the effects of dissipation.

Thus, the motion can be conveniently treated as an initial value problem with an inequality constraint on axis transfer time, that is, axis transfer time ≤ transfer to apogee time. Simulation of motion can then be appropriately accomplished on a digital computer.

A comparison of viscous and inertial forces acting on the fluid slug in the damper loop determines whether motion in the ring is laminar or turbulent. The small angle (perturbation of stable spin state) case corresponds to laminar flow, because this situation is associated with high viscous and low inertial forces along the tangential direction. *Small angle theory* permits the fluid to be modeled as a closed ring, illustrated in Figure 4.8a. Energy is dissipated by cyclic forming and degeneration of surface waves. For values of θ greater than about 0.2 rad, a *large angle* situation must be assumed, and turbulent flow models are used. In such cases the fluid forms a slug, depicted in Figure 4.8b. Motion of this slug with respect to the ring causes energy dissipation through shear stress acting on the wall of the tube, as in elementary pipe flow,

$$\tau_o = \frac{f}{4}\left(\frac{\rho v^2}{2}\right) \tag{4.6}$$

where τ_o is the shear stress, f is the Darcy-Wiesbach resistance coefficient, ρ is the density of the damper fluid, and v is the speed of the fluid slug relative to the ring. End effects are of negligible significance, but cross section diameter is restricted to be small compared with the ring radius in order to simplify integration of shear stress over the wetted area of the ring. The retarding

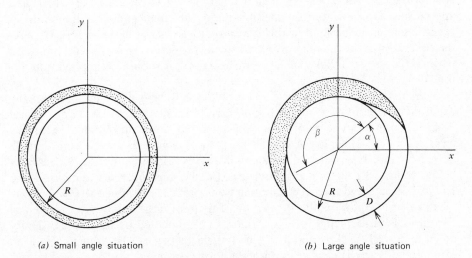

(a) Small angle situation   (b) Large angle situation

FIGURE 4.8 Fluid configurations of interest.

torque about the ring axis of symmetry becomes

$$N_z = -\tau_o(R^2 P\beta)$$

where N_z is the moment due to viscous shear in the annular ring, R is the mean radius of the viscous ring in Figure 4.8, P is the circumference of the ring cross section, and β is the angular span of the fluid slug in the ring. Therefore, the energy disipation rate is

$$\dot{T} = N_z\dot{\alpha} = -\tau_o R^2 P\beta\dot{\alpha} \tag{4.7}$$

with $\dot{\alpha}$ as the angular speed of the slug relative to the ring. Remembering that Reynolds number is $R_n = RD\dot{\alpha}/\nu$, where D is the diameter of the ring cross section and ν is the fluid kinematic viscosity, permits the accepted definition of turbulent flow as $2000 < R_n < 10^5$ in smooth round pipes. This flow regime is assumed here, allowing the resistance coefficient to be approximated by an empirical relation of Blasius,

$$f = \frac{0.316}{R_n^{1/4}} \tag{4.8}$$

Combining expressions (4.6) to (4.8) gives the energy dissipation rate in terms of physical properties

$$T = -\left(\frac{0.316\rho P\beta R^4}{8R_n^{1/4}}\right)\dot{\alpha}^3 = -0.0395\left[\frac{\rho P\beta R^{15/4}}{(D/\nu)^{1/4}}\right]\dot{\alpha}^{11/4} \tag{4.9}$$

This analytical formulation yields good correlation with experimental results and is used for all values of θ above $10°$. As θ is reduced below $10°$ and approaches approximately $5°$, neither large nor small angle theory holds. A realistic model of the fluid in this transition phase is not yet available, because the fluid exhibits both laminar and turbulent properties over varying transition periods. To insure validity of large angle theory, it should be applied only for $\theta \geq 10°$.

A great deal of insight into the problem can be gained by summarizing experimental results related to spin axis reorientation. Consider a typical AMAPS configuration with initial spin about its minor axis and θ perturbed away from $90°$. The rate of energy dissipation is very low for a considerable period of time, because the average fluid slug velocity is low. As θ decreases below about $75°$, this velocity increases on the average; thus, increasing the energy loss rate and further promoting transformation of spin axes. In fact, the mean value of θ is a parabolic function of time for large values. The reorientation process, as depicted in Figure 4.9, can be modeled by large angle theory down to $\theta = 10°$. Below about $5°$ small-angle theory can be applied. Here an exponential decay is caused by the fluid loop. However, this period of time is shortened by added dissipation from vehicle flexibility and liquid sloshing in

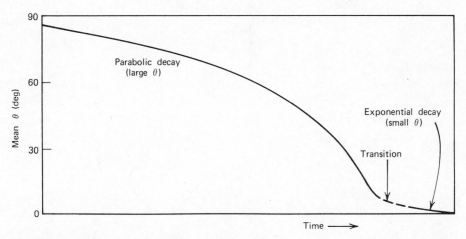

FIGURE 4.9 Axis transformation phases.

propellant tanks. Thus, the major portion of time for spin axis transformation is required to reduce the mean value of θ to about 10°.

General equations of motion for a rigid body are presented in Section 2.3.4. Formulation of dynamics must account for the damper mechanism while maintaining constant angular momentum of the system. Figure 4.10 shows the set of Euler angles and coordinates used in this analysis. Note that I_1, I_2, I_3 are the principal inertias of the rigid body about the x, y, z axes, respectively, and I_x^f, I_y^f, I_z^f are the moments of inertia of the fluid slug about these axes. Since the damper fluid is very small compared to the payload, its products of inertia and center of mass movement effects are ignored. Angular momentum components for the complete system are

$$h_x = (I_1 + I_x^f)\omega_x, \ h_y = (I_2 + I_y^f)\omega_y, \ h_z = I_3\omega_z + I_z^f(\omega_z + \dot{\alpha})$$

Note that the AMAPS configuration requires $I_3 > I_1 > I_2$. Components of angular velocity can be expressed in terms of Euler angles and rates by applying transformations analogous to those in Section 1.4.3. Moments of inertia for the fluid slug can be replaced by

$$I_x^f = \left(\frac{\rho SR^3}{4}\right)[2\beta - \sin 2(\alpha + \beta) + \sin 2\alpha] \tag{4.10a}$$

$$I_y^f = \left(\frac{\rho SR^3}{4}\right)[2\beta + \sin 2(\alpha + \beta) - \sin 2\alpha] \tag{4.10b}$$

$$I_z^f = \rho S\beta R^3 \tag{4.10c}$$

where α is the angular position of the fluid slug in the ring and S is the cross sectional area of the viscous ring, $\pi D^2/4$. The differential equations of motion

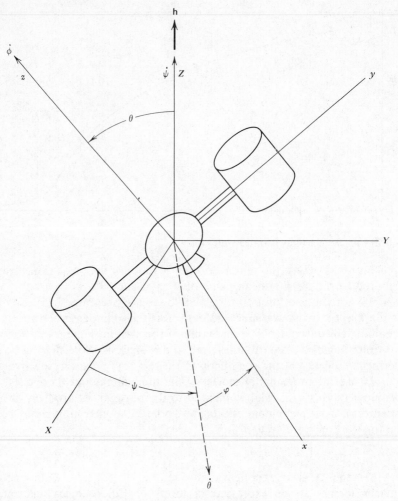

FIGURE 4.10 Coordinates system used for AMAPS modeling.

are obtained for the rigid body and fluid slug by applying Euler's moment equations. The payload and fluid are connected through shear stress, thus, equal and opposite torques act about z on both the rigid body and the slug. These equations yield set

$$(I_1 + I_x^f)(\ddot{\psi} \sin \theta \sin \phi + \dot{\psi}\dot{\theta} \cos \theta \sin \phi$$
$$+ \dot{\psi}\dot{\phi} \sin \theta \cos \phi + \ddot{\theta} \cos \phi - \dot{\theta}\dot{\phi} \sin \phi)$$
$$+ \omega_x \dot{I}_x^f + \omega_y \omega_z I_3 + \omega_y I_z^f(\omega_z + \dot{\alpha})$$
$$- \omega_z \omega_y (I_2 + I_y^f) = 0 \tag{4.11a}$$

$$(I_2 + I_y^f)(\ddot{\psi} \sin \theta \cos \phi + \dot{\psi}\dot{\theta} \cos \theta \cos \phi$$
$$- \dot{\psi}\dot{\phi} \sin \theta \sin \phi - \ddot{\theta} \sin \phi - \dot{\theta}\dot{\phi} \cos \phi)$$
$$+ \omega_y \dot{I}_y^f + \omega_z \omega_x (I_1 + I_x^f) - \omega_x \omega_z I_3$$
$$- \omega_z I_z^f (\omega_z + \dot{\alpha}) = 0 \tag{4.11b}$$

$$I_3(\ddot{\phi} + \ddot{\psi} \cos \theta - \dot{\psi}\dot{\theta} \sin \theta) + I_z^f(\ddot{\phi} + \ddot{\psi} \cos \theta$$
$$- \dot{\psi}\dot{\theta} \sin \theta + \ddot{\alpha}) + \omega_x \omega_y (I_2 + I_y^f)$$
$$- \omega_y \omega_x (I_1 + I_x^f) = 0 \tag{4.11c}$$

$$I_z^f(\ddot{\phi} + \ddot{\alpha} + \ddot{\psi} \cos \theta - \dot{\psi}\dot{\theta} \sin \theta) - \left(\frac{\omega_x^2}{2}\right)\frac{\partial I_x^f}{\partial \alpha}$$

$$- \left(\frac{\omega_y^2}{2}\right)\frac{\partial I_y^f}{\partial \alpha} + \left[\frac{0.316 \rho P \beta R^{15/4}}{8(D/\nu)^{1/4}}\right]\dot{\alpha}^{7/4} = 0 \tag{4.11d}$$

The first three equations guarantee conservation of angular momentum while the last one supplies energy dissipation. This method provides greater accuracy than the energy sink approach, because fluid slug motion is not restricted for the sake of getting a solution *quickly*. For instance, an energy sink approach would require an assumption for $\dot{\alpha}$, such as $\dot{\alpha} = \dot{\phi}$. This particular restriction has been used with questionable results. Set (4.11) can be applied to any specified situation of this type. Integration should be carried out for specified accuracy until error accumulation exceeds bounded values. The large number of integration steps required to accurately simulate rotational motion may severely limit the length of real time motion simulation. However, results for short time simulations can provide a basis for determining whether or not axis transformation will be completed in the apogee transfer interval.

4.3 ATTITUDE DETERMINATION

A basic requirement associated with attitude maneuvers of spacecraft is determination of attitude before and after execution of such maneuvers. This is usually done with respect to a set of inertial directions. There are logically two parts to this discussion. The first involves an introduction to sensor technology and the second part presents the associated geometry and sequences required to measure attitude angles.

4.3.1 Sensors

There are a limited number of ways to sense angles and directions from space. Although there are techniques which require cooperation with ground stations, devices considered here are mounted within the craft and require no inputs from ground controllers. In general, an orbiting satellite is most easily able to sense the direction of the central body about which it moves. Typically, two nonparallel directions are needed to uniquely measure orientation. Thus, a second reference may be needed. The sun is usually an excellent choice, but another star, such as Canopus or Polaris is also a possibility. If continuous attitude determination is required, the sun may not be acceptable, however, because the earth and sun may become colinear twice per orbit for many satellites. Stars which are well out of the equatorial plane provide reference directions which are never colinear with earth for low and medium inclination orbits. Earth sensors actually detect the horizon and determine the local vertical direction. There are three types of practical horizon sensors consisting of scanning, balanced radiation, and edge tracking devices. Their particular application depends on required accuracy, orbit parameters, type of satellite stabilization, power available, operational procedures, and many other factors. Examples of popular sensing techniques may be cited. One type of horizon sensor uses two scanning beams separated by an angle which is less than that subtended by the earth. The sensor is centered about the earth's center in nominal orientation. Each beam scans across the earth in a straight line. One scans across the upper half-disk and one across the lower half-disk. Attitude is determined by comparing the time between the first horizon crossing and a center reference point to the time from the reference to the second horizon crossing. Proper processing of data from both scan beams provides attitude measurements about two of the spacecraft axes. If the satellite is spinning, then this technique can be applied with static sensors which take advantage of the spin to scan earth once per revolution. Each of the scanning type sensors incorporates one resonant oscillating mirror per axis, two telescopes per mirror, and two detectors per telescope. For normal operation, information from two detectors is averaged to minimize system errors. If the sun is in a detector field-of-view, it is identified by an integral sun sensor which inhibits detector output, and the remaining detector is used with the output scale factor adjusted. Time is measured by counting pulses from the satellite reference clock. To guarantee the correct ratio of pulses per degree, the rate of scan is controlled. Another type of earth sensor uses thermocouples as detectors and operates on the *radiation balance* principle. It is intended for use on nonspinning satellites. Detectors are positioned such that their fields of view are tangent to the earth's limb with the earth centered, as shown in Figure 4.11. Such a system is actually considered to be semi-static, because a mirror having two-axis rotational capability is used to position the earth image in order to null detector outputs. The mirror position is then the output quantity which provides attitude information.

Several sun sensors are available for use. They can be classified as either

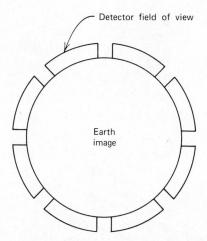

FIGURE 4.11 Detector pattern of a semi-static horizon sensor.

analog or digital types. For example, an analog sensor may use a quad silicon cell detection structure which provides a current through a resistor load in direct sunlight. Each pair of cells is connected to produce a polarity-sensitive output and a zero-output signal when cells are illuminated equally. Digital sun sensors encode the sun angle as digital numbers. For example, light passes through a slit and forms an illuminated image of that slit on a binary coded pattern. The image location depends on the angle of incidence, which is related to attitude. A silicon photocell is located behind each column of the code pattern. Figure 4.12 illustrates the processing of this light into digital information. In general, sun sensors have been very successful on both spinning and nonspinning spacecraft.

The star (other than the sun) used most commonly for attitude determination by spacecraft is Canopus. It is easier to acquire than Polaris since it is the brightest in the southern sky. However, Canopus is not typically used by synchronous satellites because it lies approximately 52° below the equator. This would require a large conical field of view to keep it in sight as the satellite circles earth. Accuracy also suffers in this case. Polaris is 89.1° above the equator, thus, providing an accurate source for attitude measurement. There is only a limited variety of star trackers available, and these are typically complex and relatively massive.

4.3.2 Cone Intercept Method

It was pointed out that attitude determination generally requires the measurement of two angles. This can be done by the instantaneous sensing of two

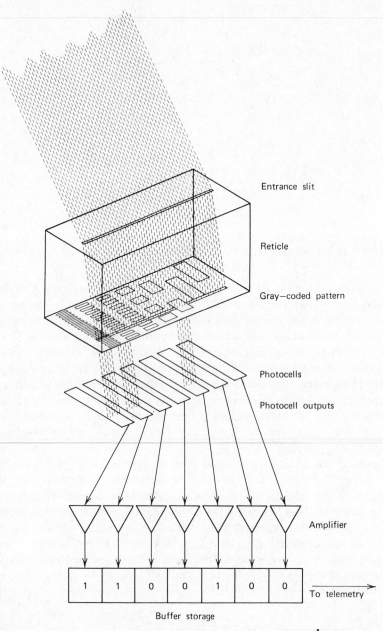

Entrance slit

Reticle

Gray—coded pattern

Photocells

Photocell outputs

Amplifier

| 1 | 1 | 0 | 0 | 1 | 0 | 0 |

To telemetry

Buffer storage

FIGURE 4.12 Schematic of a digital sun sensor.

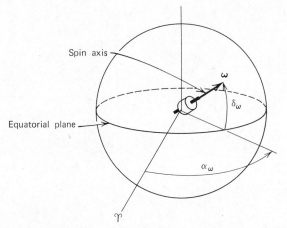

FIGURE 4.13 Right ascension and declination of spin axis.

independent directions, or in some special cases, a time history of one sensor output. The *cone intercept* method employs simultaneous detection of two directions and is the most widely used approach. One situation of importance in which attitude determination is critical to mission success occurs during the transfer orbit maneuvers of a high altitude payload to be placed into circular orbit by an apogee kick motor. Typically, the spacecraft is spin-stabilized and has earth and sun sensors onboard. The attitude coordinates normally used for these missions are illustrated in Figure 4.13 and are defined as the spin axis right ascension α_ω measured in the equatorial plane from the first point of Aries and spin axis declination δ_ω measured from the equatorial plane. The two independent measurements required to establish attitude are the angle between the sun and ω, θ_s and the angle between nadir and ω, θ_e. Each sensor individually generates a cone as illustrated in Figure 4.14. For example, the earth sensor provides all possible positions of ω that satisfy the measured value of θ_e. This locus is just the cone about the nadir direction. The sun sensor provides the other cone with half angle θ_s. Thus, ω must simultaneously satisfy both angles and be on the intersection of the two cones. In general, there are two intersections, the *real* and *image* spin axes. The ambiguity in axes is resolved by taking two additional measurements at some later time. The true solution remains unchanged but the image value will change. Of course, it is implicitly assumed that the spin axis does not change orientation between measurements.

The cone angles θ_s and θ_e can be related to spin axis angles α_ω and δ_ω through application of spherical trigonometry. Refer to Figure 4.15 which depicts relative positions of sun and spin axis in a satellite centered frame. A

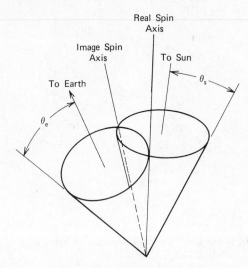

FIGURE 4.14 Attitude determination using the intersection of two cones.

two-step application of proper identities leads to

$$\theta_s = \cos^{-1}\left[\sin \delta_\omega \sin \delta_s + \cos \delta_\omega \cos \delta_s \cos (\alpha_\omega - \alpha_s)\right] \qquad (4.12)$$

where α_s, δ_s are the right ascension and declination of the sun, respectively. A similar approach to determining θ_e yields

$$\theta_e = \cos^{-1}\left[-\sin \delta_\omega \sin \delta_R - \cos \delta_\omega \cos \delta_R \cos (\alpha_\omega - \alpha_R)\right] \qquad (4.13)$$

where α_R, δ_R are the right ascension and declination of the satellite position, respectively. In principle, simultaneous solution of these two expressions will give two sets of values for α_ω and δ_ω, corresponding to the real and image axes. In actual operations the desired values of α_ω and δ_ω will be specified, and curves of deviations for measured values of θ_s and θ_e would be at hand. Thus, slight attitude corrections will be implemented over two or more orbits until correct values of cone angles are attained.

As an example, consider the synchronous orbit injection used as an illustration in Section 3.3. If the launch date is June 21, then $\alpha_s = 90°$, $\delta_s = 23.5°$ (summer solstice). Assume apogee occurs at $\alpha_R = 30°$ at which point $\delta_R = 0°$. The situation is depicted in Figure 4.16. To change planes and circularize the correct values for the spin axis are $\alpha_\omega = 120°$, $\delta_\omega = 24.4°$. Of course, the ground controller will not press the apogee kick motor ignition button unless he has confirmed these values for α_ω and δ_ω. Therefore, he will check telemetry for θ_s and θ_e about 12 min before apogee. This corresponds to $\alpha_R = 28.63°$ and $\delta_R = 0.74°$ with other quantities essentially the same as at apogee.

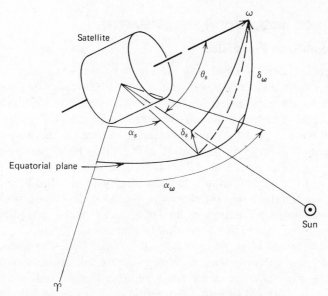

FIGURE 4.15 Relation between spin axis and sun direction.

Applying equations (4.12) and (4.13) yields $\theta_s = 27.38°$, $\theta_e = 89.03°$. If these values are achieved within his specified tolerances, then the kick motor is fired at apogee.

Notice in Figure 4.14 that if the earth and sun directions are parallel, the cones do not intersect, and attitude is indeterminate. Fortunately, this situation occurs only a small percentage of the time. If it is required that a spacecraft be ready for launch at any time, then alternate attitude determination methods may be employed.

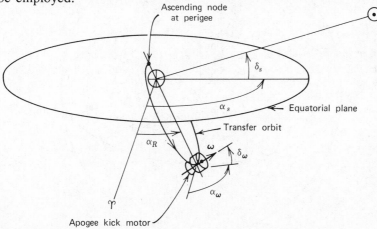

FIGURE 4.16 Synchronous injection geometry.

4.4 ATTITUDE ACQUISITION REQUIREMENTS

4.4.1 Rhumb Line Precession

Active reorientation of a spinning spacecraft momentum vector is sometimes called a spin precession maneuver and is very often associated with attitude acquisition of a newly injected satellite. This maneuver is accomplished through a series of torque pulses as described in Section 4.1. These pulses are easily timed with respect to a fixed inertial reference via the sensors. Thus, it would appear advantageous to employ a fixed reference phasing angle. The associated precession path is a *rhumb line*, that is, a line on the surface of a sphere having a constant angular relationship with the local meridian. For example, consider the situation shown in Figure 4.17. A sun sensor and an earth sensor supply information on angles θ_s and ξ of the spacecraft spin axis with respect to the sun oriented coordinate system, X_s, Y_s, and Z_s. The desired spin axis orientation is given by values $\theta_s(f)$ and $\xi(f)$. The simplest logical implementation of the maneuver is to apply a series of torque impulses about a constant body axis. This is accomplished by using the sun sensor output signal to initiate a time delay of (ψ plus a correction for thruster position)$/\omega$, where ω is the satellite spin rate. Thus, the impulse train at constant ψ causes the spin

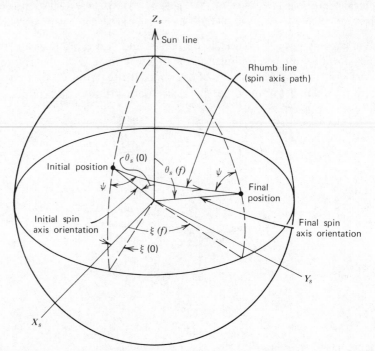

FIGURE 4.17 Rhumb line precession geometry.

axis to describe a rhumb line. The equation for the rhumb line angle ψ is

$$\cot \psi = \frac{\log_e \left[\tan(\theta_s(f)/2)/\tan(\theta_s(0)/2)\right]}{\xi(f) - \xi(0)} \tag{4.14}$$

and the angle through which the spin axis moves during this maneuver is

$$\sigma_1 = \begin{cases} |[\theta_s(f) - \theta_s(0)] \sec \psi| & \text{if } \phi \neq 90° \\ |[\xi(f) - \xi(0)] \sin \theta_s| & \text{if } \psi = 90° \end{cases} \tag{4.15}$$

From a consideration of fuel consumption, the most economical way of precessing the spin axis is along a great circle route. The precession angle would then be

$$\sigma_2 = \cos^{-1} \{\cos \theta_s(0) \cos \theta_s(f) + \sin \theta_s(0) \sin \theta_s(f) \cos [\xi(f) - \xi(0)]\} \tag{4.16}$$

However, great circle precession would necessitate a variable delay time and the advantage in fuel consumption is generally not sufficient to justify the increased complexity associated with this latter method.

4.4.2 Acquisition Sequence

It is instructive to present a rather basic attitude acquisition sequence for satellites launched individually. Consider a body-stabilized synchronous or other high altitude payload depicted in Figure 4.18 in stowed configuration. It

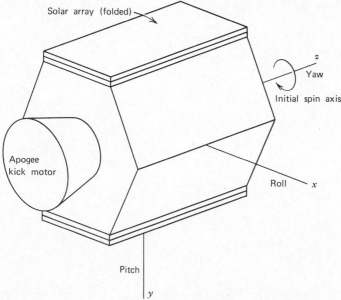

FIGURE 4.18 Assumed satellite configuration for acquisition sequence.

is spin-stabilized during apogee transfer. No internal momentum device is assumed in operation during initial acquisition (such a case is presented in Chapter 9). A logical sequence of events is illustrated in Figure 4.19 and outlined in Table 4.2. For this sequence, initial despin can be executed effectively in a matter of seconds with yo-yo devices or thrusters. A residual spin of 1° per second is left to provide stability while initial conditions for

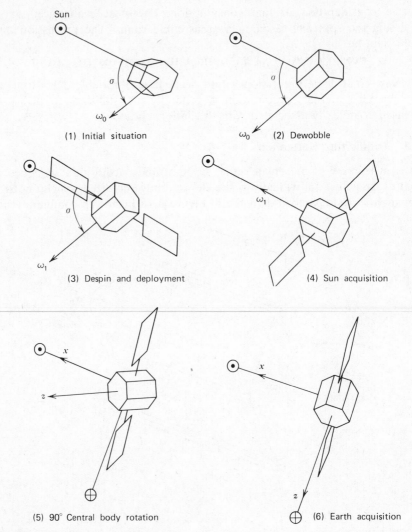

(1) Initial situation

(2) Dewobble

(3) Despin and deployment

(4) Sun acquisition

(5) 90° Central body rotation

(6) Earth acquisition

FIGURE 4.19 Sequence of events.

further maneuvers are established. Rate gyros (discussed in Chapter 5) are essential to determine wobble state, control the despin maneuver, and establish an earth acquisition spin rate about the x-axis. Once the earth enters the sensor field of view, despin should commence to capture successfully, because angular rates drop below the gyro sensitivity range, requiring earth acquisition to be completed before earth leaves this field of view. Although solar pressure and gravity gradient torques will be active during acquisition maneuvers, attitude thruster torques will always be much greater than these torques. The type of attitude control system for nominal on station operation will dictate the need for a star tracker. One is included in the sequence to illustrate its initialization if used.

TABLE 4.2   Basic Acquisition Sequence

Event	Associated Hardware	Remarks
Apogee kick	Apogee motor	Spin stabilized mode
Initial despin	Yo-yos or reaction motors	Reduce spin rate to about 1°/s; may cause wobble
Dewobble	Sun sensors and rate gyros, torque from attitude thrusters	May be done passively through dissipation mechanisms if the z-axis is the major principal axis
Despin	Rate gyros, attitude thrusters	Reduce rate to low value
Panel deployment	Deployment motors	Full power not available unless σ is near zero
Sun acquisition	Sun sensors, attitude thrusters	Bring σ to zero
90° central body rotation	Sun sensors, attitude thrusters, panel despin motors	Bring x-axis into sun line
Earth acquisition	Earth sensor, rate gyro, attitude thrusters	Continuous panel despin starts
Star acquisition	Yaw sensor, earth and sun sensors; attitude thrusters or momentum device	Bring y-axis to orbit normal position

EXERCISES

4.1 A new thruster is being studied for use in momentum precession of spinning satellites. Its thrust profile over an interval Δt is depicted in the figure.

 (a) Determine the expression for the ratio of torque impulse to thrust impulse for pulsed operation of width Δt on a satellite spinning at rate ω.

 (b) Construct a plot analogous to Figure 4.3 and discuss the attributes of this thrust profile vs a constant thrust level.

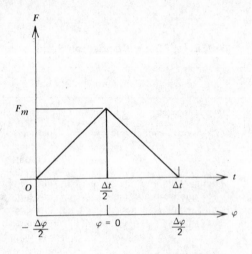

EXERCISE 4.1

4.2 Syncom satellite used precession thrusters over 60° intervals of rotation to reorient itself. Assuming a constant thrust level over these intervals, what percentage of propellant was wasted due to loss of effectiveness as the vehicle rotates?

4.3 It is necessary to precess the angular momentum vector of a spinning satellite through 10°. Assume values of $h = 1000$ N·m·s, $\omega = 4$ rad/s, and $F = 5$ N. The precession thruster is located at a distance of 1 m from the spin axis and the interval of thrust is 0.4 sec.

 (a) How many impulses are required to perform this maneuver successfully?

 (b) How long will this maneuver take to complete?

4.4 A symmetrical satellite with principal moments of inertia $I_1 = I_2 = 40$ N·m·s^2 and $I_3 = 20$ N·m·s^2 is purposely launched such that it is spinning about the 1-axis with $\omega_1(0) = 10$ rad/s. A device which adds energy was incorporated such that the rate of energy addition is

$$\dot{T} = +5 \cos \theta \text{ N·m/s}$$

where θ is the nutation angle.

(a) How long will it take for spin axes to change from 1 to 3, after θ is perturbed away from 90?

(b) What is the final value of ω_3?

4.5 Carry out the steps in the derivation of set (4.11) for the discrete parameter application to large angle reorientation.

4.6 Derive the half-cone angle expressions, equations (4.12) and (4.13), associated with the cone intercept method of attitude determination.

4.7 Carry out the example of Section 4.3.2 for a launch date of July 31 and calculate θ_s, θ_e at $\alpha_R = 26°$.

4.8 Derive the rhumb line equation given by expression (4.14) and show that the spin axis moves through angle σ_1 given by equation (4.15).

4.9 The angular momentum of a spinning satellite is to be precessed along a rhumb line such that its final orientation values are $\theta_s(f) = 60°$, $\xi(f) = 90°$.

(a) If $\theta_s(0) = 30°$ and $\xi(0) = 30°$, determine the rhumb line angle ψ and the angle through which the spin axis is precessed, σ_1.

(b) If this maneuver were done along a great circle route, what value would the angle σ_2 have?

REFERENCES

Kaplan, M. H., and N. M. Beck, Jr., "Attitude Dynamics and Control of an Apogee Motor Assembly with Paired Satellites," *Journal of Spacecraft and Rockets*, Vol. 9, *No.* 6. June 1972, pp. 410–415.

Likins, P. W., and H. K. Bouvier, "Attitude Control of Nonrigid Spacecraft," *Astronautics and Aeronautics*, Vol. 9, *No.* 5, May 1971, pp. 64–71.

Likins, P. W. "Effects of Energy Dissipation on the Free Body Motions of Spacecraft," TR 32–860, July 1966, Jet Propulsion Lab., Pasadena, Calif.

Thomson, W. T., and G. S. Reiter, "Attitude Drift of Space Vehicles," *Journal of the Astronautical Sciences*, Vol. 7, *No.* 2, 1960, pp. 29–34.

Vectors, Hughes Aircraft Company, Vol. 15, Summer/Fall 1973, pp. 7–10.

Williams, D. D., "Torques and Attitude Sensing in Spin-Stabilized Synchronous Satellites," in *Torques and Attitude Sensing in Earth Satellites*, ed. by S. F. Singer, Academic Press, 1964, pp. 159–174.

Attitude Control Devices

Most space missions require that the vehicle incorporate directional references and methods for applying attitude control torques. Chapter 4 discusses sensors which provide directional information based on an outside source, such as sun and earth sensors. An autonomous device which provides an inertially oriented direction reference through the property of momentum stiffness is generally called a *gyro* or *gyroscope*. Such a device may be used as an instrument which supplies attitude information to an automatic control system, or it may generate control torques by *momentum exchange* and *precession* techniques. Other attitude actuator choices include mass movement devices, thrusters, magnetic torquers, and passive devices. An introduction to gyroscopic instruments is presented, followed by a discussion of momentum exchange techniques. Stability criteria for dual-spin configurations are developed by using both a *discrete parameter* and an *energy sink* approach. Other methods of attitude control actuation included are mass movement, magnetic torquing, and gravity gradient schemes.

5.1 GYROSCOPIC INSTRUMENTS

Foucault is credited with originally defining the *gyroscope* in 1852 as a device possessing a large amount of angular momentum. Since torque-free motion implies an inertially fixed magnitude and direction of momentum, the gyroscope provides an excellent attitude reference. Applying a specified torque to a spinning body causes a change in the angular momentum vector in a predictable manner. These properties of gyros lead to several practical applications, including measurement of attitude angles and rates.

5.1.1 The Basic Gyroscope

A typical configuration of the basic gyroscope is depicted in Figure 5.1. Massless gimbal supports and no bearing friction are assumed for the moment.

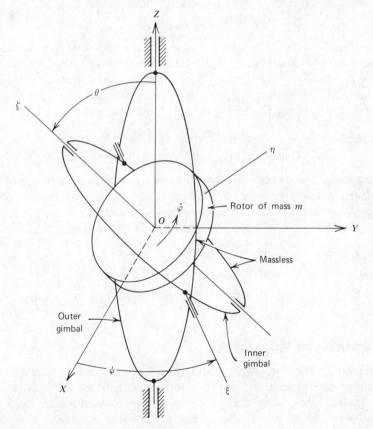

FIGURE 5.1 Basic gyro configuration.

Since the rotor is symmetrical, the two important angles are ψ and θ, which correspond to gimbal angles. To understand the motion of such a gyro, assume the center of mass is displaced from the gimbal center along ζ a distance l. If gravity acts along $-Z$ the rotor weight will cause a torque about ξ. Resulting motion is described by using the node axis frame ξ, η, ζ which do not rotate with the rotor but are attached to the gimbals. Define rotation rate of ξ about Z as *precession* $\dot{\psi}$, and rotation rate of ζ about ξ as *nutation* $\dot{\theta}$. Euler's equations, set (2.71) may be used here in modified form. Since bodies of revolution have the same inertia about any transverse axis, the equations of motion may be written about any set of axes in which the transverse axes are rotating at an arbitrary rate and one axis is parallel to the axis of symmetry. Thus, the gimbal axes are selected here and the associated equations become

$$M_\xi = \dot{h}_\xi + h_\zeta \omega_\eta - h_\eta \omega_\zeta \qquad (5.1a)$$

$$M_\eta = \dot{h}_\eta + h_\xi \omega_\zeta - h_\zeta \omega_\xi \qquad (5.1b)$$

$$M_\zeta = \dot{h}_\zeta + h_\eta \omega_\xi - h_\xi \omega_\eta \qquad (5.1c)$$

These are known as *Euler's modified equations* for bodies of revolution. Let I_1, I_2, I_3 be the principal moments of inertia of the rotor about axes centered at O with $I_1 = I_2$. Motion of the gimbals is of primary interest here. Thus, the rates ω_ξ, ω_η, ω_ζ and momentum components h_ξ, h_η, h_ζ must be expressed in gimbal angles and rates. Referring to Figure 5.1, these become

$$\omega_\xi = \dot{\theta}, \qquad \omega_\eta = \dot{\psi} \sin \theta, \qquad \omega_\zeta = \dot{\psi} \cos \theta \qquad (5.2)$$

$$h_\xi = I_1 \dot{\theta}, \qquad h_\eta = I_1 \dot{\psi} \sin \theta, \qquad h_\zeta = I_3(\dot{\phi} + \dot{\psi} \cos \theta) \qquad (5.3)$$

Comparing $\boldsymbol{\omega}$ components with set (1.26) indicates the convenience of gimbal coordinates. No reference to ϕ is required. However, $h_\zeta = I_3 \omega_3$ (not $I_3 \omega_\zeta$), because all momentum about the symmetry axis must be taken into account. Expressions (5.1) now give the gyroscope equations in terms of gimbal rates,

$$M_\xi = \dot{h}_\xi + h_\zeta \dot{\psi} \sin \theta - h_\eta \dot{\psi} \cos \theta \qquad (5.4a)$$

$$M_\eta = \dot{h}_\eta + h_\xi \dot{\psi} \cos \theta - h_\zeta \dot{\theta} \qquad (5.4b)$$

$$M_\zeta = \dot{h}_\zeta + h_\eta \dot{\theta} - h_\xi \dot{\psi} \sin \theta \qquad (5.4c)$$

5.1.2 Motion of a Spinning Top

The general motion of a basic gyro in massless gimbals is identical to that of the spinning top in Figure 5.2. Therefore, treatment of this device is very helpful in later describing the properties of gyro instruments. Noting that $M_\xi = mgl \sin \theta$, $M_\eta = M_\zeta = 0$ and using expressions (5.3) leads to a set of differential equations describing the precession and nutation of the top due to

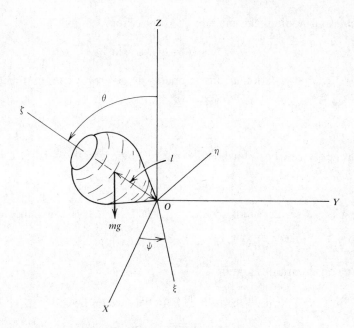

FIGURE 5.2 Spinning top pivoting above a fixed point.

its own weight,

$$mgl \sin \theta = I_1\ddot{\theta} + I_3(\dot{\phi} + \dot{\psi} \cos \theta)\dot{\psi} \sin \theta - I_1\dot{\psi}^2 \sin \theta \cos \theta \qquad (5.5a)$$

$$0 = I_1 \frac{d}{dt}(\dot{\psi} \sin \theta) + I_1\dot{\theta}\dot{\psi} \cos \theta - I_3\dot{\theta}(\dot{\phi} + \dot{\psi} \cos \theta) \qquad (5.5b)$$

$$0 = I_3 \frac{d}{dt}(\dot{\phi} + \dot{\psi} \cos \theta) \qquad (5.5c)$$

where the last equation gives

$$\omega_3 = \dot{\phi} + \dot{\psi} \cos \theta = \text{constant} = n \qquad (5.6)$$

Instead of integrating equations (5.5a) and (5.5b) it is more instructive and equally enlightening to treat the motion of this gyro through conservation principles. Since gravity is a conservative force, total energy is constant. Furthermore, since $M_Z = 0$ angular momentum about Z is conserved. Applying transformation set (1.26) to this situation yields

$$\omega_1 = \dot{\theta} \cos \phi + \dot{\psi} \sin \theta \sin \phi$$

$$\omega_2 = -\dot{\theta} \sin \phi + \dot{\psi} \sin \theta \cos \phi$$

$$\omega_3 = n$$

by squaring and adding, the magnitude of $\boldsymbol{\omega}$ is represented as

$$\omega^2 = \omega_1{}^2 + \omega_2{}^2 + n^2 = \dot{\theta}^2 + \dot{\psi}^2 \sin^2 \theta + n^2 \tag{5.7}$$

Remembering that rotational kinetic energy is given by equation (2.65), it is written here as

$$T_{\text{rot}} = \tfrac{1}{2} I_1 (\dot{\theta}^2 + \dot{\psi}^2 \sin^2 \theta) + \tfrac{1}{2} I_3 n^2$$

Potential energy can be referred to $\theta = 90°$, which gives a general form

$$U = mgl \cos \theta$$

Combining these results leads to total energy E, which is constant,

$$E = \tfrac{1}{2} I_1 (\dot{\theta}^2 + \dot{\psi}^2 \sin^2 \theta) + \tfrac{1}{2} I_3 n^2 + mgl \cos \theta \tag{5.8}$$

The other conservation equation is

$$h_Z = h_\xi \cos \theta + h_\eta \sin \theta = \text{constant}$$

Application of expressions (5.3) and (5.6) leads to

$$h_Z = I_3 n \cos \theta + I_1 \dot{\psi} \sin^2 \theta$$

which is solved for $\dot{\psi}$,

$$\dot{\psi} = \frac{h_Z - I_3 n \cos \theta}{I_1 \sin^2 \theta} \tag{5.9}$$

Equations (5.8) and (5.9) may be combined to eliminate $\dot{\psi}$,

$$E - \frac{I_3 n^2}{2} = \frac{I_1 \dot{\theta}^2}{2} + \frac{(h_Z - I_3 n \cos \theta)^2}{2 I_1 \sin^2 \theta} + mgl \cos \theta \tag{5.10}$$

Two relevant observations may be made at this point. Since $\theta \le 90°$ and $I_1 > 0$, the right-hand side of equation (5.10) is positive, requiring that

$$E - \frac{I_3 n^2}{2} > 0$$

Furthermore, expressions (5.10) represents a differential equation, which may be solved, in principle, for $\theta(t)$. Then $\psi(t)$ would be obtainable from form (5.9), and $\phi(t)$ from $\dot{\phi} = n - \dot{\psi} \cos \theta$. However, equation (5.10) is highly nonlinear. Fortunately, it need not be solved to describe the motion of interest.

Define a new set of constants for convenience,

$$a = \frac{2}{I_1}\left(E - \frac{I_3 n^2}{2}\right)$$

$$w = \frac{2mgl}{I_1}$$

$$k = \frac{h_Z}{I_1}$$

$$p = \frac{I_3 n}{I_1}$$

and substitute into equation (5.10),

$$a \sin^2 \theta = \dot{\theta}^2 \sin^2 \theta + (k - p \cos \theta)^2 + w \cos \theta \sin^2 \theta$$

Then introduce a new variable, $u = \cos \theta$, so that

$$\dot{u} = -\dot{\theta} \sin \theta$$

which leads to

$$\dot{u}^2 = (a - wu)(1 - u^2) - (k - pu)^2 \qquad (5.11)$$

It is possible to integrate this in terms of elliptic functions, but careful interpretation of expression (5.11) and other known properties of motion permits a general description of the gyro attitude history. If the right-hand side of equation (5.11) is thought of as a functional of u, then

$$f(u) = (a - wu)(1 - u^2) - (k - pu)^2 \qquad (5.12)$$

This may be qualitatively plotted after noting that at $u = \pm 1.0$, $f(u) < 0$. Furthermore, physical limits on θ imply that $0 \le u \le 1$ is the range of interest. Also, since $f(u) = \dot{u}^2$, $f(u)$ must become positive in this range of interest. Figure 5.3 illustrates the resulting curve of $f(u)$ vs u. The physically permissible limits of u are u_1 and u_2 corresponding to the bounds on positive values of $f(u)$ between $\theta = 0°$ and $\theta = 90°$. Remembering that $f(u) = \dot{\theta}^2 \sin^2 \theta$ and noting $f(u_1) = f(u_2) = 0$, leads to the requirement that $\dot{\theta} = 0$ at both u_1 and u_2. If gyro motion is described by the trajectory of its spin axis on a unit sphere, then $u_1 = \cos \theta_1$ and $u_2 = \cos \theta_2$ represent two bounding circles. A typical spin axis path is shown in Figure 5.4. This motion can be further defined for a given set of inertia and momentum data by considering changes of sign of $\dot{\psi}$ between θ_1 and θ_2. Rewriting equation (5.9) in terms of the new constants gives

$$\dot{\psi} = \frac{k - pu}{1 - u^2} \qquad (5.13)$$

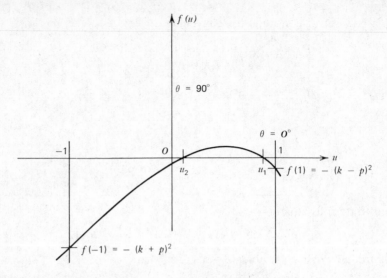

FIGURE 5.3 Qualitative plot of gyro nutation limits.

Noting that $u < 1$ between u_1 and u_2, the sign of $\dot{\psi}$ depends only on the relative sizes of k and pu. If, for example, $k > pu_1$, then $\dot{\psi} > 0$ for all values of θ between θ_1 and θ_2, because $u_1 > u_2$. This is the situation depicted in Figure 5.4. If $k = pu_1$, then $\dot{\psi} = 0$ at $\theta = \theta_1$ and $\dot{\psi} > 0$ for all values of $\theta > \theta_1$. This motion is illustrated in Figure 5.5. As a final example, consider the case in which $k = pu_i$ where $u_2 < u_i < u_1$. Then $\dot{\psi}$ changes sign between θ_1 and θ_2 as depicted in Figure 5.6.

The type of motion which the gyro or top exhibits is purely a function of initial conditions. For example, assume that at $t = 0$, $\theta = \theta_0$ and $\dot{\theta} = \dot{\psi} = 0$. This

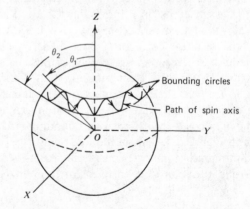

FIGURE 5.4 Depiction of spin axis trajectory of a gyro or top.

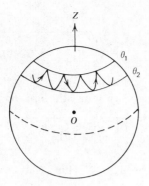

FIGURE 5.5 Cusp motion of gyro or top.

leads to constant values

$$h_Z = I_3 n \cos \theta_0$$

$$E - \tfrac{1}{2} I_3 n^2 = mgl \cos \theta_0$$

Substituting into expressions (5.9) and (5.10), respectively, leads to

$$\dot{\psi} = \frac{p(\cos \theta_0 - \cos \theta)}{\sin^2 \theta} \tag{5.14}$$

$$\dot{\theta}^2 = (\cos \theta_0 - \cos \theta)\left[w - \frac{p^2}{\sin^2 \theta}(\cos \theta_0 - \cos \theta) \right] \tag{5.15}$$

Since $\theta = 0$ at $t = 0$, the gyro must be initially at one of the limiting circles corresponding to θ_1 or θ_2. Equation (5.15) indicates which one is physically acceptable. If θ_0 were equal to θ_2, then $(\cos \theta_0 - \cos \theta)$ is negative and the right-hand side must be negative. Since this is not permissible, $\theta_0 = \theta_1$. The value of θ_2 can be determined from equation (5.15) by setting $\dot{\theta} = 0$ and solving

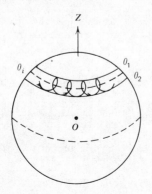

FIGURE 5.6 Precession rate reversal case.

for cos θ:

$$\cos\theta = \frac{p^2}{2w} \pm \sqrt{1 - \frac{p^2}{w}\cos\theta_1 + \left(\frac{p^2}{2w}\right)^2}$$

The correct value can be decided upon by testing the $+$ sign and constructing an inequality. Noticing that $p^2/2w$ must be less than 1 for this case,

$$\sqrt{1 - \frac{p^2}{w}\cos\theta_1 + \left(\frac{p^2}{2w}\right)^2} > \sqrt{1 - \frac{p^2}{w} + \left(\frac{p^2}{2w}\right)^2} = 1 - \frac{p^2}{2w}$$

This leads to the condition, $\cos\theta > 1$. Thus, the correct value is

$$\cos\theta_2 = \frac{p^2}{2w} - \sqrt{1 - \frac{p^2}{w}\cos\theta_1 + \left(\frac{p^2}{2w}\right)^2}$$

The resulting motion was depicted in Figure 5.5. Solutions such as those shown in Figures 5.4 and 5.6 require different initial conditions. For example, if at $t = 0$, $\theta = \theta_0$, $\dot{\theta} = 0$, $\dot{\psi} = \dot{\psi}_0 > 0$, then the motion illustrated in Figure 5.4 would be expected.

To illustrate the use of results for a spinning top, consider a case where the following quantities are observed at one instant,

$$\dot{\psi} = 0, \qquad \dot{\phi} = 100\,\text{rad/s}, \qquad \dot{\theta} = 2\,\text{rad/s}, \qquad \theta = 30°$$

The top is a large one with

$$I_1 = 1\,\text{N·m·s}^2, \qquad I_3 = 2\,\text{N·m·s}^2$$
$$m = 1\,\text{kg}, \qquad l = 0.2\,\text{m}$$

To determine θ_1, θ_2, and the shape of the spin axis path on the unit sphere, start with expression (5.6) which indicates that $n = 100\,\text{rad/s}$. Then forms (5.9) and (5.10) yield

$$h_Z = I_3 n\cos\theta_0 = 173.21\,\text{N·m·s}$$

$$E - \frac{I_3 n^2}{2} = \frac{I_1 \dot{\theta}_0^2}{2} + mgl\cos\theta_0 = 3.697\,\text{N·m}$$

Substitute these forms into expression (5.10), and solve for $\dot{\theta}^2$, to get

$$\dot{\theta}^2 = (\cos\theta_0 - \cos\theta)\left[w - \frac{p^2}{\sin^2\theta}(\cos\theta_0 - \cos\theta)\right] + \dot{\theta}_0^2$$

where $p = 200\,\text{rad/s}$ and $w = 3.92\,\text{s}^{-2}$. By trial and error methods the limiting values of θ are

$$\theta_1 = 29.43°$$
$$\theta_2 = 30.58°$$

which are very close because of the high spin rate, and stiffness of this top. Applying equation (5.9) gives the corresponding values of $\dot{\psi}$ as

$$\dot{\psi}_1 = -4.08 \text{ rad/s}$$

$$\dot{\psi}_2 = 3.95 \text{ rad/s}$$

Therefore, the spin axis follows the form of Figure 5.6 in a very narrow band between θ_1 and θ_2.

In many applications of gyroscopic instruments and gyrotorquing devices, *steady precession* is required. The conditions for such motion can be established by considering the curve of $f(u)$ shown in Figure 5.3. If this curve is tangent to the $f(u) = 0$ line then $u_1 = u_2$ and $\theta_1 = \theta_2$. This situation corresponds to constant nutation angle, θ_s. Expression (5.9) then implies a constant precession rate, $\dot{\psi}_s$. The associated initial conditions are $\theta = \theta_s$, $\dot{\theta} = 0$, $\dot{\psi} = \dot{\psi}_s$ at $t = 0$. However, the value of $\dot{\psi}_s$ which satisfies the physical situation is determined by specializing moment equation (5.5a),

$$mgl \sin \theta_s = I_3 n \dot{\psi}_s \sin \theta_s - I_1 \dot{\psi}_s^2 \sin \theta_s \cos \theta_s \qquad (5.16)$$

which becomes a quadratic in $\dot{\psi}_s$,

$$\dot{\psi}_s^2 - \left(\frac{I_3 n}{I_1 \cos \theta_s}\right) \dot{\psi}_s + \frac{mgl}{I_1 \cos \theta_s} = 0$$

The two possible solutions are

$$\dot{\psi}_s = \frac{I_3 n}{2 I_1 \cos \theta_s} \pm \sqrt{\left(\frac{I_3 n}{2 I_1 \cos \theta_s}\right)^2 - \frac{mgl}{I_1 \cos \theta_s}} \qquad (5.17)$$

An observation of importance is that $\dot{\psi}_s$ is real only if the quantity in the radical is not negative. Thus, steady precession is possible only if

$$n^2 \geq \frac{4 I_1 mgl \cos \theta_s}{I_3^2} \qquad (5.18)$$

Assuming this condition is satisfied, then either the $+$ or $-$ sign in equation (5.17) is mathematically valid. In order to explain the situation, a qualitative plot of $M_\xi = mgl \sin \theta_s$ vs $\dot{\psi}_s$ is shown in Figure 5.7. Note that for any given value of M_ξ, below $M_{\xi_{max}}$, two values of $\dot{\psi}_s$ are possible. Values above $\dot{\psi}_p$ are called *fast precession* rates and below correspond to *slow precession* rates. At $M_{\xi_{max}}$ equality in condition (5.18) holds and $\dot{\psi}_s = \dot{\psi}_p = I_3 n / 2 I_1 \cos \theta_s$. This also leads to

$$M_{\xi_{max}} = \frac{I_3^2 n^2}{4 I_1} \tan \theta_s \qquad (5.19)$$

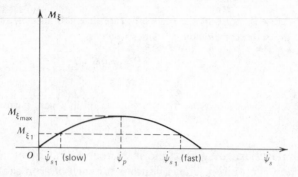

FIGURE 5.7 Range of steady precession values.

If the moment due to gravity is greater than this, steady precession is not possible unless n is increased for given gyro inertias and θ_s. In actual practice fast precession is very difficult to attain because of the high values of kinetic energy required. Thus, slow precession is assumed in all cases, and equation (5.17) should assume the form

$$\dot{\psi}_s = \frac{I_3 n}{2I_1 \cos \theta_s} - \sqrt{\left(\frac{I_3 n}{2I_1 \cos \theta_s}\right)^2 - \frac{mgl}{I_1 \cos \theta_s}} \qquad (5.20)$$

A special case of steady precession occurs when $\theta_s = 0$. Then, according to expression (5.18), the required spin rate is

$$n \geq \frac{2}{I_3} \sqrt{mgl I_1}$$

in order to maintain a *sleeping top*. Thus, a steady, small perturbing torque would induce steady precession at a small value of θ_s. This principle is used when launching sounding rockets and small missiles. A high spin rate is induced to allow passive stabilization for short flights. If the attitude is perturbed away from the velocity vector, a torque due to drag results. The vehicle reacts by precessing steadily around the velocity vector. Thus, no course change results.

A steadily precessing gyro may experience a disturbance in both precession and nutation. It is important to anticipate the resulting effects on nominal motion. A disturbance can be modeled by small variations of nutation angle and precession rate, θ_d and $\dot{\psi}_d$, respectively. Therefore, general values of these parameters become

$$\theta = \theta_0 + \theta_d, \qquad \dot{\psi} = \dot{\psi}_0 + \dot{\psi}_d$$

Note that $\dot{\theta} = \dot{\theta}_d$, $\ddot{\psi} = \ddot{\psi}_d$, since θ_0 and $\dot{\psi}_0$ are assumed constant. Limiting the analysis to first order effects permits several simplifications, such as

$$\dot{\theta}\dot{\psi} = \dot{\theta}_d(\dot{\psi}_0 + \dot{\psi}_d) \cong \dot{\theta}_d\dot{\psi}_0$$

$$\sin\theta = \sin\theta_0 \cos\theta_d + \cos\theta_0 \sin\theta_d \cong \sin\theta_0 + \theta_d \cos\theta_0 \Bigg\}$$

$$\cos\theta \cong \cos\theta_0 - \theta_d \sin\theta_0$$

$$(5.21)$$

First consider the η-axis motion given by equation (5.5b). Incorporating these identities and ignoring second and higher order terms in the disturbance parameters leads to

$$I_1 \sin\theta_0 \frac{d\dot{\psi}_d}{dt} = (I_3 n - 2I_1\dot{\psi}_0 \cos\theta_0)\frac{d\theta_d}{dt}$$

This can be integrated directly, assuming that $\dot{\psi}_d$ is zero when θ_d is zero,

$$I_1\dot{\psi}_d \sin\theta_0 = (I_3 n - 2I_1\dot{\psi}_0 \cos\theta_0)\theta_d \qquad (5.22)$$

Next, substitute identities (5.21) into equation (5.5a) to obtain the disturbed motion about ξ,

$$I_1\ddot{\theta}_d - I_1(\dot{\psi}_0^2 + 2\dot{\psi}_0\dot{\psi}_d)(\sin\theta_0 \cos\theta_0 - \theta_d \sin^2\theta_0 + \theta_d \cos^2\theta_0)$$
$$+ I_3 n(\dot{\psi}_0 + \dot{\psi}_d)(\sin\theta_0 + \theta_d \cos\theta_0) = mgl(\sin\theta_0 + \theta_d \cos\theta_0) \quad (5.23)$$

Since the motion of interest is that which is superimposed on nominal steady precession, extract an expression in which $\ddot{\theta}_d = 0$, $\dot{\psi}_d = 0$, $\dot{\psi} = \dot{\psi}_0$, and $\theta = \theta_0$ from equation (5.23),

$$-I_1\dot{\psi}_0^2 \cos\theta_0 + I_3 n\dot{\psi}_0 = mgl$$

which is equivalent to form (5.16). Now subtract this from expression (5.23) to describe the perturbed motion with respect to steady precession,

$$I_1\ddot{\theta}_d - I_1(2\dot{\psi}_0\dot{\psi}_d \sin\theta_0 \cos\theta_0 - \dot{\psi}_0^2\theta_d \sin^2\theta_0 + \dot{\psi}_0^2\theta_d \cos^2\theta_0)$$
$$+ I_3 n(\dot{\psi}_0\theta_d \cos\theta_0 + \dot{\psi}_d \sin\theta_0) = mgl\theta_d \cos\theta_0 \quad (5.24)$$

Eliminating $\dot{\psi}_d$ by combining this with expression (5.22) gives a differential equation for disturbed nutation,

$$I_1^2\ddot{\theta}_d + [(I_3 n)^2 - 4I_1 mgl \cos\theta_0 + I_1^2\dot{\psi}_0^2 \sin^2\theta_0]\theta_d = 0 \qquad (5.25)$$

This has the form

$$\ddot{\theta}_d + \omega_\xi^2\theta_d = 0$$

where

$$\omega_\xi = \frac{\sqrt{(I_3 n)^2 - 4I_1 mgl \cos\theta_0 + I_1^2\dot{\psi}_0^2 \sin^2\theta_0}}{I_1} \qquad (5.26)$$

Thus, the spin axis nods with frequency $\omega_\xi/2\pi$. If n is very large, which is typical of gyros, then expression (5.26) can be simplified to

$$\omega_\xi \cong \frac{I_3 n}{I_1}$$

which indicates that the nutation rate is larger than the spin rate because $I_3 > I_1$. Precession effects may be determined from equation (5.22) rewritten as

$$\dot{\psi}_d = \left(\frac{I_3 n - 2I_1 \dot{\psi}_0 \cos \theta_0}{I_1 \sin \theta_0}\right) \theta_d \qquad (5.27)$$

Define λ by

$$\lambda^2 = \frac{I_3 n - 2I_1 \dot{\psi}_0 \cos \theta_0}{I_1 \sin \theta_0}$$

to rewrite equation (5.27) as

$$\dot{\psi}_d = \lambda^2 \theta_d \qquad (5.28)$$

Differentiate twice and note that $\ddot{\theta}_d = -\omega_\xi^2 \theta_d$, to obtain

$$\dddot{\psi}_d = -\lambda^2 \omega_\xi^2 \theta_d$$

Reusing equation (5.28) allows elimination of θ_d,

$$\dddot{\psi}_d + \omega_\xi^2 \dot{\psi}_d = 0 \qquad (5.29)$$

Therefore, the frequency of precession rate is the same as that of the nutation angle, and the period of $\dot{\psi}_d$ oscillation equals the period of θ_d. Thus, a disturbance to a steadily precessing gyro or top results in bounded oscillations about the steady state. This is *stable* motion. If damping is present in the inner gimbal support about ξ, then these oscillations will dissipate and steady precession would return. This is *asymptotic stability*.

5.1.3 Gimbal Effects

Unfortunately, most of the gyro devices in use have mechanical gimbals with finite masses and moments of inertia. Effects of these gimbals on gyro motion may be detrimental and should be anticipated. Consider the case in which the gyro center of mass coincides with the gimbal geometric center, with nomenclature defined in Figure 5.1. Gimbal rates about each of the gimbal degrees of freedom are ω_ξ, ω_ζ, and $\dot{\psi}$. Appropriate principal moments of inertia for the inner and outer gimbals are defined as:

$$I_\xi^i = \text{moment of inertia of inner gimbal about } \xi$$

$$I_\eta^i = \text{moment of inertia of inner gimbal about } \eta$$

$$I_\zeta^i = \text{moment of inertia of inner gimbal about } \zeta$$

$$I_Z^o = \text{moment of inertia of outer gimbal about } Z$$

Formulation of the appropriate equations of motion is accomplished in several steps. The combined inertias of gyro and inner gimbal are

$$I_\xi = I_1 + I_\xi^i$$
$$I_\eta = I_1 + I_\eta^i$$
$$I_\zeta = I_\zeta^i$$

Note that I_ζ includes only I_ζ^i because the gyro is unaffected by gimbal rotating about ζ. The moment of inertia of gyro and inner gimbal about the Z axis is required to include outer gimbal effects. Using formula (2.62) to determine inertia about the axis of rotation yields

$$I_Z = I_\xi l_{Z\xi}^2 + I_\eta l_{Z\eta}^2 + I_\zeta l_{Z\zeta}^2 + I_Z^o$$

where $l_{Z\xi} = 0$, $l_{Z\eta} = \sin\theta$, $l_{Z\zeta} = \cos\theta$. With these substitutions, I_Z becomes,

$$I_Z = (I_1 + I_\eta^i)\sin^2\theta + I_\zeta^i \cos^2\theta + I_Z^o \qquad (5.30)$$

The angular momentum components about the gimbal axes are

$$h_\xi = (I_1 + I_\xi^i)\dot\theta \qquad (5.31a)$$

$$h_\eta = (I_1 + I_\eta^i)\dot\psi \sin\theta \qquad (5.31b)$$

$$h_\zeta = I_\zeta^i \dot\psi \cos\theta + I_3(\dot\phi + \dot\psi \cos\theta) \qquad (5.31c)$$

Since the outer gimbal can only rotate about Z and the inner gimbal about ξ, it is convenient to write the equations of motion about these two axes. This is accomplished by transforming the components of momentum from ξ, η, ζ axes to a new set, ξ', η', ζ' defined in Figure 5.8. Transformation equations are

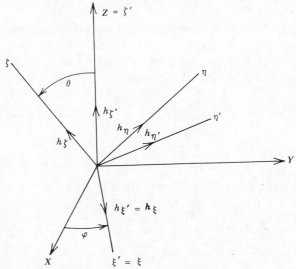

FIGURE 5.8 Gimbal coordinate transformation.

given by

$$\xi' = \xi$$
$$\eta' = \eta \cos \theta - \zeta \sin \theta$$
$$\zeta' = \eta \sin \theta + \zeta \cos \theta$$

Thus, the components of momentum become

$$h_{\xi'} = h_\xi \tag{5.32a}$$

$$h_{\eta'} = h_\eta \cos \theta - h_\zeta \sin \theta \tag{5.32b}$$

$$h_{\zeta'} = h_\eta \sin \theta + h_\zeta \cos \theta + I_z°\dot{\psi} \tag{5.32c}$$

The two appropriate equations of motion are now obtained by transforming set (5.4) into the new coordinates, and remembering to include the outer gimbal:

$$M_{\xi'} = M_\xi = \dot{h}_\xi + h_\zeta \dot{\psi} \sin \theta - h_\eta \dot{\psi} \cos \theta$$

$$M_{\zeta'} = M_Z = M_\eta \sin \theta + M_\zeta \cos \theta + I_z°\ddot{\psi}$$

$$= (\dot{h}_\eta + h_\xi \dot{\psi} \cos \theta - h_\zeta \dot{\theta}) \sin \theta$$

$$+ (\dot{h}_\zeta + h_\eta \dot{\theta} - h_\xi \dot{\psi} \sin \theta) \cos \theta + I_z°\ddot{\psi}$$

Combining these with set (5.32) permits reduction to

$$\left. \begin{array}{c} M_\xi = \dot{h}_{\xi'} - h_{\eta'} \dot{\psi} \\ \\ M_Z = \dot{h}_{\zeta'} \end{array} \right\} \tag{5.33}$$

Condition (2.76) for the free motion of an axisymmetric body is assumed to hold here. Thus, $\dot{\phi} + \dot{\psi} \cos \theta = n = $ constant. This should be valid since no torque is applied to the rotor and gimbal effects are expected to be small. Combining sets (5.31) and (5.32) with (5.33) leads to the desired equations of motion,

$$M_\xi = (I_1 + I_\xi^i)\ddot{\theta} + I_3 n\dot{\psi} \sin \theta + [I_\zeta^i - (I_1 + I_\eta^i)]\dot{\psi}^2 \sin \theta \cos \theta \tag{5.34}$$

and

$$M_z = (I_1 + I_\eta^i)(\ddot{\psi} \sin^2 \theta + 2\dot{\psi}\dot{\theta} \sin \theta \cos \theta) - (I_3 n + I_\zeta^i \dot{\psi} \cos \theta)\dot{\theta} \sin \theta$$

$$+ (I_\zeta^i \ddot{\psi} \cos \theta - I_\zeta^i \dot{\psi}\dot{\theta} \sin \theta) \cos \theta + I_z°\ddot{\psi}$$

$$= [(I_1 + I_\eta^i) \sin^2 \theta + I_\zeta^i \cos^2 \theta + I_z°]\ddot{\psi} + 2(I_1 + I_\eta^i - I_\zeta^i)\dot{\psi}\dot{\theta} \sin \theta \cos \theta$$

$$- I_3 n\dot{\theta} \sin \theta$$

or

$$M_z = \frac{d}{dt}(I_Z\dot{\psi}) - I_3 n\dot{\theta} \sin \theta \tag{5.35}$$

It is implicit in the development of these equations that $M_\xi = M_Z = 0$. There are cases in which gimbal bearing friction is important. To study these one cannot assume $n = $ constant, and a third equation for M_ζ must be added. However, this last equation is simply

$$M_\zeta = I_3 \frac{d}{dt}(\dot\phi + \dot\psi \cos \theta)$$

Differential equations (5.34) and (5.35) are generally not solvable analytically, because they are highly nonlinear. As an illustration of this difficulty and the effects of gimbal mass, a typical situation is presented. Start by setting $M_\xi = M_Z = 0$ with the gimbals at rest at $t = 0$, and $\theta(0) = \theta_0$. An impulsive moment is applied about ξ such that $\dot\theta(0) \to \alpha$. Note that $\dot\psi(0)$ is zero, because this moment is applied about ξ and a finite time is required for coupling into Z. No assumptions can yet be made with respect to the magnitude of $\dot\psi$, because this is the object of the analysis. Thus, equation (5.35) becomes

$$d(I_Z \dot\psi) = I_3 n \sin \theta \, d\theta$$

which can be integrated between limits corresponding to $t = 0$ and some later time

$$I_Z \dot\psi = -I_3 n(\cos \theta - \cos \theta_0) \tag{5.36}$$

Oscillations about the ξ-axis are assumed small, permitting the use of

$$\theta = \theta_0 + \theta_d$$

and the approximations of set (5.21) regarding θ, in addition to

$$\sin \theta \cos \theta \cong \sin \theta_0 \cos \theta_0 + \theta_d(\cos^2 \theta_0 - \sin^2 \theta_0)$$

Making the appropriate substitutions in expression (5.30), I_Z becomes

$$I_Z = I_Z(0) + 2\theta_d(I_1 + I_\eta{}^i - I_\zeta{}^i) \sin \theta_0 \cos \theta_0 \tag{5.37}$$

where

$$I_Z(0) = (I_1 + I_\eta{}^i) \sin^2 \theta_0 + I_\zeta{}^i \cos^2 \theta_0 + I_Z{}^o$$

Equation (5.36) can now be written as

$$I_Z(0)\dot\psi - (I_3 n \sin \theta_0)\theta_d + 2\theta_d \dot\psi(I_1 + I_\eta{}^i - I_\zeta{}^i) \sin \theta_0 \cos \theta_0 = 0 \tag{5.38}$$

Next, equation (5.34) becomes

$$(I_1 + I_\xi{}^i)\ddot\theta_d + (I_3 n \sin \theta_0)\dot\psi + \{(I_3 n \cos \theta_0)\theta_d \dot\psi - (I_1 + I_\eta{}^i - I_\zeta{}^i)[\sin \theta_0 \cos \theta_0$$
$$+ \theta_d(\cos^2 \theta_0 - \sin^2 \theta_0)]\dot\psi^2\} = 0 \tag{5.39}$$

These two equations represent the disturbed motion of the gyro. However, they are both nonlinear in their last terms because products of θ_d and $\dot\psi$ are retained. Perturbation techniques have been applied to this problem and an approximate solution obtained. Details are not presented here but are readily available in the literature. Results indicate that gimbal precession, $\dot{\psi}$ can be thought of as small, superimposed periodic oscillations of frequencies ω and 2ω where

$$\omega^2 = \frac{(I_3 n \sin \theta_0)^2}{I_Z(0)(I_1 + I_\xi^i)}$$

plus a constant drift, $\dot{\psi}_s$ given as

$$\dot{\psi}_s = -\frac{\alpha^2 I_3 n}{2\omega^2 I_Z^2(0) \sin \theta_0}[(I_1 + I_n^i - I_\xi^i) \sin^3 \theta_0 \cos \theta_0 - I_Z(0) \sin \theta_0 \cos \theta_0]$$

Thus, the outer gimbal oscillates and drifts in the direction opposite to spin. This drift is known as *gimbal walk*, and it is accompanied by a high speed nutation oscillation, which is not associated with a net drift in θ.

5.1.4 Basic Gyro Instruments

Two-degree-of-freedom gyros, the earliest type used, have found applications in attitude instruments, fire control systems, and satellite stabilization. As higher accuracy requirements developed related technology advanced with respect to the problems of gimbals. For example, when inner and outer gimbals become aligned ($\theta = 0°$ or $180°$) they can no longer provide two degrees of freedom. One way to avoid this situation is to add an outer gimbal which insures orthogonal alignment of the two inner gimbals. Another improvement is to float the rotor gimbals to remove bearing friction. This results in the greater accuracy required for inertial navigation. Progress has been made to the point where two-degree-of-freedom gyros have been produced without gimbals. This is accomplished by suspending the rotor in a fluid, electrostatic field, or magnetic field. The latter two methods are still awaiting widespread application. All such gyros must have nutation damping with respect to inertial space to insure asymptotically stable performance.

The single-degree-of-freedom gyro has only an inner gimbal. Such gyros operate by nulling applied torques about an output axis rather than by geometrical reorientation as with the two-degree-of-freedom version. A typical single-degree-of-freedom gyro is illustrated in Figure 5.9. There are three types of these gyros. The *rate gyro* uses an elastic restraining torque, for example, a spring. Thus, steady deflection about the output axis allows a sensor to indicate a constant inertial angular velocity about the input axis. An *integrating gyro* uses a damper to restrain output axis motion. Thus, if the spring in Figure 5.9 is removed, this becomes an integrating gyro. Deflection of the rotor relative to the case is a measure of the change of angular attitude about the input axis.

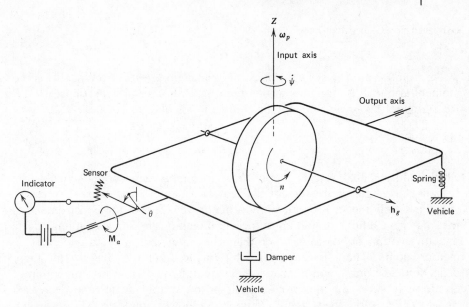

FIGURE 5.9 Single-degree-of-freedom rotor used as a rate gyro.

Finally, an *unrestrained* gyro is essentially an integrating gyro with very low damping.

A major contribution to the practical understanding of the gyro is a simplified vector model introduced by Dr. Charles Draper of M.I.T. The classical treatment is tedious to apply, thus, the simplified approach is summarized briefly. Generally, the solution of motion has transient and forced parts of the response. The forced solution results in precession, while transients are damped out for practical gyros. Therefore, the precessional solution is the only one of interest, and it can be written as

$$\boldsymbol{\omega}_p \times \mathbf{h}_g = \mathbf{M}_a \qquad (5.40)$$

where $\boldsymbol{\omega}_p$ is the angular velocity of the gyro momentum vector, $\mathbf{h}_g$ is the angular momentum of the gyro, and $\mathbf{M}_a$ is the applied torque. Equation (5.40) can be interpreted physically: angular momentum precesses in an attempt to align itself with the applied torque. This is also equivalent to $\mathbf{M}_a = \dot{\mathbf{h}}_g$, since $|\mathbf{h}_g|$ is constant here. Rate gyro performance can be quickly obtained by applying expression (5.40) to Figure 5.9. If the vehicle to which the rate gyro is mounted turns about the input axis at a rate $\dot{\psi}$, then the torque produced about the output axis is $I_3 n\dot{\psi}$. Notice that the gimbal is deflected through θ until a restraining torque is generated by the spring. The situation is then equivalent to the spacecraft applying torque $\mathbf{M}_a$ in order to generate precession $\dot{\psi}$. Thus equation (5.40) reduces to

$$I_3 n \dot\psi = k \delta l$$

where k and δ are the spring stiffness and deflection, respectively, and l is the moment arm of the spring about the output axis. For small deflections, $\theta = \delta / l$. Rearranging this gives gimbal deflection in terms of vehicle rate,

$$\theta = \frac{I_3 n \dot\psi}{k l^2} \tag{5.41}$$

which indicates that rate is proportional to gimbal deflection. Actually, the response of a rate gyro is essentially that of a damped oscillator. Expression (5.41) is simply the steady state part. A major weakness of these instruments is that they have a finite output angle for a finite input rate. Thus, coupling effects due to rotation about axes other than the input axis may cause errors in measurements. There is an upper limit on accurate use due to this finite deflection. The unit could be made *stiffer* by designing in a low momentum to elastic force ratio, but this would result in low output angles and decreased signal to noise ratio. Thus, there are upper and lower bounds on stiffness when using elastic restoring forces. Nevertheless, rate gyros offer moderate accuracy with a minimum amount of physical equipment. Early designs were supported mechanically with ball bearings and a linkage indicator gave direct readings. Both air and electric driven rotors were available. Inexpensive aircraft rate gyros still employ this type of design. More recently, fluid supports have been introduced to improve accuracy and life.

In the case of an integrating gyro there is no elastic restraint. If a viscous damper is employed with c as the coefficient of viscous damping, then the torque $I_3 n \dot\psi$ is balanced by $cl\,\dot\delta$. Thus, equation (5.40) becomes

$$I_3 n \dot\psi = cl\,\dot\delta = cl^2 \dot\theta \tag{5.42}$$

which can be integrated directly to give a change in θ, corresponding to a change in precession angle ψ,

$$\Delta\theta = \frac{I_3 n}{cl^2} \int \dot\psi \, dt = \frac{I_3 n}{cl^2} \Delta\psi \tag{5.43}$$

Therefore, deflection of the gimbal $\Delta\theta$ is proportional to vehicle rotation about the input axis $\Delta\psi$. The damping coefficient is usually greater than that for a rate gyro to limit gimbal deflection, which in turn limits cross-coupling effects. In many designs damping is accomplished through the fluid support of the gimbal. Equation (5.42) indicates integrating gyros are simple first-order devices. Equation (5.43) implies that this device acts like a bevel gear train with a gear ratio of $I_3 n / cl^2$ between input and output axes.

5.1.5 The Gyrocompass

Return to a two-degree-of-freedom gyro of importance, the gyrocompass. This is an instrument which indicates a given inertial reference direction at all times. It must periodically be updated because of drift arising from gimbal effects and other perturbations. The application of greatest use is in aircraft, where this instrument maintains a reference to the earth's north (usually magnetic north because a magnetic compass is used for setting the gyro). However, this application introduces a dilemma: the gyro wants to maintain an inertial direction, but the earth is rotating. This situation is depicted in Figure 5.10. If the gyro spin axis is aligned with the *local north* direction $-\mathbf{i}$, then it will point at north for only a short period of time. Since **i** rotates with the earth and the gyro axis is inertially fixed these two directions are aligned only once per day. The solution is to introduce a steady precession rate into the gyro which is just enough to maintain the $-\mathbf{i}$ direction. Figure 5.11 illustrates a way of doing this. A small torque is applied about ξ, which can be thought of as a small weight, w hanging from the inner gimbal. The requirement is to precess this gyro about **k** at a rate

$$\dot{\psi} = \Omega \sin \lambda \tag{5.44}$$

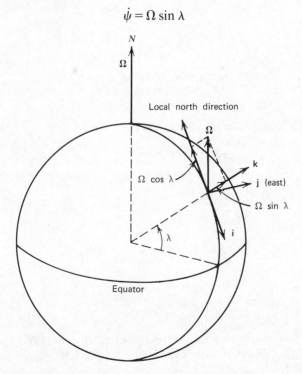

FIGURE 5.10 Nomenclature for gyrocompass.

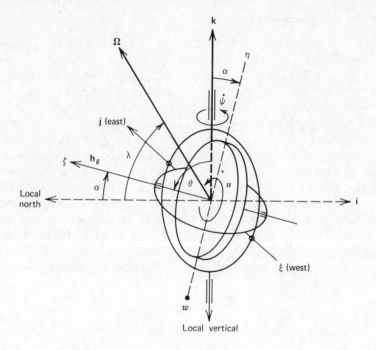

FIGURE 5.11 Aircraft gyrocompass geometry.

which is the component of $\boldsymbol{\Omega}$ about the local vertical. By initially setting the gyro at an angle α above the local horizon, a torque and a precession rate can be induced. Since the rotor spin rate n is assumed large, equation (5.40) can be used to relate α to $\dot{\psi}$. The applied moment $wl \sin \alpha$ must be balanced by $\boldsymbol{\omega}_p \times \mathbf{h}_g$, or

$$wl \sin \alpha = I_3 n \omega_\eta$$

Here

$$\omega_\eta = \Omega(\sin \lambda \cos \alpha - \cos \lambda \sin \alpha)$$

which leads to

$$\tan \alpha = \frac{I_3 n \Omega \sin \lambda}{wl + I_3 n \Omega \cos \lambda} \tag{5.45}$$

This gives the required value of α to maintain north direction if at latitude λ. Thus, each aircraft gyrocompass is set for a given latitude. Since high accuracy is not essential for most applications, gyrocompasses in the U.S. are set for an appropriate latitude of between 25 and 48 degrees.

5.2 MOMENTUM EXCHANGE TECHNIQUES

Gyroscopic instruments are typically so small with respect to the vehicle in which they are used that torques which they produce have no effect on the vehicle itself. Thus, they must direct actuators which apply large torques to the spacecraft or aircraft. When aerodynamic forces are available, ailerons, rudders, or elevators can provide these control torques. In space, however, the environment does not permit the use of conventional control surfaces, and other methods have been developed. The technique of interest in this section is the use of *angular momentum exchange*. Such methods utilize redistribution of momentum among the various subsystems of a spacecraft. Thus, if large enough gyros are used in a vehicle, their torques may be able to provide direct control of attitude. There are several types of momentum exchange concepts and devices in use today. Selection for a particular application is a function of the mission requirements, spacecraft size, and other constraints. Several of the devices and techniques are now presented. Their integration into attitude control systems is considered in Chapter 6.

5.2.1 Spin Stabilization

The simplest way to apply momentum directly to attitude control is to allow the whole spacecraft to spin about a single axis. Of course, this severely constraints the design and mission applications, but is inherently stable and essentially passive. In other words, a large spinning body is gyroscopically stiff. Design constraints include the requirement that the vehicle be configured to permit spin about its major principal axis due to the stability criterion of Section 2.5. Such designs have typically been axisymmetric, such as the first successful synchronous communications satellite, Syncom II, described in Section 4.1. Small perturbing torques may precess the momentum vector slowly away from the desired orbit-normal direction. Periodic attitude corrections are carried out with small thrusters. This precession correction technique was also discussed in Section 4.1.

5.2.2 Internal Moving Parts

Most missions cannot be accomplished with *simple spinners* such as Syncom II. Thus, other momentum exchange techniques have been developed. Spacecraft employing these methods fall into a large class of dynamical systems sometimes referred to as bodies with internal moving parts. Vehicles with internal gyros represent a more specialized category, because the spacecraft center of mass does not move when attitude torques are applied. Consider first the general case of a body with internal moving mass. Such effects on attitude motion of a space vehicle were first investigated separately by Roberson and Grubin. Both developed the equations of motion for a rigid body carrying an arbitrary number of rigid bodies. Roberson chose the composite center of mass of the system as a reference point for the equations of motion. This formulation leads to time varying moments of inertia of the main vehicle (or platform), because the reference point is moving relative to the vehicle as the masses move inside. Grubin circumvented this complication by choosing the center of mass of the platform as the reference point. Then, moments of inertia of the main vehicle are constant relative to the body fixed coordinate system. The resulting equations of motion are not simpler than those of Roberson, but the platform center of mass seems more natural. Furthermore, motion of an internal mass may be simply expressed relative to the vehicle axes. General equations of motion using Grubin's approach are derived here. Then specialized forms are extracted to handle simple rotor and gyro devices. Section 5.3 will present attitude stabilization schemes using other types of moving masses.

In the early days of space flight, Grubin pointed out that Euler's angular momentum equation need not be restricted to cases of origin fixed or attached at the accelerating center of mass. In fact, the origin can be arbitrary and have arbitrary motion, provided an additional term is included in the equation. Start with classical equation (2.69)

$$\mathbf{M}_o = \frac{d\mathbf{h}_o}{dt}$$

referred to the center of mass, point O in Figure 5.12. Introduce a translational equation

$$\mathbf{F} = m\mathbf{a}_o$$

where $\mathbf{F}$ is an external applied force, m is the total system mass, and $\mathbf{a}_o$ is the acceleration of the mass center. Consider an arbitrary point A with arbitrary motion. The transformations of momentum and torque about A referred to O are

$$\mathbf{h}_o = \mathbf{h}_A - \mathbf{r}_A m \times \dot{\mathbf{r}}_A \qquad (5.46a)$$

$$\mathbf{M}_o = \mathbf{M}_A - \mathbf{r}_A \times \mathbf{F} \qquad (5.46b)$$

$$\mathbf{a}_o = \mathbf{a}_A + \ddot{\mathbf{r}}_A \qquad (5.46c)$$

where $\mathbf{r}_A$ is the position vector of O with respect to A. Combining these equations yields

$$\mathbf{M}_A - \mathbf{r}_A \times m(\mathbf{a}_A + \ddot{\mathbf{r}}_A) = \frac{d}{dt}(\mathbf{h}_A - m\mathbf{r}_A \times \dot{\mathbf{r}}_A)$$

or

$$\mathbf{M}_A = \dot{\mathbf{h}}_A + \mathbf{S}_A \times \mathbf{a}_A \qquad (5.47)$$

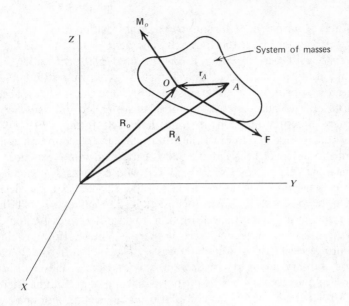

FIGURE 5.12 Motion referred to an arbitrary point.

where $S_A = m\mathbf{r}_A$, the static moment of the body with respect to point A, and $\mathbf{a}_A$ is the absolute acceleration of point A. Thus, for a rigid body with 1, 2, 3 as principal axes and origin at an arbitrary point the *generalized Euler equations* become

$$M_1 = I_1\dot{\omega}_1 + (I_3 - I_2)\omega_2\omega_3 + (S_2 a_3 - S_3 a_2) \tag{5.48a}$$

$$M_2 = I_2\dot{\omega}_2 + (I_1 - I_3)\omega_3\omega_1 + (S_3 a_1 - S_1 a_3) \tag{5.48b}$$

$$M_3 = I_3\dot{\omega}_3 + (I_2 - I_1)\omega_2\omega_1 + (S_1 a_2 - S_2 a_1) \tag{5.48c}$$

5.2.3 Dual Spinners

There is a special case in which internal motion takes place that is of great interest to the spacecraft designer. This is the situation when an internal gyro or rotor is used to redistribute momentum by either changing its speed or through the phenomenon of gyrotorquing. Such techniques include the use of momentum wheels to maintain bias momentum for stiffness, control moment gyros to apply control torques when needed, and reaction wheels which nominally maintain zero momentum but change speed to apply torques about their spin axes. These will be considered in connection with control system design. The particular technique of interest here has evolved from the simple spinner idea used originally on Explorer I. One primary limitation of these satellites is that no oriented sensors or antennas can be employed, because all parts rotate together about the spin axis. The next logical step in the evolution of such spacecraft was to combine an oriented platform and a rotor. This concept appeared to maintain the advantage of gyroscopic stiffness and permit an oriented platform for scientific instruments, antennas, etc. Spacecraft with large rotors and despun (oriented) platforms are called *dual spinners*. Most of the commercial communications satellites launched in the 1970's will have this configuration. In fact, satellites of the International Telecommunications Satellite Organization (INTELSAT) have been of the dual spin type since 1968. The first of these was known as INTELSAT III, whose spin axis coincided with its major principal axis. Beginning in 1971, the INTELSAT IV designs were launched. One of these is shown in Figure 5.13. Unlike the INTELSAT III, which was *oblate* in its inertia distribution, this satellite was *prolate*. That is, it spins about its minor principal axis of inertia. The platform is despun and earth-pointing, while the rotor maintains gyroscopic stiffness. Of course, the major axis stability condition of Section 2.5 for rigid bodies with dissipation prompted the question of stability of these dual spinners. Early designs, such as INTELSAT III, had relatively small platforms or despun antennas and rotors spun about the major vehicle axis. Stability criteria were assumed identical to those of a simple spinner, and flight performance demonstrated that these were essentially stable. However, launch vehicle shroud constraints limited rotor diameters. The major axis stability condition effectively limited spinning spacecraft sizes. Nevertheless, larger communications satellites were in demand in the mid-1960's.

The U.S. Air Force decided to try and bypass this limitation by orbiting an experimental dual-spinner whose rotor spun about the vehicle minor axis. This was to be the Tactical Communications Satellite (TACSAT). The primary problem was to devise appropriate stability criteria for such a configuration. Hughes Aircraft Company did develop a set of conditions which were related to energy dissipation mechanisms in the platform and rotor. Briefly stated, *the addition of dissipating devices on the despun portion of a dual spinner would offset the destabilizing effect of dissipation in the rotor.* Furthermore, a large nutation damper mounted on the platform would provide nutational stability about the minor axis. Hughes Aircraft Company was able to patent this concept and labeled it *Gyrostat*. This development meant that the platform size was no longer limited by inertia ratio constraints. The rotor need not be rigid, but it must not dissipate energy at a higher rate than the platform in order to be stable. Arguments concerning stability of dual spinners were not rigorous at that time, but the Air Force decided to go ahead with TACSAT. It was launched in February 1969. The spacecraft performed successfully and the INTELSAT IV series followed.

Several rigorous stability arguments were considered by Likins, Pringle, and

FIGURE 5.13 INTELSAT IV, the first commercial prolate dual spinner (Hughes Aircraft Company photo).

others in the late 1960's. The fact that the major spin axis criterion does not apply to dual spinners had been independently discovered by V. D. Landon and A. J. Iorillo. However, dissemination of early results was delayed by an unfortunate rejection of Landon's first paper in 1962. The problem was treated by evaluating the force-torque balance in the dampers which tend to stabilize motion. Landon and B. Stewart of RCA later published (1964) an *energy sink* approximation which had been previously applied to simple spinners. This work was limited to a fully axisymmetric dual spinner with a massless energy dissipation *sink* in one of the bodies. At the same time Iorillo was developing the Gyrostat concept at Hughes and in 1965 presented arguments extending knowledge to axisymmetric satellites with damping in both bodies. This extension made the Gyrostat possible. Rigorous and energy sink stability arguments will now be presented for dual-spinners with dampers.

The rigorous stability analysis follows a discrete parameter approach to describing the damper. Consider a dual-spin satellite, illustrated in Figure 5.14, with an asymmetric platform P attached to a rotor R which is axisymmetric. A platform damper is assumed to be of the form of a spring-mass-dashpot. Stability results should be independent of the type of damper, provided that damping takes place for any nutational motion. However, time constants will vary. The rotor is centered on z and is permitted rotation about z only. The

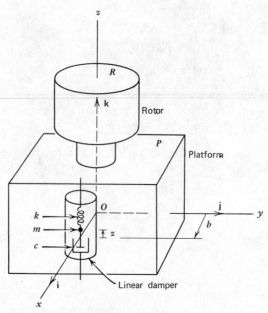

FIGURE 5.14 Dual spinner with platform damper.

damper is centered on the x axis and mass movement is parallel to z at distance b. Spring has constant k, and dashpot has damping constant c. The vehicle (including rotor and platform) center of mass is at point O when $z = 0$, and I_x, I_y, I_z are the principal moments of inertia of the entire spacecraft when $z = 0$. In general $(z \neq 0)$, the inertia tensor is

$$\mathbf{I} = \begin{bmatrix} I_x + mz^2 & 0 & -mbz \\ 0 & I_y + mz^2 & 0 \\ -mbz & 0 & I_z \end{bmatrix} \tag{5.49}$$

The vector equation of motion is given by expression (5.47), when the undeformed $(z = 0)$ center of mass position is taken as the reference point. Thus, O and A coincide when $z = 0$. For this case angular momentum about the reference point is expressed without subscript as

$$\mathbf{h} = \mathbf{I} \cdot \boldsymbol{\omega} - mb\dot{z}\mathbf{j} + I_z^R \Omega \mathbf{k} \tag{5.50}$$

where I_z^R is the rotor moment of inertia about the bearing axis and Ω is used here as the angular velocity of the rotor with respect to the platform. Note that $\boldsymbol{\omega}$ is the angular velocity of the platform. Also,

$$\mathbf{r}_A = \mu z \mathbf{k}$$

and the static moment is

$$\mathbf{S} = (m + M_P + M_R)\mu z \mathbf{k}$$

where M_P, M_R are the platform and rotor masses, respectively, and μ is defined as

$$\mu = \frac{m}{m + M_P + M_R} \tag{5.51}$$

This leads to a simple form of $\mathbf{S}$,

$$\mathbf{S} = mz\mathbf{k} \tag{5.52}$$

The absolute acceleration of the reference point is just $-\ddot{\mathbf{r}}_A$, thus,

$$\mathbf{a} = -\mu \frac{d^2}{dt^2}(z\mathbf{k})$$

which gives

$$\mathbf{a} = -\mu[(2\omega_y\dot{z} + \dot{\omega}_y z + \omega_x\omega_z z)\mathbf{i} - (2\omega_x\dot{z} + \dot{\omega}_x z - \omega_y\omega_z z)\mathbf{j} + (\ddot{z} - \omega_y^2 z - \omega_x^2 z)\mathbf{k}] \tag{5.53}$$

Combining expressions (5.52) and (5.53) gives the correction term

$$\mathbf{S} \times \mathbf{a} = (-m\mu\dot{\omega}_x z^2 + m\mu\omega_y\omega_z z^2 - 2m\mu\omega_x z\dot{z})\mathbf{i} + (-m\mu\dot{\omega}_y z^2 - m\mu\omega_x\omega_z z^2 - 2m\mu\omega_y z\dot{z})\mathbf{j}$$

Carrying out the operations indicated in equation (5.50) yields

$$\mathbf{h} = [(I_x + mz^2)\omega_x - mb\omega_z z]\mathbf{i} + [(I_y + mz^2)\omega_y - mb\dot{z}]\mathbf{j}$$
$$+ [-mb\omega_x z + I_z\omega_z + I_z{}^R\Omega]\mathbf{k} \tag{5.54}$$

Situations of interest exclude outside torques, thus, $\mathbf{M}_A = 0$. The Euler equations of motion for a dual spinner with platform damping, as described above, become

$$I_x\dot{\omega}_x - \omega_y\omega_z(I_y - I_z) + I_z{}^R\Omega\omega_y + m(1-\mu)\dot{\omega}_x z^2 - m(1-\mu)\omega_y\omega_z z^2$$
$$+ 2m(1-\mu)\omega_x z\dot{z} - mb\dot{\omega}_z z - mb\omega_x\omega_y z = 0 \tag{5.55}$$

$$I_y\dot{\omega}_y - \omega_z\omega_x(I_z - I_x) - I_z{}^R\Omega\omega_x + m(1-\mu)\dot{\omega}_y z^2 + m(1-\mu)\omega_x\omega_z z^2$$
$$+ 2m(1-\mu)\omega_y z\dot{z} - mb\ddot{z} + mb\omega_x{}^2 z - mb\omega_z{}^2 z = 0 \tag{5.56}$$

$$I_z\dot{\omega}_z - \omega_x\omega_y(I_x - I_y) + I_z{}^R\dot{\Omega} + mb\omega_y\omega_z z - 2mb\omega_x\dot{z} - mb\dot{\omega}_x z = 0 \tag{5.57}$$

These three equations in five unknowns, ω_x, ω_y, ω_z, Ω, and z will require two additional relationships associated with wheel torque and damper force balance. Since $I_z{}^R\dot{\omega}_z{}^R = T$ and $\omega_z{}^R = \omega_z + \Omega$, one of these equations is

$$I_z{}^R(\dot{\omega}_z + \dot{\Omega}) = T \tag{5.58}$$

where T is the magnitude of torque applied about the rotor axis. Bearing friction is ignored here. The acceleration-force balance of the damper mass gives

$$m(1-\mu)\ddot{z} + c\dot{z} + kz - m(1-\mu)(\omega_x{}^2 + \omega_y{}^2)z + mb\omega_x\omega_z - mb\dot{\omega}_y = 0 \tag{5.59}$$

which is obtained in the moving frame centered at A. Set (5.55) to (5.59) constitutes a complete description of attitude motion for this type of dual spinner. The first three of these are a statement of conservation of momentum. Equation (5.58) can represent changes of energy due to variation of rotor and platform speeds, and form (5.59) accounts for energy dissipation. If the rotor happens to be a small momentum wheel, then these equations represent the motion of a *fixed-gimbal, bias-momentum* satellite with damping. These are used in the design of automatic attitude control systems for such vehicles.

One special solution of interest corresponds to nominal orientation and operation on orbit for a typical dual-spin communications satellite. Thus,

$$\left.\begin{array}{l} \omega_z = \omega_P = \text{const.} \\ \Omega = \Omega_R = \text{const.} \\ \omega_x = \omega_y = z = 0 \\ T = 0 \end{array}\right\} \tag{5.60}$$

is a solution giving nominal attitude motion. In most practical cases ω_P is just the orbital rate or mean motion for a circular orbit. This permits continuous earth-pointing of the platform which may consist of several high-gain antennas. Stability of the motion described by set (5.60) can be tested by perturbing this nominal solution, which results in linearized equations of motion. These are then treatable by linear techniques. A perturbed form of nominal motion is established by defining variables as follows:

$$\omega_x = \omega_x$$
$$\omega_y = \omega_y$$
$$\omega_z = \omega_P + \omega_P{}^d$$
$$z = z$$
$$\Omega = \Omega_R + \Omega_R{}^d$$

where ω_P and Ω_R are constants, and ω_x, ω_y, $\omega_P{}^d$, z, and $\Omega_R{}^d$ are small quantities. These are substituted into the equations (5.55) to (5.59) to yield a new set of linear equations in the perturbed variables,

$$I_x\dot{\omega}_x + [(I_z - I_y)\omega_P + I_z{}^R\Omega_R]\omega_y = 0 \qquad (5.61a)$$

$$I_y\dot{\omega}_y - [(I_z - I_x)\omega_P + I_z{}^R\Omega_R]\omega_x - mb\ddot{z} - mb\omega_P{}^2 z = 0 \qquad (5.61b)$$

$$I_z\dot{\omega}_P{}^d + I_z{}^R\dot{\Omega}_R{}^d = 0 \qquad (5.61c)$$

$$I_z{}^R(\dot{\omega}_P{}^d + \dot{\Omega}_R{}^d) = 0 \qquad (5.61d)$$

$$m(1 - \mu)\ddot{z} + c\dot{z} + kz + mb\omega_P\omega_x - mb\dot{\omega}_y = 0 \qquad (5.61e)$$

Expressions (5.61c) and (5.61d) may be ignored, because they represent perturbed motion about the z-axis. Such motion is not of interest in considering stability. To simplify the remaining expressions define

$$\lambda_1 = \frac{[(I_z - I_y)\omega_P + I_z{}^R\Omega_R]}{I_x} = \frac{[I_z{}^P\omega_P + I_z{}^R\omega_R - I_y\omega_P]}{I_x}$$

$$\lambda_2 = \frac{[(I_z - I_x)\omega_P + I_z{}^R\Omega_R]}{I_y} = \frac{[I_z{}^P\omega_P + I_z{}^R\omega_R - I_x\omega_P]}{I_y}$$

where $\Omega_R = \omega_R - \omega_P$, ω_R is the constant part of $\omega_z{}^R$, and $I_z{}^P$ is the platform moment of inertia about the rotor axis. The angular momentum magnitude is nominally

$$h = I_z{}^P\omega_P + I_z{}^R\omega_R$$

and it is assumed unchanged by the perturbation. Applying this to the definitions of λ_1 and λ_2 leads to

$$\lambda_1 = \frac{h - I_y\omega_P}{I_x} \qquad \lambda_2 = \frac{h - I_x\omega_P}{I_y} \qquad (5.62)$$

For convenience define

$$p = \sqrt{\frac{k}{m}}, \qquad \beta = \frac{c}{m}, \qquad \zeta = \frac{z}{b}, \qquad \delta = \frac{mb^2}{I_y} \qquad (5.63)$$

so that equations (5.61a), (5.61b), and (5.61e) become

$$\dot{\omega}_x + \lambda_1 \omega_y = 0 \qquad (5.64a)$$

$$\dot{\omega}_y - \lambda_2 \omega_x - \delta\ddot{\zeta} - \delta\omega_P^2 \zeta = 0 \qquad (5.64b)$$

$$(1-\mu)\ddot{\zeta} + \beta\dot{\zeta} + p^2\zeta + \omega_P\omega_x - \dot{\omega}_y = 0 \qquad (5.64c)$$

Routh's method for testing stability of a linear system can now be implemented. It consists of establishing the characteristic equation of set (5.64) and arranging its coefficients in a special array. This characteristic equation is obtained by first converting set (5.64) into a set of simultaneous, linear algebraic equations by taking Laplace transforms,

$$s\omega_x(s) + \lambda_1\omega_y(s) = 0$$

$$s\omega_y(s) - \lambda_2\omega_x(s) - \delta s^2\zeta(s) - \delta\omega_P^2\zeta(s) = 0$$

$$(1-\mu)s^2\zeta(s) + \beta s\zeta(s) + p^2\zeta(s) + \omega_P\omega_x(s) - s\omega_y(s) = 0$$

Here s is the Laplace variable and initial values are taken as zero for the purpose of evaluating stability. Then the determinant of coefficients

$$\begin{vmatrix} s & \lambda_1 & 0 \\ -\lambda_2 & s & (-\delta s^2 - \delta\omega_P^2) \\ \omega_P & -s & [(1-\mu)s^2 + \beta s + p^2] \end{vmatrix} = 0$$

gives the characteristic equation as

$$s^4(1-\mu-\delta) + s^3\beta + s^2[p^2 - \delta\omega_P^2 + \lambda_1\lambda_2(1-\mu) - \lambda_1\delta\omega_P]$$
$$+ s\beta\lambda_1\lambda_2 + (\lambda_1\lambda_2 p^2 - \lambda_1\delta\omega_P^3) = 0 \quad (5.65)$$

Routh's criteria for asymptotic stability consist of the following conditions applied to equation (5.65):

1. All coefficients must be nonzero and of the same sign. This is a necessary condition.
2. Necessary and sufficient conditions for asymptotic stability are established by constructing and testing an array as described below.

For a general form of characteristic equation,

$$a_n s^n + a_{n-1} s^{n-1} + \cdots + a_1 s + a_o = 0$$

arrange the coefficients in array form,

$$
\begin{array}{cccc}
a_n & a_{n-2} & a_{n-4} & \cdots \\
a_{n-1} & a_{n-3} & a_{n-5} & \cdots \\
b_1 & b_2 & b_3 & \cdots \\
c_1 & c_2 & c_3 & \cdots \\
d_1 & d_2 & d_3 & \cdots \\
\cdots & \cdots & \cdots & \cdots
\end{array}
$$

where

$$
b_1 = \frac{a_{n-1}a_{n-2} - a_n a_{n-3}}{a_{n-1}}, \qquad b_2 = \frac{a_{n-1}a_{n-4} - a_n a_{n-5}}{a_{n-1}}, \cdots
$$

$$
c_1 = \frac{b_1 a_{n-3} - a_{n-1} b_2}{b_1}, \qquad c_2 = \frac{b_1 a_{n-5} - a_{n-1} b_3}{b_1}, \cdots
$$

$$
d_1 = \frac{c_1 b_2 - b_1 c_2}{c_1}, \qquad d_2 = \frac{c_1 b_3 - b_1 c_3}{c_1}, \cdots
$$

$$
e_1 = \frac{d_1 c_2 - c_1 d_2}{d_1}, \qquad \cdots
$$

This array terminates when all remaining terms in the first column are zero. Now the stability test consists of inspecting terms in the first column only. If all signs are alike the motion is asymptotically stable. Note that if the first term of any row is zero while others in that row are not, replace the zero with a very small number to complete the array. All signs in the first column must still be alike.

Applying this test to expression (5.65) leads to the following results:

1. Since, for typical systems $\delta \ll 1$, and $\mu \ll 1$, the only coefficient constraint is

$$
\lambda_1 \lambda_2 > 0 \tag{5.66}
$$

2. The Routhian array becomes

$$
\begin{array}{ccc}
(1-\mu-\delta) & [p^2 - \delta\omega_P{}^2 + \lambda_1\lambda_2(1-\mu) - \lambda_1\,\delta\omega_P] & (\lambda_1\lambda_2 p^2 - \lambda_1\,\delta\omega_P{}^3) \\[4pt]
\beta & \beta\lambda_1\lambda_2 & 0 \\
(p^2 - \delta\omega_P{}^2 + \delta\lambda_1\lambda_2 - \delta\lambda_1\omega_P) & (\lambda_1\lambda_2 p^2 - \lambda_1\,\delta\omega_P{}^3) & 0 \\[4pt]
\left(\dfrac{\delta\beta\lambda_1(\omega_P - \lambda_2)(\omega_P{}^2 - \lambda_1\lambda_2)}{p^2 - \delta\omega_P{}^2 + \delta\lambda_1\lambda_2 - \delta\lambda_1\omega_P} \right) & 0 & 0 \\[4pt]
[\lambda_1(\lambda_2 p^2 - \delta\omega_P{}^3)] & 0 & 0 \\[4pt]
0 & 0 & 0
\end{array}
$$

Since $\beta > 0$, all signs in the first column must be $+$. Therefore, the necessary and sufficient conditions for asymptotic stability of a dual spinner with platform damping become

$$\lambda_1\lambda_2 > 0 \tag{5.67a}$$

$$1 - \mu - \delta > 0 \tag{5.67b}$$

$$\beta > 0 \tag{5.67c}$$

$$p^2 - \delta(\omega_P^2 + \omega_P\lambda_1 - \lambda_1\lambda_2) > 0 \tag{5.67d}$$

$$\lambda_1(\omega_P - \lambda_2)(\omega_P^2 - \lambda_1\lambda_2) > 0 \tag{5.67e}$$

$$\lambda_1(\lambda_2 p^2 - \delta\omega_P^3) > 0 \tag{5.67f}$$

remembering that $\delta > 0$ by definition. All of the conditions in (5.67) must be satisfied. Real dampers require $p > 0$, and practical systems dictate that several of these inequalities are trivial. Take the limiting case as $m \to 0$ (and also $\mu \to 0$, $\delta \to 0$). The only nontrivial conditions left are

$$\lambda_1\lambda_2 > 0 \tag{5.68a}$$

$$\lambda_1(\omega_P - \lambda_2)(\omega_P^2 - \lambda_1\lambda_2) > 0 \tag{5.68b}$$

With the help of equations (5.62), the second of these may be written as

$$\frac{\lambda_1 h[\omega_P(I_x + I_y) - h]^2}{I_x I_y^2} > 0$$

and since $I_x I_y^2 > 0$ as is the squared term, this reduces to $\lambda_1 h > 0$. However, h can be established as positive by sign convention, leaving

$$\lambda_1 > 0$$

Comparing this with condition (5.68a) indicates the only other condition is

$$\lambda_2 > 0$$

Thus, all stability conditions reduce to

$$h - I_y\omega_P > 0, \qquad h - I_x\omega_P > 0 \tag{5.69}$$

Three important observations can now be made for special situations:

(a) In many applications the platform is completely despun, such that $\omega_P = 0$. Conditions (5.69) then requires only that $h > 0$, which is true by sign convention. It should be emphasized that this condition is trivial for damping on the platform only. Note that this result does not specify any inertia relationships.

(b) Now assume the rotor is despun and the platform with damper is spinning. Stability requires that

$$I_z^P - I_y > 0, \qquad I_z^P - I_x > 0$$

which is analogous to the major axis spin rule. Thus, the damper location is critical to stability.

(c) If both rotor and platform spin at different rates then conditions (5.69) become

$$\left. \begin{array}{l} I_z^R \omega_R + \omega_P (I_z^P - I_y) > 0 \\ I_z^R \omega_R + \omega_P (I_z^P - I_x) > 0 \end{array} \right\} \tag{5.70}$$

which indicate a minimum rotor speed is required for stability. For example, ω_R and ω_P could have opposite signs such that these conditions are not satisfied, even though I_z may be the major principal moment of inertia. Thus, it is conceivable that a dual spinner could be unstable while spinning about its major axis.

The rigorous treatment of dual-spin stability has provided much insight into conditions appropriate for many of the present communications satellites. However, the analysis just completed is very restrictive in terms of damping, and other meaningful approaches need to be attempted in order to develop more generalized conditions for stability. Thus, an energy sink approach developed by Iorillo is now offered to handle a dual spinner with dissipation in both rotor and platform. Energy and momentum considerations are applied to arrive quickly at stability conditions. The dual spin satellite to be treated is illustrated in Figure 5.15. For this analysis both rotor and platform are assumed axisymmetric and the only relative motion may be about z. Assume the rotor and platform are uncoupled and no torques are applied. Then the system angular momentum magnitude can be written as

$$h^2 = (I_z^R \omega_z^R + I_z^P \omega_z^P)^2 + (I_\eta \omega_\eta)^2$$

where I_η is the transverse moment of inertia of the spacecraft and ω_η is the transverse angular velocity. The total rotational kinetic energy is

$$2T = I_\eta \omega_\eta^2 + I_z^R (\omega_z^R)^2 + I_z^P (\omega_z^P)^2$$

Angular momentum is conserved and energy dissipation is accounted for through $\dot{T}$. Thus, differentiating the two preceding expressions yields

$$0 = (I_z^R \omega_z^R + I_z^P \omega_z^P)(I_z^R \dot{\omega}_z^R + I_z^P \dot{\omega}_z^P) + I_\eta^2 \omega_\eta \dot{\omega}_\eta \tag{5.71a}$$

$$\dot{T} = I_\eta \omega_\eta \dot{\omega}_\eta + I_z^P \omega_z^P \dot{\omega}_z^P + I_z^R \omega_z^R \dot{\omega}_z^R \tag{5.71b}$$

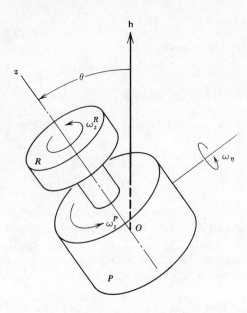

FIGURE 5.15 Dual-spin configuration for energy sink approach.

Combining these to eliminate $\omega_\eta \dot\omega_\eta$ leads to

$$\dot{T} = -I_z^R \lambda_R \dot\omega_z^R - I_z^P \lambda_P \dot\omega_z^P \tag{5.72}$$

where λ_R and λ_P are defined as

$$\lambda_R = \lambda_o - \omega_z^R$$

$$\lambda_P = \lambda_o - \omega_z^P$$

with

$$\lambda_o = \frac{I_z^P \omega_z^P + I_z^R \omega_z^R}{I_\eta}$$

The quantities ω_η and λ_o are taken as positive without loss of generality, and nutation angle θ is measured from the direction of the positive angular momentum vector. Note that $\dot{T}$ is the rate at which work is done by nonconservative forces in the rotor and platform, that is, the rate of energy dissipation. Each component contributes to this rate and $\dot{T}$ can be written as the sum of two parts

$$\dot{T} = \dot{T}_R + \dot{T}_P$$

with all quantities being negative in this expression. Since rotor and platform are assumed uncoupled, reaction torques which tend to change angular rates can be identified from equation (5.72) as

$$
\left.\begin{aligned}
I_z^{P}\dot{\omega}_z^{P} &= -\frac{\dot{T}_P}{\lambda_P} \\[2mm]
I_z^{R}\dot{\omega}_z^{R} &= -\frac{\dot{T}_R}{\lambda_R}
\end{aligned}\right\} \tag{5.73}
$$

Combining expression (5.71b) and (5.73) gives the transverse rate equation

$$
I_\eta \omega_\eta \dot{\omega}_\eta = \lambda_o\left(\frac{\dot{T}_P}{\lambda_P} + \frac{\dot{T}_R}{\lambda_R}\right) \tag{5.74}
$$

The energy sink argument may now be applied to this situation. Dissipation is a consequence of nonrigid elements moving within the system, but inertia and momentum variations associated with this motion are assumed negligible. It is the sustained cyclic forces acting throughout the spacecraft that cause this dissipation with fundamental frequencies λ_P and λ_R, the nutation frequencies with respect to the platform and rotor, respectively. The criterion for stability is simply that

$$
\dot{\omega}_\eta < 0
$$

Applying this to equation (5.74) implies stability when one of the following sets of conditions holds:

$$
\text{(a)} \quad \lambda_P > 0 \quad \text{and} \quad \lambda_R > 0 \tag{5.75a}
$$

$$
\text{(b)} \quad \lambda_P > 0,\, \lambda_R < 0 \quad \text{and} \quad \left|\frac{\dot{T}_P}{\lambda_P}\right| > \left|\frac{\dot{T}_R}{\lambda_R}\right| \tag{5.75b}
$$

$$
\text{(c)} \quad \lambda_P < 0,\, \lambda_R > 0 \quad \text{and} \quad \left|\frac{\dot{T}_P}{\lambda_P}\right| < \left|\frac{\dot{T}_R}{\lambda_R}\right| \tag{5.75c}
$$

Equations (5.73) and (5.74) indicate that a decrease in transverse rate must be accompanied by changes in spin rates about the z axis. If conditions (5.75) are met then the nutation angle, θ will decrease because $\tan\theta = h_\eta/h_z$ or

$$
\tan\theta = \frac{I_\eta \omega_\eta}{I_z^{P}\omega_z^{P} + I_z^{R}\omega_z^{R}} = \frac{\omega_\eta}{\lambda_o}
$$

As ω_η decreases and body spin rates change, nutation frequencies also change. If this frequency becomes zero in one of the bodies before θ reaches zero, energy dissipation will stop in that body and it then acts as a rigid body. As an example, assume a prolate dual spinner with a very slowly rotating platform. Then condition (5.75b) would have to be satisfied for stability. For realistic inertia values this requires that the platform have much more dissipation than the rotor. If this condition is not satisfied until θ reaches zero, then some residual nutation angle will persist. Thus, when the condition

$$\omega_z{}^P = \lambda_o$$

is reached, platform damping goes to zero, which means that all elements in the platform experience static forces. This implies the satellite should be designed to insure that as nutation is damped the nutation frequency of the platform λ_P must be kept nonzero. Otherwise, the satellite will find some equilibrium nutation angle greater than zero. Thus, the existence of damping on both rotor and platform complicates the problem of insuring asymptotic stability in nutation.

5.3 MASS MOVEMENT TECHNIQUES

The preceding section dealt with spinning inertias used to maintain a desired orientation. Large attitude and momentum changes were not considered. However, many spacecraft require reorientation and momentum adjustment during their lifetime in orbit. Mass movement devices, called *yo-yos*, have been used to adjust spin rate on several single-body satellites which were spun up during the launch phase for stability reasons. Their operational objectives required them to despin or, at least, reduce momentum. Mass movement techniques may also be useful in some reorientation maneuvers. Although thrust devices are very popular for such applications, there are a few situations in which energy dissipation can be used effectively to reorient a semirigid body. The example of Section 4.2.2 is a case in point. A mass movement scheme may be effective in handling the pointing ambiguity problem through active control. Spin adjustment and reorientation maneuvers which employ moving masses are emphasized in this section.

5.3.1 Yo-Yo Devices

Consider the situation in which deployment of instruments, booms, or antennae is required and the launch vehicle has to spin up the payload for stability during orbit insertion maneuvers. An adjustment of spin rate is required to bring the satellite into an operational state. One very reliable method of doing this is with expendable masses attached to the satellite by light cords. These are initially wrapped around the body and spin axis as shown in Figure 5.16. If two masses are used, no unbalanced forces occur about the center of mass. This device is known as a *yo-yo* and it absorbs a prescribed percentage of the initial angular momentum. When the masses are released centrifugal force pulls them away from the satellite as illustrated in Figure 5.17.

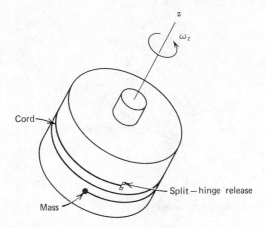

FIGURE 5.16 Yo-yo spin adjustment configuration.

The cords unwind to their full length. At this point the split-hinge release device permits both masses to escape, carrying away part or all of the system angular momentum. To determine the performance of these yo-yos consider first the situation in which a single mass m on a cord of length l is released as soon as the full length is unwound tangentially. This will have the same effect on spin rate as two masses of size $m/2$ on cords of length l. The appropriate analysis is carried out using the model in Figure 5.18. Conservation of angular momentum is applied directly in this case. I_z is the inertia of the spacecraft, excluding m, about z. Thus, as the cord unwinds the moment of inertia about z of the spacecraft-plus-mass system increases, resulting in a decrease of ω_z. As illustrated, the satellite is assumed axisymmetric and of radius R. Axes x and y are body fixed and 1 and 2 are allowed to rotate such that 1 passes through the tangent point of the cord. Initially, at $\phi = 0$, 1 and x coincide and m is in contact with the satellite. The position of m at any time t is referred to the vehicle center of mass O by

$$\mathbf{r} = -R\phi\mathbf{j} + R\mathbf{i} \tag{5.76}$$

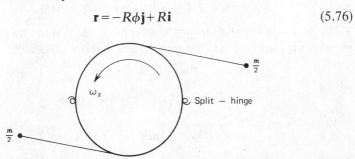

FIGURE 5.17 Yo-yo unwinding process.

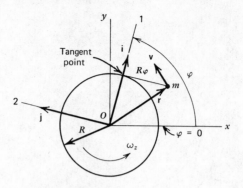

FIGURE 5.18 Yo-yo model for analysis.

and the absolute velocity of m is obtained by applying equation (1.31),

$$\mathbf{v} = [\dot{\mathbf{r}}]_b + \boldsymbol{\omega} \times \mathbf{r}$$

However, the proper $\boldsymbol{\omega}$ for m is

$$\boldsymbol{\omega} = (\omega_z + \dot{\phi})\mathbf{k}$$

which gives

$$\mathbf{v} = R\phi(\omega_z + \dot{\phi})\mathbf{i} + R\omega_z\mathbf{j} \tag{5.77}$$

The angular momentum of m about O is

$$\mathbf{h}_m = \mathbf{r} \times m\mathbf{v}$$

After inserting expressions (5.76) and (5.77) this becomes

$$\mathbf{h}_m = mR^2[\omega_z + \phi^2(\omega_z + \dot{\phi})]\mathbf{k}$$

Therefore, the total angular momentum is the sum of this and $I_z\omega_z\mathbf{k}$,

$$\mathbf{h} = \{I_z\omega_z + mR^2[\omega_z + \phi^2(\omega_z + \dot{\phi})]\}\mathbf{k} \tag{5.78}$$

Since there is no applied torque from outside the system, this total momentum is constant. Furthermore, the total kinetic energy of satellite and mass is

$$T = \tfrac{1}{2}I_z\omega_z^2 + \tfrac{1}{2}mv^2$$

where

$$v^2 = \dot{R}^2[\omega_z^2 + \phi^2(\omega_z + \dot{\phi})^2]$$

Thus

$$T = \tfrac{1}{2}I_z\omega_z^2 + \tfrac{1}{2}mR^2[\omega_z^2 + \phi^2(\omega_z + \dot{\phi})^2] \tag{5.79}$$

There is no mechanism in these yo-yos for dissipation. Hence, kinetic energy is also constant.

Performance of the device clearly depends on initial conditions. Consider the typical situation in which $\phi = 0$ and $\omega_z = \omega_{z0}$ at $t = 0$. These initial conditions lead to

$$\mathbf{h} = (I_z + mR^2)\omega_{z0}\mathbf{k}, \qquad T = \tfrac{1}{2}(I_z + mR^2)\omega_{z0}^2$$

Equating these to expressions (5.78) and (5.79), respectively, gives

$$c(\omega_{z0} - \omega_z) = \phi^2(\omega_z + \dot\phi) \qquad (5.80a)$$
$$c(\omega_{z0}^2 - \omega_z^2) = \phi^2(\omega_z + \dot\phi)^2 \qquad (5.80b)$$

where

$$c = \frac{I_z}{mR^2} + 1$$

Now divide equation (5.80b) by (5.80a) to get

$$\omega_{z0} + \omega_z = \omega_z + \dot\phi$$

or

$$\omega_{z0} = \dot\phi = \text{constant} \qquad (5.81)$$

Thus

$$\phi = \omega_{z0}t \qquad (5.82)$$

where t is the time to unwind a length ϕR. This result implies that the cord unwinds at a constant rate, ω_{z0}. Substituting form (5.82) into (5.80a) yields the satellite spin rate as a function of time,

$$\omega_z = \omega_{z0}\left(\frac{c - \omega_{z0}^2 t^2}{c + \omega_{z0}^2 t^2}\right) \qquad (5.83)$$

It is also instructive to express ω_z as a function of cord length l. Since $\phi = l/R = \omega_{z0}t$, equation (5.83) becomes

$$\omega_z = \omega_{z0}\left(\frac{cR^2 - l^2}{cR^2 + l^2}\right) \qquad (5.84)$$

Typically, ω_z is specified and the corresponding cord length is to be determined. Rearranging equation (5.84) gives the length for specified ω_z as

$$l = R\sqrt{c\left(\frac{\omega_{z0} - \omega_z}{\omega_{z0} + \omega_z}\right)} \qquad (5.85)$$

If all spin is to be removed, then $\omega_z = 0$, and expression (5.85) is simply

$$l = R\sqrt{c} = \sqrt{R^2 + \frac{I_z}{m}} \qquad (5.86)$$

Remember that the result given by this expression is based on cord release as it tangentially unwinds to length l. Notice that the required length for complete despin is independent of initial spin rate. Therefore, a value of ω_{z0} need not be specified for missions requiring elimination of all angular momentum. Furthermore, the time to completely despin is

$$t = \frac{l}{R\omega_{z0}} \qquad (5.87)$$

which is inversely proportional to ω_{z0}.

To illustrate application of these yo-yo performance equations, consider a situation in which a satellite is to be completely despun. Assume its spin axis moment of inertia is $I_z = 200 \text{ N·m·s}^2$ and initial spin rate is $\omega_{z0} = 5$ rad/s. A single yo-yo mass of 4 kg is used at a radius of $R = 1$ m, and the required cord length and time to despin are to be determined. Result (5.86) can be employed directly to give the cord length,

$$l = 7.14 \text{ m}$$

which will require just over one waist circuit for wrapping. Expression (5.87) gives the despin time as

$$t = 1.43 \text{ s}$$

Two observations should be made before leaving this topic. If the yo-yo cannot be wrapped in the center of mass plane, then two equal masses, on opposite sides, should be used to minimize attitude perturbation during unwinding. In many cases the masses are released when the cord is parallel to the radial direction rather than along the tangent to the drum. The preceding results are nevertheless valid in describing the unwinding process up to the point where the cord begins rotating about the split-hinge. The important performance parameter, length of cord for complete despin, is easily obtained through reapplication of momentum and energy conservation laws,

$$l = R(\sqrt{c} - 1)$$

where the cord is released radially.

5.3.2 Control of Reorientation Ambiguity

The use of internal energy dissipation for large angle reorientation was discussed through a specific application in Section 4.2.2. During this presentation of the AMAPS concept, the existence of a pointing ambiguity was exposed, that is, a passive reorientation of this type could result in either positive or negative spin about the major body axis. This uncertainty is the result of uncontrolled transition from spin with nutation and precession about the minor axis of a semirigid body to spin with similar oscillations about the major axis. Prediction of transition dynamics for a given body depends on initial spin conditions and precise modeling of the energy dissipation characteristics, which is generally not possible to the accuracy required. Therefore, final spin sense is unpredictable and can be controlled only through an active system. A moving mass device which could do this is now discussed. This concept uses the displacement of internal masses to guarantee crossing the proper separatrix, as described in Section 2.5. The *polhode of dissipation* for a large angle reorientation was illustrated in Figure 2.21. Separatrices represent boundaries between major and minor axis rotational motion. As the polhode crosses one of these two boundaries, motion is transformed from spin with wobble about the minor axis to spin with wobble about the positive or negative major axis, depending on which separatrix is crossed. Thus, the ambiguity of concern here is associated with the sign of $\boldsymbol{\omega} \cdot \mathbf{k}$ after a separatrix is crossed.

Remembering that $\mathbf{i}$, $\mathbf{j}$, $\mathbf{k}$ are units vectors along 1, 2, 3 principal body axes and nutation angle θ is measured between $\mathbf{k}$ and $\mathbf{h}$, the analytical interpretation of polhode motion can be handled through the use of θ. From Section 2.3.6, separatrices correspond to an energy level of $h^2/2I_2$. Thus, it is apparent that only the relationship between energy level (i.e., θ) and the sign of $\boldsymbol{\omega} \cdot \mathbf{k}$ are required to control the ambiguity, and no detailed solution of motion is needed. Begin by evaluating the limits of θ as the polhode crosses a separatrix. These are obtained for the two regions of interest, that is, *before* and *after* separatrix crossing. After this crossing the energy level is $T < h^2/2I_2$, and the corresponding upper and lower limits are obtained by noting that

$$\sin^2 \theta = \frac{h_{12}{}^2}{h^2} \tag{5.88}$$

where $h_{12}{}^2 = I_1{}^2 \omega_1{}^2 + I_2{}^2 \omega_2{}^2$. The upper limit, θ_u corresponds to $\omega_1 = 0$,

$$\sin^2 \theta_u = \frac{I_2(2I_3T - h^2)}{h^2(I_3 - I_2)} \tag{5.89}$$

and the lower limit, θ_l occurs when $\omega_2 = 0$,

$$\sin^2 \theta_l = \frac{I_1(2I_3T - h^2)}{h^2(I_3 - I_1)} \tag{5.90}$$

Before separatrix crossing the energy level is $T > h^2/2I_2$. Since ω_1 is always nonzero in this region, $\mathbf{k}$ oscillates in inertial space such that the limits of θ are supplements of each other, that is, $\theta_u = \pi - \theta_l$. These limiting values are then obtained from

$$\left.\begin{array}{c} \dfrac{\pi}{2} - \theta_l \\[2ex] \theta_u - \dfrac{\pi}{2} \end{array}\right\} = \sin^{-1} \sqrt{\frac{2I_1 I_3 T}{h^2(I_1 - I_3)} - \frac{I_3}{I_1 - I_3}} \tag{5.91}$$

To nondimensionalize results, define

$$I_{13} = \frac{I_1}{I_3}, \; I_{12} = \frac{I_1}{I_2}, \; T^* = \frac{T}{T_{\max}}$$

where $T_{\max} = h^2/2I_1$. The limits of θ are summarized as follows:
Before separatrix crossing

$$\left.\begin{array}{l} \theta_u = \dfrac{\pi}{2} + \sin^{-1}\sqrt{\dfrac{1-T^*}{1-I_{13}}} \\[3mm] \theta_l = \dfrac{\pi}{2} - \sin^{-1}\sqrt{\dfrac{1-T^*}{1-I_{13}}} \end{array}\right\} \tag{5.92}$$

After separatrix crossing

$$\left.\begin{array}{l} \theta_u = \sin^{-1}\sqrt{\dfrac{T^*-I_{13}}{I_{12}-I_{13}}} \\[3mm] \theta_l = \sin^{-1}\sqrt{\dfrac{T^*-I_{13}}{1-I_{13}}} \end{array}\right\} \tag{5.93}$$

Remembering that $I_1 < I_2 < I_3$, the profile of nutational motion as energy dissipates can be described by Figure 5.19. Separatrix crossing is shown as a critical point in attitude reorientation after which nutational amplitude decreases to establish spin about the positive or negative major axis, that is, a final nutation angle of 0° or 180°, respectively.

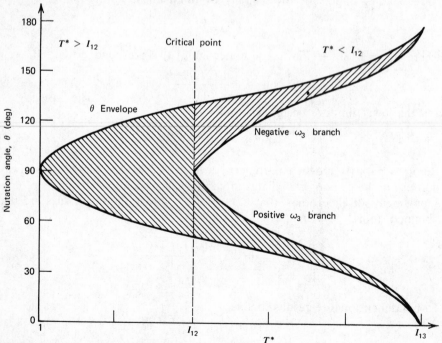

FIGURE 5.19 General nutational envelope.

From the above considerations it is apparent that the critical point in the motion is when energy satisfies $T^* = I_{12}$. Nutation angle limits or the sign of ω_3 defines the motion thereafter. Since the sign of ω_3 is easily measurable, this is used as a control input. In addition, magnitudes of $\boldsymbol{\omega}$ components are employed. Active control is engaged as T^* equals I_{12}. That is, moving internal masses are proposed in order to lower the inertia ratio I_{12} at the critical point if $\omega_3 < 0$. This would have the effect of extending separatrix crossing to the other side where ω_3 is positive. Then these masses are returned to their initial positions, bringing I_{12} back to its original value. Spinning up an internal mass about an axis can increase mechanical energy and alter the distribution of angular momentum. This has essentially the same effect as the moving mass technique. Reaction jets can also be used for this application, and both energy and angular momentum would be controlled. However, application of torques will generally effect the final orientation and magnitude of the angular momentum vector, unless corrective torques are applied later. Selection of control elements depends on the particular situation at hand.

A moving mass concept is depicted in Figure 5.20. Activation of the device depends on the sign of ω_3 as T reaches the critical value. If ω_3 is negative at

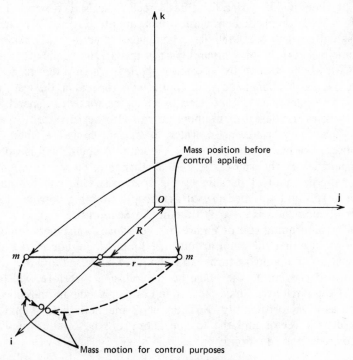

FIGURE 5.20 Example moving mass configuration.

this point, the masses are quickly brought together, as shown. Resulting changes in moments of inertia cause a shift in energy state such that separatrix crossing is delayed. When ω_3 becomes positive; the masses return to their original positions. This drives the polhode across the proper separatrix and $\boldsymbol{\omega}$ proceeds in the proper direction. It is desirable to have high dissipation of energy near the separatrices in order to avoid problems with sensor noise levels and false indications of crossings. On the other hand, control masses must be sized so that T^* does not again reach I_{12} before ω_3 changes to a positive value.

5.4 MAGNETIC TORQUERS

In the mid-1960s Harold Perkel conceived of a three axis attitude control system utilizing a single, fixed-gimbal momentum wheel and magnetic torquers for controlling transverse momentum components. *Stabilite* is the name given to this concept of controlling pitch, roll, and yaw such that pitch is maintained parallel to the orbit normal and yaw parallel to the local vertical (nadir). Thus, roll and yaw axes rotate once per orbit about the orbit normal. Originally, this concept was applied to the TIROS weather satellite program, but has been used on the later ITOS series. The basic technique is incorporated into the synchronous RCA domestic communications satellite. Figure 5.21 illustrates a weather satellite which uses Stabilite. The single momentum wheel is the chief active element and is the only moving part in this attitude control system. Yaw and roll axes are controlled by *magnetic torquing*, which refers to an interaction with the earth's magnetic field to provide torque in the yaw and roll directions. The pitch axis is controlled by changing the wheel speed. Periodically a net excess or deficit of momentum can also be corrected by magnetic torquing. Reaction jets are incorporated as backup control elements. Figure 5.22 illustrates the placement of control elements. A simple coil is wound with its plane normal to the spin axis, and a current is passed through this to produce a dipole moment parallel to the spin axis. One coil is sufficient to control both roll and yaw because of gyroscopic coupling between these two axes. This phenomenon is discussed further in Section 6.2.

The roll/yaw torquing coil of Figure 5.22 produces a magnetic dipole when current flows in either direction around the loop. Magnitude of this dipole is proportional to the ampere-turns and the area enclosed by the coil. The direction will be normal to the plane of the coil, and therefore, will lie in the direction of the pitch axis. Since the dipole has magnitude and direction it may be defined as a vector quantity **D**. The earth's magnetic field **B** may also be represented as a vector and the torque acting on the satellite may be represented by a vector equation:

$$\mathbf{M} = \mathbf{D} \times \mathbf{B} \tag{5.94}$$

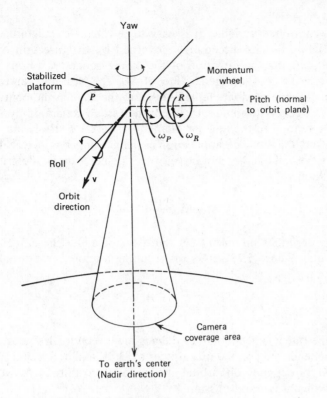

FIGURE 5.21 Schematic of Stabilite concept.

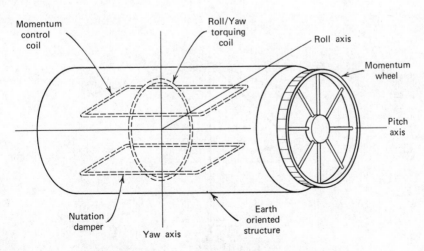

FIGURE 5.22 Stabilite control elements for high inclination orbits.

Since **D** is a controlled variable, it is easy to see that the magnitude of **M** is infinitely variable, but its direction is governed by the direction of **B**. This restriction on defining the direction of the dipole-generated torque has been overcome by using an averaged value of **B**. The attitude control system requires that the generated dipole be parallel to the angular momentum stored in the inertia wheel. With this requirement the torque generated by the dipole will always be normal to the angular momentum vector **h**. If the magnitude of **M** is small, and the time of application is long compared to the free-body precession period, then the motion of the satellite can be described by a familiar equation,

$$\mathbf{M} = \frac{d\mathbf{h}}{dt}$$

which merely describes the motion of the **h** vector under the influence of the magnetic torque. Figure 5.23 shows the resulting motion, **h** rotates about **B** at the rate

$$|\boldsymbol{\omega}| = \left| \frac{DB}{h} \right| \tag{5.95}$$

Notice that the rate is independent of the angle between the **h** vector and the **B** vector, but depends only on the magnitudes of **D**, **B**, and **h**. Specific application of these devices depends on orbital inclination and altitude as well as the spacecraft configuration and mission.

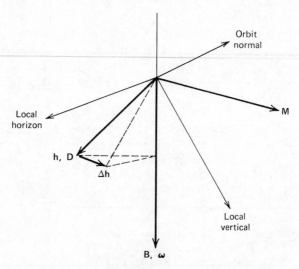

FIGURE 5.23 Motion induced by magnetic torquing.

5.5 GRAVITY GRADIENT STABILIZATION

In addition to those active control methods already mentioned in this chapter, there are a limited number of passive stabilization techniques available. Spin stabilization is, of course, one popular method previously cited, which is independent of altitude and orbital shape. However, no consistent pointing of the spin axis is guaranteed and periodic thrusting is typically required. Use of gravity gradients to maintain earth pointing of an antenna or other instruments has been successful and is appropriate for restricted missions and low eccentricity orbits. This principle can be explained simply by considering the attitude motion of a dumbbell satellite. Refer to Figure 5.24, and note that the center of mass follows a circular orbit. A deflection away from the local vertical causes a restoring torque to be generated by the imbalance of forces acting on equal masses m_1 and m_2. The centrifugal force acting on 1 is greater than the gravitational force on it, because these two forces are equal only at the center of mass. The opposite is true of mass 2, gravity is stronger than centrifugal force. Thus, a net torque is created which forces the masses toward a local vertical orientation.

A more general case is depicted in Figure 5.25. If **G** is the gravity produced torque, then the equation of motion is simply

$$\mathbf{G} = \frac{d\mathbf{h}}{dt}$$

where **h** is the angular momentum of body B about its center of mass O given

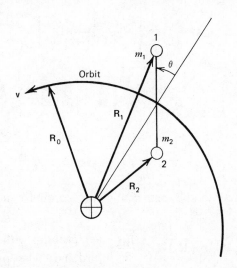

FIGURE 5.24 Gravity gradient torque principle.

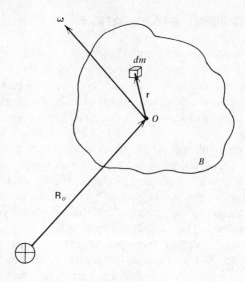

FIGURE 5.25 General gravity gradient model.

by equation (2.37). Using the triple vector product identity,

$$\mathbf{h} = \int_B [r^2\boldsymbol{\omega} - (\mathbf{r} \cdot \boldsymbol{\omega})\mathbf{r}] \, dm$$

The gravity torque has the form

$$\mathbf{G} = \int_B \mathbf{r} \times \left[-\frac{\mu_\oplus(\mathbf{R}_o + \mathbf{r}) \, dm}{|\mathbf{R}_o + \mathbf{r}|^3} \right] \tag{5.96}$$

Noting that

$$|\mathbf{R}_o + \mathbf{r}|^{-3} = R_o^{-3}\left[1 + \frac{2(\mathbf{r} \cdot \mathbf{R}_o)}{R_o^2} + \frac{r^2}{R_o^2}\right]^{-3/2} = \frac{1}{R_o^3}\left[1 - \frac{3(\mathbf{r} \cdot \mathbf{R}_o)}{R_o^2} + O\!\left(\frac{r^2}{R_o^2}\right)\right]$$

and

$$\int_B \mathbf{r} \, dm = 0$$

because **r** is referred to the center of mass, permits equation (5.96) to be rewritten as

$$\mathbf{G} = \frac{3\mu_\oplus}{R_o^5} \int_B (\mathbf{r} \cdot \mathbf{R}_o)(\mathbf{r} \times \mathbf{R}_o) \, dm + \frac{\mu_\oplus}{R_o} \times \begin{array}{l} \text{(vector terms with magnitudes of} \\ \text{third and higher orders in } r/R_o) \end{array}$$

$$\tag{5.97}$$

This is balanced by the rate of **h**,

$$\frac{d\mathbf{h}}{dt} = \int_B [r^2\dot{\boldsymbol{\omega}} - (\mathbf{r} \cdot \dot{\boldsymbol{\omega}})\mathbf{r}] \, dm + \int_B (\mathbf{r} \cdot \boldsymbol{\omega})(\mathbf{r} \times \boldsymbol{\omega}) \, dm \tag{5.98}$$

The question of stability about the local vertical is prominent. However, the associated equilibrium orientation must first be determined. Assume **I**, **J**, **K** are unit vectors along the outward local vertical, the orbit path, and the orbit normal, respectively. If the only body angular rate is the orbital rate, then equation (5.98) gives

$$\frac{d\mathbf{h}}{dt} = \dot{\theta}^2 (I_{yz}\mathbf{I} - I_{xz}\mathbf{J}) \tag{5.99}$$

where I_{yz}, I_{xz} are products of inertia about the x, y, z axes which are instantaneously parallel to **I**, **J**, **K**, respectively. Note that $\dot{\theta}^2 = \mu_\oplus/R_o^3$ and expression (5.97) becomes

$$\mathbf{G} = \frac{3\mu_\oplus}{R_o^3} (I_{xz}\mathbf{J} - I_{xy}\mathbf{K}) \tag{5.100}$$

Therefore, equilibrium requires $I_{xy} = I_{xz} = I_{yz} = 0$, that is, the body must have its principal axes aligned with the **I**, **J**, **K** frame.

Stability of motion about this aligned position can now be investigated. A small orbital eccentricity will be included for generality. Euler's equations are written here as

$$I_x \frac{d}{dt} \omega_x + (I_z - I_y)\omega_y\omega_z = \frac{3\mu_\oplus}{R^5} (I_z - I_y)R_yR_z \tag{5.101a}$$

$$I_y \frac{d}{dt} \omega_y + (I_x - I_z)\omega_z\omega_x = \frac{3\mu_\oplus}{R^5} (I_z - I_x)R_zR_x \tag{5.101b}$$

$$I_z \frac{d}{dt} \omega_z + (I_y - I_x)\omega_x\omega_y = \frac{3\mu_\oplus}{R^5} (I_y - I_x)R_xR_y \tag{5.101c}$$

where I_x, I_y, I_z are the principal moments of inertia, and R_x, R_y, R_z are the components of the satellite position **R** (subscript dropped) along principal body directions. Note that these do not necessarily coincide with **I**, **J**, **K** directions. If **i**, **j**, **k** are the body fixed unit vectors along x, y, z, suppose that there is a small misalignment. Referring to Figure 5.26,

$$\begin{bmatrix} \mathbf{I} \\ \mathbf{J} \\ \mathbf{K} \end{bmatrix} = \begin{bmatrix} \mathbf{i} \\ \mathbf{j} \\ \mathbf{k} \end{bmatrix} - \begin{bmatrix} 0 & \psi_z & -\psi_y \\ -\psi_z & 0 & \psi_x \\ \psi_y & -\psi_x & 0 \end{bmatrix} \begin{bmatrix} \mathbf{i} \\ \mathbf{j} \\ \mathbf{k} \end{bmatrix}$$

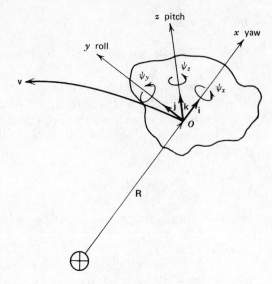

FIGURE 5.26 Attitude geometry for stability analysis.

where

$$\psi_x = \text{small yaw to the left}$$
$$\psi_y = \text{small roll to the right}$$
$$\psi_z = \text{small pitch down}$$

Notice that

$$\mathbf{R} = R\mathbf{I} = R\mathbf{i} - \psi_z R\mathbf{j} + \psi_y R\mathbf{k}$$

and

$$\boldsymbol{\omega} = (\dot{\psi}_x - \dot{\theta}\psi_y)\mathbf{i} + (\dot{\psi}_y + \dot{\theta}\psi_x)\mathbf{j} + (\dot{\psi}_z + \dot{\theta})\mathbf{k}$$

For small eccentricity two approximations can be useful,

$$\dot{\theta} \cong n[1 + 2e \cos nt_p]$$
$$R \cong a[1 - e \cos nt_p]$$

where t_p is the time since periapsis passage. The first of these permits

$$\ddot{\theta} \cong -2n^2 e \sin nt_p$$

Neglecting squares and products of small quantities, set (5.101) becomes

$$I_x(\ddot{\psi}_x - n\dot{\psi}_y) + (I_z - I_y)(n\dot{\psi}_y + n^2\psi_x) = 0 \tag{5.102a}$$
$$I_y(\ddot{\psi}_y + n\dot{\psi}_x) + (I_x - I_z)(n\dot{\psi}_x - n^2\psi_y) = 3n^2(I_x - I_z)\psi_y \tag{5.102b}$$
$$I_z(\ddot{\psi}_z - 2n^2 e \sin nt_p) = -3n^2(I_y - I_x)\psi_z \tag{5.102c}$$

The small pitching z motion is uncoupled from the small roll-yaw motions, and is stable provided $I_y > I_x$. The general solution for the pitching motion, assuming stability, including both natural oscillation and forced motion due to eccentricity, is of the form

$$\psi_z = \frac{2e \sin nt_p}{3\left(\dfrac{I_y - I_x}{I_z}\right) - 1} + A \cos\left[nt_p \sqrt{3\left(\frac{I_y - I_x}{I_z}\right)} + \alpha \right] \qquad (5.103)$$

Define k_P as

$$k_P = \frac{I_y - I_x}{I_z}$$

and pitch resonance occurs at $k_P = \frac{1}{3}$. The coupled roll-yaw motion satisfies

$$\ddot{\psi}_x + k_Y n^2 \psi_x - (1 - k_Y)n\dot{\psi}_y = 0$$
$$(1 - k_R)n\dot{\psi}_x + \ddot{\psi}_y + 4k_R n^2 \psi_y = 0 \qquad (5.104)$$

where k_Y and k_R are defined as

$$k_Y = \frac{I_z - I_y}{I_x}, \qquad k_R = \frac{I_z - I_x}{I_y}$$

Stability of small roll-yaw motion is evaluated by assuming a solution of the form

$$\psi_x = Be^{snt}, \qquad \psi_y = Ce^{snt}$$

and inserting it into set (5.104) to obtain a quartic characteristic equation,

$$s^4 + s^2(1 + 3k_R + k_Y k_R) + 4k_Y k_R = 0 \qquad (5.105)$$

For stability the roots must be purely imaginary. Thus, the conditions for roll-yaw stability are

$$k_Y k_R > 0 \qquad (5.106a)$$

$$1 + 3k_R + k_Y k_R > 4\sqrt{k_Y k_R} \qquad (5.106b)$$

The first of these implies that the pitch axis must be the minor or major principal axis for roll-yaw stability. Pitch stability requires that roll inertia be greater than yaw inertia. Therefore, only two orientations are permissible, when $I_{pitch} > I_{roll} > I_{yaw}$ or $I_{roll} > I_{yaw} > I_{pitch}$. However, condition (5.106b) restricts these choices to limited ranges of inertia ratios. In general, it is best to design such that $I_{pitch} > I_{roll} > I_{yaw}$. Figure 5.27 summarizes the zones of stability for various combinations of inertia ratios. Pitch stability requires that $k_R > k_Y$, thus, the upper-left region is eliminated. Condition (5.106b) eliminates the

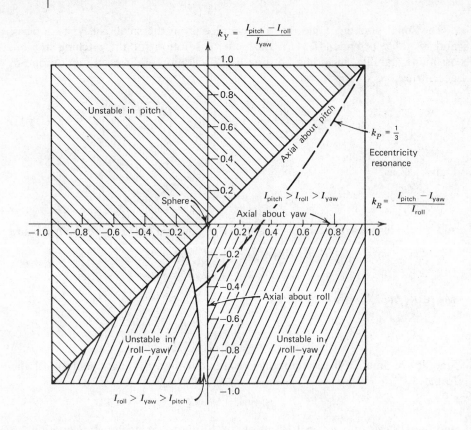

FIGURE 5.27 Gravity gradient stability regions.

bottom-left region and condition (5.106a) accounts for the lower-right quadrant. Notice that each axis corresponds to a special inertia shape, for example, the $k_Y = 0$ axis is associated with a body of revolution whose axis is aligned in the yaw direction.

Assuming a stable orientation is established, the restoring torque generated by small attitude errors is calculated from equation (5.100). Approximate I_{xz} and I_{xy} by

$$I_{xz} = \psi_y I_x + \psi_y I_z$$
$$I_{xy} = \psi_z I_x + \psi_z I_y$$

(5.107)

Motion about the yaw direction cannot be effectively controlled by the gravity gradient technique. Fortunately, perturbing torques about this axis are generally much smaller than those about roll and pitch.

EXERCISES

5.1 Derive the equations of motion for the precessing top with $\theta = 90° =$ constant, and determine the precession rate $\dot{\psi}$ from these equations.

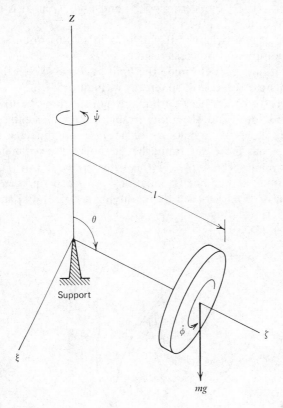

EXERCISE 5.1

5.2 Consider a spinning top of the type shown in Figure 5.2. It is observed instantaneously to have rates

$$\dot{\psi} = 0, \qquad \dot{\phi} = 50 \, \text{rad/s}, \qquad \dot{\theta} = 2 \, \text{rad/s}$$

when

$$\psi = 120°, \qquad \theta = 30°, \qquad \phi = 90°$$

Its physical properties are

$$I_1 = 1 \, \text{N·m·s}^2, \qquad I_3 = 1.5 \, \text{N·m·s}^2$$
$$m = 1 \, \text{kg}, \qquad l = 0.2 \, \text{m}$$

   (a)   Determine the limits of nutation, θ_1 and θ_2
   (b)   Sketch the spin axis path on a unit sphere of the type used in Figure 5.4.

5.3   A large symmetric top of the kind in Figure 5.2 has properties $I_1 = 1\ \text{N}\cdot\text{m}\cdot\text{s}^2$, $I_3 = 2\ \text{N}\cdot\text{m}\cdot\text{s}^2$, $mg = 7\ \text{N}$, and $l = 0.97\ \text{m}$. It is initially spun up with $\theta = 0$, $\dot{\phi} = 2\ \text{rad/s}$.
   (a)   Determine whether it will remain in this orientation or perform nutation and precession maneuvers.
   (b)   If it nutates, determine the limit of θ and sketch the motion of the spin axis on a unit sphere as in Figure 5.4.

5.4   A bright student notices during a demonstration of a toy gyroscope that both the rotor and the outer frame rotate in steady precession. He wonders how the equations of motion are effected by this frame. Assume that the outer frame has a spherically symmetric inertia of I^o and its spin speed is $\dot{\phi}^o$. If the rotor spins at rate $\dot{\phi}$, write the new momentum and moment equations for steady precession. Assume a total mass of m. Proceed by modifying set (5.3) and specializing equations (5.4).

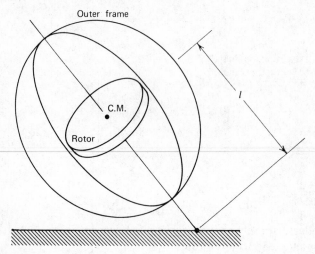

Outer frame

C.M.

Rotor

l

EXERCISE 5.4

5.5   Consider the two-degree-of-freedom gimbal system of Figure 5.1. Instead of an axisymmetric rotor, however, assume an asymmetrical rigid body with distinct principal axes, 1, 2, 3. The 3-axis is aligned with the gimbal ζ-axis. Derive the three equations of motion about the gimbal axles, ξ', ζ', and ζ as defined in Figure 5.8. Note that *Euler's modified equations* do not apply here.

5.6 Mr. Sat A. Lite, having survived his duel of Chapter 1 (which is, hopefully, not a surprise to the reader), now has another irate husband after him. This time the duel is more civilized. They decide on a 3 hour race along the equator in Brazil using two identical Beechcraft Bonanzas from Boston. For navigation only their gyrocompasses can be used as directional references (no magnetic compasses are allowed), which are set just before takeoff.

(a) Sat realizes that the precession of these gyros is adjusted for the latitude of Boston ($\lambda = 45°$). Assuming he has his tool kit along, how could he adjust his gyrocompass to eliminate any errors due to this latitude ($\lambda = 0°$)?

(b) How much error will his opponent have in his direction at the end of 3 hours, if he does not adjust his gyro?

5.7 Since the Moon has no magnetic field, the Lunar Roving Vehicle (LRV) used on Apollo 15 did not have a magnetic compass for navigation. However, a gyrocompass was used. The latitude of the landing site was 30°N. Assume the following data:

$$I_3 = 0.34 \text{ N·m·s}^2$$
$$I_1 = 0.20 \text{ N·m·s}^2$$
$$\dot{\phi} = 1000 \text{ rad/s}$$
$$mgl = 0.011 \text{ N·m}$$

(a) What is the required precession rate to eliminate the effect of the 28-day lunar rotation cycle?

(b) If $M_\xi = mgl \sin \alpha$, what value of α will insure this precession rate?

(c) M_ξ is the result of a weight mg. Would the precession rate be the same on earth if the LRV returned with the astronauts?

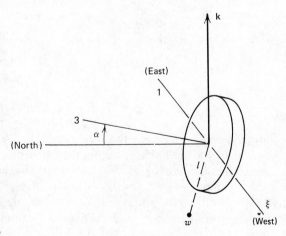

EXERCISE 5.7

5.8 Derive equation (5.59) by application of relative motion concepts in Section 1.5.1.

5.9 Given the characteristic equation of a dynamic system

$$s^4 + 8s^3 + ks^2 + 80s + 100 = 0$$

Use Routh's method to determine the range of k values which keep the system stable.

5.10 A huge space station of diameter 6 m is spinning out of control about its major principal axis at a rate of $\omega_0 = 10.0$ rad/s. The inertia tensor for this vehicle is

$$\mathbf{I} = \begin{bmatrix} 4 & 0 & 0 \\ 0 & 4 & 0 \\ 0 & 0 & 5 \end{bmatrix} \times 10^4 \, \text{N·m·s}^2$$

excluding the yo-yo device. A 1000 kg mass is attached to the emergency yo-yo device and is released tangentially. If the line is 7 m long, (a) what will be the final spin rate of the station? (b) how long will it take for the yo-yo to unwind? (c) what is the magnitude of angular deceleration half-way through this interval of time?

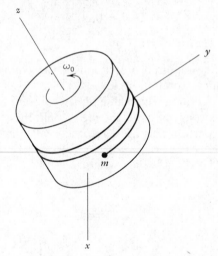

EXERCISE 5.10

5.11 As the ultimate defensive strategy the Department of Defense proposes to use a yo-yo device to completely despin the earth just after launch of ICBMs against the United States. Neglecting gravity and the atmosphere, and assuming $I_\oplus = 9.5 \times 10^{37}$ N·m·s^2, $R_\oplus = 6378$ km, $\omega_0 = 7.28 \times 10^{-5}$ rad/s, and a yo-yo mass of 4.4×10^5 kg, determine the required cord length and time to stop the world, assuming a tangential release.

5.12 The satellite of radius 1 m is initially spun up about its 2-axis at the rate $\omega_0 = 100$ rad/s. An energy dissipation device has been included in the satellite. Also, inertias about the 1, 2, 3 body axes are:

$$\mathbf{I} = \begin{bmatrix} 20 & 0 & 0 \\ 0 & 20 & 0 \\ 0 & 0 & 40 \end{bmatrix} \text{N·m·s}^2$$

(a) What will be the spin rate about the 3-axis after the energy dissipator has done its job?

(b) A radially released yo-yo despin device is also included in the satellite. If the yo-yo mass is 20 kg, how long a string is required to completely despin the satellite?

(c) The spin axis ambiguity discussed in Sections 4.2.2 and 5.3.2 may lead to *negative* spin about the major axis. What would happen if the yo-yo is released anyway?

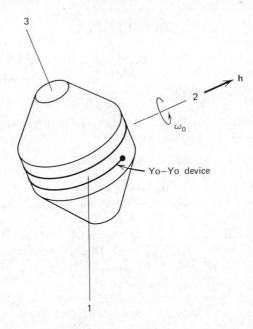

EXERCISE 5.12

5.13 Show that the upper and lower limits of nutation angle *after* separatrix crossing correspond to $\omega_1 = 0$ and $\omega_2 = 0$, respectively. Then derive expressions (5.89) and (5.90) from (5.88).

5.14 Show that the upper and lower limits of nutation angle *before* separatrix crossings are given by expression (5.91).

5.15 For a gravity gradient stabilized satellite, show that pitch stability requires $k_R > k_Y$.

5.16 A satellite has principal moments of inertia

$$I_1 = 300 \text{ N·m·s}^2, \qquad I_2 = 400 \text{ N·m·s}^2, \qquad I_3 = 500 \text{ N·m·s}^2.$$

Determine the permissible orientations in a circular orbit for gravity gradient stabilization. Specify which axes may be aligned in the pitch, roll, and yaw directions.

5.17 Develop two sketches of weather satellites which are in a low altitude, polar orbit. The purpose of these satellites is to take cloud pictures at noon on March 21 over the Western hemisphere. Assume a dual spinner and a nonspinning configuration. Show orbit orientations, sun position, solar cells or arrays, and momentum devices.

REFERENCES

DeBra, D. B., and R. H. Delp, "Rigid Body Attitude Stability and Natural Frequencies in a Circular Orbit," *Journal of the Astronautical Sciences*, Vol. 8, *No. 1*, Spring 1960, pp. 14–17.

Grubin, C., "Dynamics of a Vehicle Containing Moving Parts," *Transactions of the ASME: Journal of Applied Mechanics*, Vol. 29, September 1962, pp. 486–488.

Iorillo, A. J., "Nutation Damping Dynamics of Axisymmetric Rotor Stabilized Satellites," presented at the ASME Winter Meeting, Chicago, Nov. 1965.

Kaplan, M. H., and R. J. Cenker, "Control of Spin Ambiguity During Reorientation of an Energy Dissipating Body," *Journal of Spacecraft and Rockets*, Vol. 10, *No. 12*, Dec. 1973, pp. 757–760.

Landon, V., and B. Stewart, "Nutational Stability of an Axisymmetric Body Containing a Rotor," *Journal of Spacecraft and Rockets*, Vol. 1, *No. 6*, Nov.–Dec. 1964, pp. 682–684.

Likins, P. W., "Attitude Stability Criteria for Dual Spin Spacecraft," *Journal of Spacecraft and Rockets*, Vol. 4, *No. 12*. Dec. 1967, pp. 1638–1643.

Perkel, H., "Stabilite—A Three-Axis Attitude Control System Utilizing a

Single Reaction Wheel," *Progress in Astronautics and Aeronautics,* Vol. 19, Academic Press, 1966, pp. 375–399.

Roberson, R. E., "Torques on a Satellite Vehicle from Internal Moving Parts," *Transactions of the ASME: Journal of Applied Mechanics,* Vol. 25, June 1958, pp. 196–200.

Thomson, W. T., *Introduction to Space Dynamics,* Wiley, 1961, Chapters 5, 6, 7.

Wrigley, W., W. M. Hollister, and W. G. Denhard, *Gyroscopic Theory, Design, and Instrumentation,* Massachusetts Institute of Technology Press, 1969, Chapter 1.

Automatic
Attitude Control

The first autopilot wâs dramatically introduced to the world on June 18, 1914 in Paris. A crowd of spectators lined both sides of the Seine to view a series of aerial feats sponsored by the Concours de la Sécurité en Aéroplane (Airplane safety Competition). Silence fell upon the audience in anticipation as the Curtiss Hydroplane came into view at an altitude of 120 m. At the controls was 22-year-old Lawrence B. Sperry. He was accompanied by a French mechanic, Emile Cochin. Abruptly, the mechanic emerged from the cockpit and stepped onto the lower wing of the biplane. Then Sperry stood up in the cockpit with his hands stretched high above his head. Obviously, no one was controlling the plane at this point. Nevertheless, the aircraft continued to fly straight and level. The mechanic then shifted to the aft fuselage with no apparent change in the aircraft attitude. The crowd erupted into a roar as it realized the implications of this feat. The spectators had just seen the first public demonstration of a workable *automatic pilot* that could accommodate normal buffeting and gross shifts in center of gravity. Sperry did not win the

Grand Prix, even though his device was the most significant safety innovation demonstrated. For unknown reasons the grand prize was withdrawn, but he was awarded the second prize of 50,000 francs.

The 1914 Sperry *autopilot* used four gyros to control yaw, pitch, and roll. This is the same principle upon which modern autopilots are based. The idea of using a gyro to stabilize aircraft motion was an offshoot of ship gyro-stabilizers developed by Elmer Sperry, Lawrence's father and well-known American inventor. A significant difference, however, was that seagoing gyros were huge and used direct gyrotorquing to force a ship into stable attitudes. Aircraft autopilots use gyros to sense attitude errors, which are then transformed into torques through deflection of aerodynamic control surfaces (e.g., ailerons and elevators). Spacecraft autopilots may use gyros as torques producers or error sensors, depending on the mission. It is also possible to design spacecraft autopilots without gyros of any kind. This chapter deals with the fundamentals of automatic control and design of satellite attitude control systems. Linear control theory is presented to the extent appropriate. Performance evaluation methods and stability criteria are developed. Examples of control system design are offered to illustrate application of the principles.

6.1 LINEAR CONTROL THEORY

In order to understand and design automatic attitude control systems for spacecraft, at least a basic understanding of associated theories is required. The concepts to be presented are realistic, but simple enough to be treated by linear control theory. Therefore, elements of linear control are now offered, following the approach of R. N. Clark. This will serve as either a refresher or an introduction to the concepts. Firstly, a mathematical model of a physical system is linear if the system is linear. Of course, a linear system is one whose general motion may be described by superposition of more than one simpler motion generated by linear differential equations. The systems of interest here are characterized by second order, ordinary differential equations, many of which are linear or can be linearized for situations of concern. Thus, if a linear system exhibits a response $x_o(t)$ to a stimulus $x_i(t)$ it will exhibit a response $[x_o(t) + y_o(t)]$ to a stimulus $[x_i(t) + y_i(t)]$. Automatic control theory, as applied here, can be thought of as an array of special techniques for solving linear differential equations. In many instances, these methods lead to information about the system without actually solving the equations. Typically, the designer is looking for the influence of individual system parameters on dynamic performance of the whole system, with particular attention given to the stability problem.

6.1.1 Transfer Functions

In general, if given an input $x_i(t)$ and a complete description of the physical system, it is possible to compute the output $x_o(t)$. This concept is illustrated as a basic block diagram in Figure 6.1. No random inputs will be assumed for this discussion, but only relatively simply modeled functions are permitted. These include step functions and sine waves. A response can be derived from an input by writing the differential equation of each component, giving at least one equation per component. Input is the forcing function. These differential equations could then be solved simultaneously for the response, $x_o(t)$. However, the *transfer function* approach permits the dynamic response of several cascaded units to be calculated in a single step simply by multiplication of individual transfer functions. A *transfer function* is the ratio of the Laplace transform of the output quantity to the Laplace transform of the input, with the restriction that the initial conditions appearing in the transformed differential equations are all zero. Thus

$$\text{T.F.} = \frac{\mathscr{L}(\text{Output})}{\mathscr{L}(\text{Input})}$$

Consider a simple electric circuit, illustrated in Figure 6.2. This is referred to as an *R-L circuit*, and its associated differential equation is

$$e = R\,i(t) + L\frac{di}{dt}$$

where e is the applied voltage, R is the resistor value, i is the loop current, and L is the inductance of the coil. This equation is obtained by summing voltages. The Laplace transform gives

$$I(s) = \frac{E}{s}\left(\frac{1/L}{s + R/L}\right)$$

where E is a step input in voltage such that

$$E(s) = \frac{E}{s}$$

If E is not specified then

$$I(s) = E(s)\left(\frac{1/L}{s + R/L}\right)$$

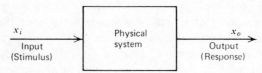

FIGURE 6.1 Basic block diagram.

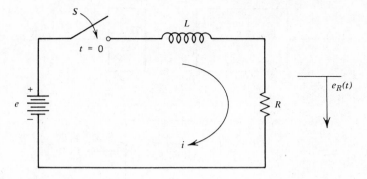

FIGURE 6.2 Simple *R-L* circuit.

and the general transfer function relating current to voltage for this type of circuit is

$$\text{T.F.} = \frac{I(s)}{E(s)} = \frac{1/L}{s + R/L} \qquad (6.1)$$

Many electrical circuits have *mechanical analogs*, whose differential equations have identical form. This is the basic principle behind analog computers. The analog of the simple *R-L* circuit is a spring-dashpot system, as shown in Figure 6.3. Its transfer function is

$$\text{T.F.} = \frac{X(s)}{F(s)} = \frac{1}{k + Bs} \qquad (6.2)$$

Note that $f(t)$ is the forcing function, k is the spring constant, and B is the dashpot constant. Referring to the circuit in Figure 6.2, if the output of the *R-L* circuit is to be $e_R(t)$, use Ohm's law to get

$$E_R(s) = RI(s) \qquad (6.3)$$

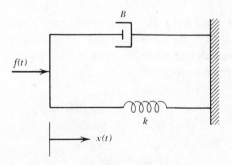

FIGURE 6.3 Simple first order mechanical system.

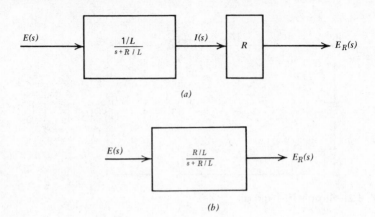

(a)

(b)

FIGURE 6.4 Block diagrams for simple *R-L* circuit.

Figure 6.4 depicts this in block diagram form and illustrates the ability to combine blocks without changing the response. A general procedure may now be outlined for deriving transfer functions of linear systems:

(a) Decide on input and output variables.
(b) Decide on which properties may be ignored, if any.
(c) Write the differential (or integrodifferential) equations.
(d) Laplace transform these equations and rearrange to give a transfer function. Take all initial conditions as zero. If nonzero initial conditions must be assumed, use the principle of superposition, that is, the response to input is calculated as though the initial conditions are zero and a separate calculation is made for the response due to initial conditions. The two responses are added to give the same result as if the input is applied at the time initial conditions exist.

One of the most important electrical components used in satellite control systems is the integrator. Its associated block diagram is illustrated in Figure 6.5. An *integrator* is a device whose output is proportional to the time integral of its input. In its most basic form,

$$x_o(t) = \int_0^t x_i(\tau)\, d\tau$$

which gives

$$X_o(s) = \frac{1}{s} X_i(s) \tag{6.4}$$

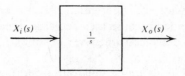

FIGURE 6.5 Integrator block symbol.

The circuit of Figure 6.6 shows an electronic integrator. Note that there are other types as well, such as the integrating gyro. An operational amplifier has a gain of $-A$. If $i_g = 0$ due to high internal impedance, then $e_o = -Ae_g$, and

$$-i_1 = i_2$$

$$i_1 = \frac{e_1 - e_g}{R}$$

$$e_o - e_g = \frac{1}{C} \int_0^t i_2 \, d\tau$$

Assuming $A \gg 1$, which implies e_g is small, results in

$$\text{T.F.} = \frac{E_o(s)}{E_1(s)} = \frac{-1/RC}{s} \tag{6.5}$$

as the transfer function of the integrator in Figure 6.6. The next logical component to introduce is the differentiator. This is a device whose output is proportional to the time derivative of the input,

$$x_o(t) = \frac{d}{dt}[x_i(t)]$$

which yields

$$\text{T.F.} = \frac{X_o(s)}{X_i(s)} = s \tag{6.6}$$

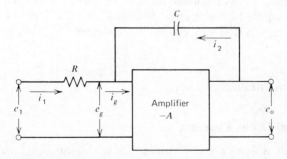

FIGURE 6.6 Integrator circuit.

For example, a tachometer is a differentiator. Unfortunately, such devices are not generally recommended because they decrease signal-to-noise ratio. Therefore, their use should be avoided whenever possible. Integrators, on the other hand, increase this important ratio.

It is now appropriate to identify standard transfer function forms for first and higher order systems. A first order system would display a transfer function whose denominator has s raised only to the first power. In general, an nth order system has a transfer function of nth order. For example, the transfer function of a first order system has the form

$$\text{T.F.} = \frac{K}{s + \alpha} \tag{6.7}$$

where K is a constant. The response or output can be obtained from the transfer function and knowledge of the input. For a unit step input,

$$X_i(s) = \frac{1}{s}$$

form (6.7) gives

$$X_o(s) = \frac{K}{s(s + \alpha)}$$

Through application of partial fraction methods this becomes

$$X_o(s) = \frac{K_1}{s} + \frac{K_2}{s + \alpha}$$

where $K_1 = -K_2 = K/\alpha$ for this case. In general, the constants K_1 and K_2 are the *residues* of $X_o(s)$ at $s = 0$ and $s = -\alpha$, respectively. This concept of residues is borrowed from complex variable theory. Results can be generalized for basic systems of interest here. Response can always be written in the form

$$X_o(s) = \frac{K N(s)}{D(s)} \tag{6.8}$$

where $N(s)$, $D(s)$ are polynomials in s. If $D(s) = (s + \alpha)(s + \beta)(s + \gamma) \cdots$ then $X_o(s)$ can be expanded into partial fractions, and the form of $x_o(t)$ is determined by the factors of $D(s)$. Hence, the equation

$$D(s) = 0$$

is called the *characteristic equation*.

6.1.2 Second Order Systems

Second order systems are very common in spacecraft dynamics. They possess several properties which may be common to all such systems. Consider the

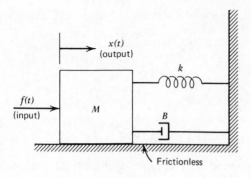

FIGURE 6.7 Second order mechanical system.

mass-spring-dashpot configuration of Figure 6.7. It has an equation of motion

$$M\ddot{x} = f(t) - kx - B\dot{x} \tag{6.9}$$

obtained by balancing forces and accelerations. Initial conditions are assumed to be $x(0) = \dot{x}(0) = 0$. The associated transfer function is easily extracted as

$$\text{T.F.} = \frac{X(s)}{F(s)} = \frac{1}{Ms^2 + Bs + k} \tag{6.10}$$

Consider the response to a step input $f(t) = Pu(t)$, where $u(t)$ is the unit step function. The denominator becomes $F(s) = P/s$. This allows expression (6.10) to be written as

$$X(s) = \frac{K_1}{s} + \frac{K_2}{s+\alpha} + \frac{K_3}{s+\beta} \tag{6.11}$$

where

$$\alpha, \beta = \frac{B}{2M} \pm \frac{1}{2M}\sqrt{B^2 - 4kM}$$

and K_1, K_2, K_3 and residues of $X(s)$ at 0, $-\alpha$, $-\beta$, respectively. There are three possible forms of response, depending on the relative sizes of B^2 and $4kM$:

1. If $B^2 > 4kM$, then α, β are real and unequal. The solution is

    $$x(t) = K_1 + K_2 e^{-\alpha t} + K_3 e^{-\beta t} \tag{6.12}$$

    where

    $$K_1 = \frac{P}{k}, \qquad K_2 = \frac{P/M}{\alpha^2 - \alpha\beta}, \qquad K_3 = \frac{P/M}{\beta^2 - \alpha\beta}$$

    The solution indicates a secular displacement K_1, and damped transient motion. This is known as the *overdamped* case.

2. If $B^2 = 4kM$, then $\alpha = \beta = B/2M$, which is real. Equation (6.11) becomes

$$X(s) = \frac{P/M}{s\left(s + \dfrac{B}{2M}\right)^2} = \frac{P/M}{s\left(s + \sqrt{\dfrac{k}{M}}\right)^2}$$

which gives a solution

$$x(t) = \frac{P}{k}\left[1 - e^{-(B/2M)t} - \frac{B}{2M}\,te^{-(B/2M)t}\right] \tag{6.13}$$

Here again there is a secular displacement and two damping terms. This situation is referred to as the *critically damped* case, because it corresponds to the lowest allowable value of damper constant B which prevents any oscillation of the mass for a given value of spring constant.

3. If $B^2 < 4kM$, then α, β are complex conjugates and equation (6.11) may be written as

$$X(s) = \frac{K_1}{s} + \frac{K_2}{s + a + ib} + \frac{K_3}{s + a - ib}$$

where α, $\beta = a \pm ib$, $a = B/2M$, $b = \sqrt{4kM - B^2}/2M$, and

$$K_1 = \frac{P}{k}$$

$$K_2 = -\frac{P}{2k}\left[1 + i\left(\frac{B}{\sqrt{4kM - B^2}}\right)\right]$$

$$K_3 = -\frac{P}{2k}\left[1 - i\left(\frac{B}{\sqrt{4kM - B^2}}\right)\right]$$

The solution is

$$x(t) = \frac{P}{k} + \frac{P}{k}\left(1 + \frac{B^2}{4kM - B^2}\right)^{1/2} e^{-(B/2M)t} \sin\left(\frac{\sqrt{4kM - B^2}}{2M}\,t + \psi\right) \tag{6.14}$$

where

$$\psi = \tan^{-1}\left(\frac{-\sqrt{4kM - B^2}}{-B}\right) \tag{6.15}$$

This is the *underdamped* case because motion is oscillatory.

All three possible motions are illustrated in Figure 6.8 with each case labeled. Notice that the underdamped solution causes *overshoot*. The solutions shown are valid for any second order system whose differential equation has form (6.9) and is subjected to a step input.

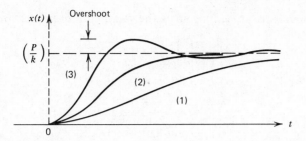

FIGURE 6.8 Possible responses of a second order system due to a step input.

Convenient parameters describing the performance of second order systems may now be defined and applied. The *damping ratio* indicates which of the three cases will occur and is defined by

$$\zeta = \frac{B}{\sqrt{4kM}} \tag{6.16}$$

Thus, $\zeta > 1$ implies motion is overdamped, $\zeta = 1$ implies motion is critically damped, and $\zeta < 1$ implies motion is underdamped. Next define the *undamped natural frequency* (i.e., $B = 0$) as

$$\omega_n = \sqrt{\frac{k}{M}} \tag{6.17}$$

and the *frequency of oscillation* as

$$\omega_o = \frac{\sqrt{4kM - B^2}}{2M} \tag{6.18}$$

which is meaningful for underdamped motion and may be expressed in terms of ω_n and ζ as

$$\omega_o = \omega_n\sqrt{1 - \zeta^2} \tag{6.19}$$

Notice that equation (6.19) indicates $\omega_n > \omega_o$. Thus, damping reduces the oscillation frequency. Substituting these new quantities into equation (6.14) gives a simpler form of underdamped motion,

$$x(t) = \frac{P}{k}\left[1 + \frac{e^{-\zeta\omega_n t}}{\sqrt{1 - \zeta^2}}\sin(\omega_o t + \psi)\right] \tag{6.20}$$

The maximum overshoot is sometimes important in designing a control system and is obtained by setting $\dot{x}(t) = 0$ and determining the time at which the first peak is reached. This procedure gives the *maximum overshoot fraction* as

$$\left[\frac{x(t)}{x(\infty)}\right]_{\max} = e^{-\pi\zeta/\sqrt{1-\zeta^2}} \tag{6.21}$$

which occurs at

$$t = \frac{\pi}{\omega_n\sqrt{1-\zeta^2}}$$

Higher order systems can be treated by generalizing the preceding results. The Laplace transform of the output can be expressed as

$$X_o(s) = \frac{K(s+Z_1)(s+Z_2)\cdots(s+Z_m)}{(s+P_1)(s+P_2)\cdots(s+P_n)} \tag{6.22}$$

where $n > m$ in typical situations. If all P_j's $(j = 1, 2, \ldots, n)$ are distinct, then this expression can take the form

$$X_o(s) = \frac{K_1}{s+P_1} + \frac{K_2}{s+P_2} + \cdots + \frac{K_n}{s+P_n}$$

where the residues are calculated from

$$\left.\begin{aligned}
K_1 &= (s+P_1)X_o(s)\big|_{s=-P_1} \\
K_2 &= (s+P_2)X_o(s)\big|_{s=-P_2} \\
&\text{-----------------------} \\
K_n &= (s+P_n)X_o(s)\big|_{s=-P_n}
\end{aligned}\right\} \tag{6.23}$$

If all P_j's $(j = 1, 2, \ldots, n)$ are not distinct, multiple roots occur. For example, if P_1 has a value which appears r times, then

$$X_o(s) = \frac{K_1}{s+P_1} + \frac{C_2}{(s+P_1)^2} + \frac{C_3}{(s+P_1)^3} + \cdots + \frac{C_r}{(s+P_1)^r} + \frac{K_2}{s+P_2} + \cdots$$

is obtained by partial fraction methods. $C_k(k = 2, 3, \ldots, r)$ is not a residue, but $K_2, K_3, \ldots,$ etc. are residues obtained from form (6.23). For $K_1, C_2, \ldots, C_r$, use

$$\left.\begin{aligned}
C_r &= (s+P_1)^r X_o(s)\big|_{s=-P_1} \\
C_{r-1} &= \frac{1}{1!}\left\{\frac{d}{ds}\left[(s+P_1)^r X_o(s)\right]\right\}_{s=-P_1} \\
&\text{--} \\
C_2 &= \frac{1}{(r-2)!}\left\{\frac{d^{(r-2)}}{ds^{(r-2)}}\left[(s+P_1)^r X_o(s)\right]\right\}_{s=-P_1} \\
K_1 &= \frac{1}{(r-1)!}\left\{\frac{d^{(r-1)}}{ds^{(r-1)}}\left[(s+P_1)^r X_o(s)\right]\right\}_{s=-P_1}
\end{aligned}\right\} \tag{6.24}$$

6.1.3 Pole-Zero Plots

The preceding discussion illustrates the analytical approach to extracting residues and solutions. It is also possible, and advantageous in later discussions to graphically determine these residues. This approach is called the *complex plane* analysis. A map of poles and zeros of a function is plotted initially. Define a *pole* of $X_o(s)$ as a value of s which makes $X_o(s) = \infty$, and a *zero* of $X_o(s)$ as a value of s which makes $X_o(s) = 0$. In general, $X_o(s)$ has form (6.22). Therefore,

$$s = -Z_1, -Z_2, \ldots, -Z_m = \text{zeros of } X_o(s)$$

$$s = -P_1, -P_2, \ldots, -P_n = \text{poles of } X_o(s)$$

Since $n > m$ for systems of interest, $X_o(\infty) = 0$. It follows that there are $(n - m)$ zeros at ∞. The term *singularity* is used to indicate either a pole or a zero. Note that $X_o(s)$ is completely determined if its poles, zeros, and K are specified. The use of a pole-zero map on the complex plane is best explained by considering an example.

Assume the case

$$X_o(s) = \frac{15(s + 1)}{s(s^2 + 5s + 6)}$$

It is obvious that only one zero occurs at $s = -1$, and the roots of the denominator must be found. They are simply determined as $s = 0$, $s = -2$, $s = -3$, which are also the poles of $X_o(s)$. These poles and zeros are plotted on the format shown in Figure 6.9 on the complex plane, $\sigma, i\omega$. Here poles are distinguished from zeros by $\times$'s and $\bigcirc$'s. Constant K is shown in a box. Double poles can be represented as double $\times$'s. If there are complex poles, then some $\times$'s will appear off the σ axis. For example, if

$$X_o(s) = \frac{3(s + 1)}{s(s^2 + 2s + 2)} = \frac{3(s + 1)}{s(s + 1 + i)(s + 1 - i)}$$

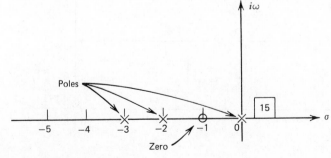

FIGURE 6.9 Pole-zero plot format.

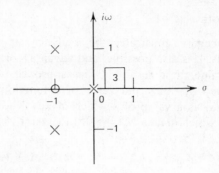

FIGURE 6.10 Pole-zero plot with complex poles.

then the pole-zero plot would be represented by Figure 6.10. Once the pole-zero plot has been constructed for a given function, residues can be determined graphically. Consider the example case in which

$$X_o(s) = \frac{18(s+1)}{s(s+3)(s+6)} = \frac{K_1}{s} + \frac{K_2}{s+3} + \frac{K_3}{s+6}$$

with its pole-zero plot shown in Figure 6.11. First draw vectors from each pole and zero to the pole whose residue is being calculated. The value of K_1 (corresponding to $s = 0$) is being determined in the figure. Vector $(\mathbf{s+1})$ has a value of $+1$, because its length is 1 and direction is along the positive real axis. It has a reference angle of zero degrees. Thus, $(\mathbf{s+1})$ is equivalent to

$$(s+1)|_{s=0}$$

Similarly, $(\mathbf{s+3})$ is equivalent to

$$(s+3)|_{s=0}$$

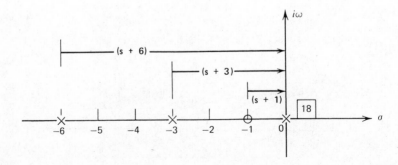

FIGURE 6.11 Graphical determination of residues.

and so on. Comparing this with set (6.23) indicates that

$$K_1 = \frac{18(s+1)}{(s+3)(s+6)} = \frac{18(+1)}{(+3)(+6)} = 1$$

In a general sense,

$$K_1 = K\left(\frac{\text{Product of vectors drawn from zeros to } s = 0}{\text{Product of vectors drawn from poles to } s = 0}\right)$$

or

$$K_1 = 18\left(\frac{\Pi Z}{\Pi P}\right)_{s=0}$$

In a similar manner K_2, which corresponds to the pole at $s = -3$, can be found,

$$K_2 = 18\left(\frac{\Pi Z}{\Pi P}\right)_{s=-3} = 18\frac{-2}{(-3)(+3)} = 4$$

and finally, $K_3 = -5$. Combining results gives

$$X_o(s) = \frac{1}{s} + \frac{4}{s+3} - \frac{5}{s+6}$$

To illustrate determination of residues when complex poles are present, consider an example in which

$$X_o(s) = \frac{s+2}{(s+1)(s+5)(s^2+3s+9)}$$

$$= \frac{K_1}{s+1} + \frac{K_2}{s+5} + \frac{K_3}{s+1.5+i2.6} + \frac{K_4}{s+1.5-i2.6}$$

The associated pole-zero plot is given in Figure 6.12, with vectors for calculating K_1 and K_2 shown. Using the *solid* vectors, K_1 is

$$K_1 = 1\left(\frac{\Pi Z}{\Pi P}\right)_{s=-1} = \frac{1}{4(2.648 @ 79.1°)(2.648 @ -79.1°)} = \frac{1}{28}$$

Note that $(2.648 @ 79.1°)$ is the polar representation of $(0.5 + i2.6)$. Using the *dotted* vectors,

$$K_2 = \frac{-3}{(-4)(4.39 @ 143.4°)(4.39 @ -143.4°)} = 0.039$$

In a similar manner K_3 is found

$$K_3 = \frac{(2.648 @ -79.1°)}{(2.645 @ -100.9°)(4.39 @ -36.6°)(5.2 @ -90°)} = -0.037 + i0.023$$

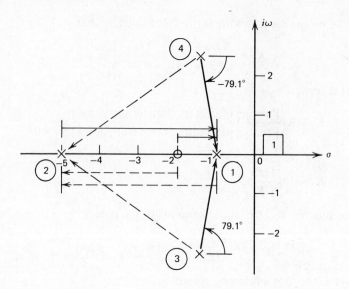

FIGURE 6.12 Pole-zero plot for calculating residues of complex poles.

and K_4 is the conjugate of K_3,

$$K_4 = -0.037 - i0.023$$

If $X_o(s)$ has one or more multiple poles this graphical procedure becomes cumbersome and is not recommended in such cases. Nevertheless, the pole-zero approach makes it possible to quickly judge relative sizes of the various residues, thus permitting assessment of importance associated with each term in $X_0(s)$. Notice that since

$$K_n = K\left(\frac{\Pi\mathbf{Z}}{\Pi\mathbf{P}}\right)_{s=-P_n}$$

the size of this residue may be made small if $\Pi\mathbf{Z}$ is small or if $\Pi\mathbf{P}$ is large. For example, consider the qualitative case shown in Figure 6.13. Here $\Pi\mathbf{Z}$ is small for $s = -P_n$, so that the residue at each of the other poles will be large compared to K_n. If a pole and a zero coincide they cancel each other, and the residue is zero. Furthermore the residue of a remote pole is small, because $\Pi\mathbf{P}$ would be large. Since second-order systems produce complex roots in conjugate pairs, if an odd number of poles plus zeros lie to the right of a pole whose residue is being determined, then that residue has a negative real part.

Consider again an underdamped second order system with step input $P\,u(t)$. Output is expressed as

$$X_o(s) = \frac{P}{s}\frac{\omega_n^2/k}{s^2 + 2\zeta\omega_n s + \omega_n^2} \tag{6.25}$$

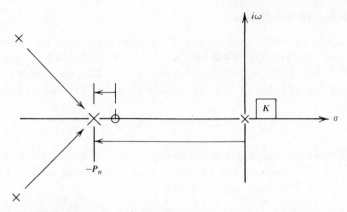

FIGURE 6.13 Illustration of small residue resulting from a close zero.

The pole-zero map of Figure 6.14 gives the relationship between ω_n, ζ, and ω_o graphically. The angle ν is defined by $\zeta = \cos \nu$ and expression (6.19) verifies

$$\omega_o = \omega_n \sin \nu$$

Residues of $X_o(s)$ are now obtained by writing

$$X_o(s) = \frac{K_1}{s} + \frac{K_2}{s + \zeta\omega_n + i\omega_n\sqrt{1-\zeta^2}} + \frac{\bar{K}_2}{s + \zeta\omega_n - i\omega_n\sqrt{1-\zeta^2}}$$

where $\bar{K}_2$ is the conjugate of K_2. To determine K_1 and K_2 use the graphical procedure,

$$K_1 = \frac{P\omega_n^2}{k}\left[\frac{1}{(\omega_n\underline{/-\nu})(\omega_n\underline{/+\nu})}\right] = \frac{P}{k}$$

$$K_2 = \frac{P/k}{2\sqrt{1-\zeta^2}}\underline{/270° - \nu}$$

(6.26)

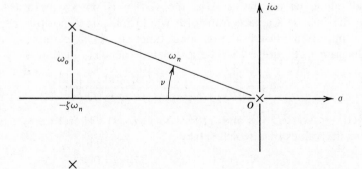

FIGURE 6.14 Pole-zero map of second order system with step input.

6.1.4 Network Synthesis

The process by which one arrives at a design for a network having a given pole-zero configuration is called *network synthesis*. In general, for given input a system will have m zeros and n poles, with $n > m$. This inequality is important because the initial value of x_o must usually be finite. The *initial value theorem* states that

$$x_o(0^+) = \lim_{s \to \infty} s\, X_o(s) \tag{6.27}$$

where (0^+) indicates that this is the limiting value of x_o as $t = 0$ is approached from $t > 0$. This implies n must be greater than m for finite $x_o(0^+)$. If $n \geq m+2$, then $x_o(0^+) = 0$. The *final value theorem* states

$$x_o(\infty) = \lim_{s \to 0} s\, X_o(s) \tag{6.28}$$

with certain restrictions. It will suffice here to limit use of this theorem to situations in which all poles of $s\, X_o(s)$ lie in the left-half plane. Unfortunately, locations of these poles are not generally given and may have to be determined before using the theorem. Unless there is a pole at $s = 0$, expression (6.28) will then give $x_o(\infty) = 0$. A system whose transfer function has one or more poles or zeros in the right-half plane is a *nonminimum phase* system. All others are *minimum phase* systems. For stability, a system must not have any poles in the right-half plane. Thus, a nonminimum phase system with only zeros in the right-half plane is stable.

A control system is *asymptotically stable* if, in response to an impulsive input applied when the system is at rest, the output approaches its initial state as time increases. The importance of linear stability analysis here is in predicting whether unstable motions will occur and in showing how to avoid such motions, rather than in computing the exact form of such responses. Stability is determined by constructing the transfer function of a given system. Once it is in factored form, stability can be determined by inspection. If any pole is in the right-half plane or on the $i\omega$-axis, the system is not asymptotically stable because the inverse Laplace transform would indicate a positive exponential power or pure oscillation in the response function. As an example consider the block diagram in Figure 6.15. The associated transfer function is

$$W(s) = \frac{C(s)}{R(s)} = \frac{N_1(s)D_2(s)}{N_1(s)N_2(s) + D_1(s)D_2(s)} \tag{6.29}$$

where $G(s) = N_1(s)/D_1(s)$ and $H(s) = N_2(s)/D_2(s)$. Therefore, the zeros of $W(s)$ are the values of s which satisfy

$$N_1(s) = 0 \quad \text{or} \quad D_2(s) = 0$$

In other words, $W(s)$ has zeros at the zeros of $G(s)$ and at the poles of $H(s)$.

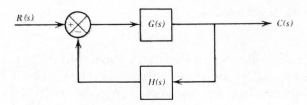

FIGURE 6.15 Example system for stability criteria.

The poles of $W(s)$ occur at the values of s which satisfy

$$N_1(s)N_2(s) + D_1(s)D_2(s) = 0 \qquad (6.30)$$

Stability may be determined by factoring this polynomial or by applying Routh's criteria, presented in Section 5.2.3. However, in the design of control systems there are usually several variables to be specified. The range of these parameters, which allows satisfaction of performance requirements while maintaining stability, is of critical importance. This situation has brought about the use of other methods for determining stability such that system performance can be studied at the same time. The *root locus* stability analysis method is one such technique in linear control theory.

6.1.5 Feedback and the Root Locus Plot

The concept of *feedback* has brought about development of automatic control theory by permitting the response of a physical system to be used for correcting its own motion. Figure 6.16 illustrates the fundamental feedback loop, the *unity feedback* system. Its transfer function is simply

$$W(s) = \frac{C(s)}{R(s)} = \frac{N(s)}{N(s) + D(s)} \qquad (6.31)$$

where $G(s) = N(s)/D(s)$. The zeros of $W(s)$ lie at the zeros of $G(s)$. Poles of $W(s)$ are the roots of $N(s) + D(s) = 0$ or

$$G(s) + 1 = 0$$

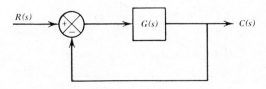

FIGURE 6.16 Unity feedback system.

This can also be interpreted as

$$G(s) = -1 \qquad (6.32)$$

which is a condition for the poles of $W(s)$. From this, the *root locus* stability analysis method can be developed. Remember that stability of the unity feedback system requires that the poles of $W(s)$ lie in the left-half plane. If $N(s)$ contains an adjustable parameter, K, which is usually called the *gain*, then every allowable value of K produces a new set of poles of $W(s)$. Thus, if K varies from zero to infinity, all possible pole positions will form a set of loci on the complex plane. Each path is the locus of a pole of $W(s)$. As long as all loci remain in the left-half plane, the system is stable. Typically, there is only a limited range of K values which correspond to sections of the loci in this half plane. Values of K which correspond to points on loci in the right-half plane and on the imaginary axis are not allowable if a stable system is required. Noting that s is complex rewrite $G(s)$ in polar form,

$$G(s) = |G(s)| @ \underline{/G(s)}$$

as depicted in Figure 6.17. This allows a more convenient form of condition (6.32) in two parts

$$\begin{aligned} |G(s)| &= 1 \\ \underline{/G(s)} &= \pm 180°, \pm 540°, \ldots \end{aligned} \qquad (6.33)$$

In general, $G(s)$ has the form

$$G(s) = \frac{K(s + Z_1) \cdots (s + Z_m)}{(s + P_1)(s + P_2) \cdots (s + P_n)} \qquad (6.34)$$

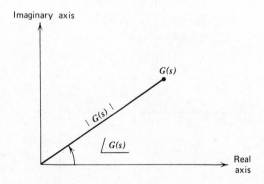

FIGURE 6.17 Nomenclature for polar form of a complex number.

which permits conditions (6.33) to become

$$|G(s)| = \frac{K|s+Z_1||s+Z_2|\cdots|s+Z_m|}{|s+P_1||s+P_2|\cdots|s+P_n|} = 1 \qquad (6.35a)$$

$$\underline{/G(s)} = \underline{/s+Z_1} + \underline{/s+Z_2} + \cdots + \underline{/s+Z_m} - \underline{/s+P_1} - \underline{/s+P_2} - \cdots - \underline{/s+P_n}$$

$$= \pm180°, \pm540°, \ldots \qquad (6.35b)$$

Implementation of the root locus method is best introduced by assuming a specific example. If

$$G(s) = \frac{K(s+2)}{s(s+1)(s+4)(s+10)}$$

the angle condition is

$$\underline{/G(s)} = \underline{/s+2} - \underline{/s} - \underline{/s+1} - \underline{/s+4} - \underline{/s+10} = \pm180°, \pm540°, \ldots$$

This can be used first to locate sections of loci on the real axis. Refer to the pole-zero plot of $G(s)$ in Figure 6.18. If $s = a$, where a is real and positive, then $\underline{/G(s)} = 0°$. This does not satisfy the angle condition, thus, any point on the positive real axis cannot be a pole of $W(s)$. If $-1 < s < 0$, then $\underline{/G(s)} = -180°$. This means that the segment of the real axis between 0 and -1 is also a segment of a locus. If $-2 < s < -1$, then $\underline{/G(s)} = 360°$, which is not satisfactory. Further testing shows that real axis segments between -2 and -4, and between -10 and $-\infty$ are on loci of poles of $W(s)$. In summary, all points on the real axis which lie to the left of an odd number of poles and zeros of $G(s)$ satisfy angle conditions (6.35b). Now consider points on the $i\omega$ axis as potential locus points. Refer to Figure 6.19 and take an arbitrary value, $s = i\omega$. The angle condition becomes

$$\underline{/G(s)} = \tan^{-1}\left(\frac{\omega}{2}\right) - 90° - \tan^{-1}\left(\frac{\omega}{1}\right) - \tan^{-1}\left(\frac{\omega}{4}\right) - \tan^{-1}\left(\frac{\omega}{10}\right)$$

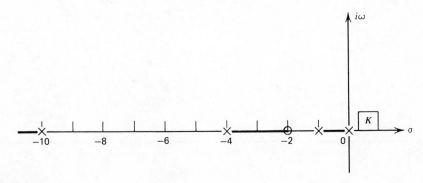

FIGURE 6.18 Loci on the real axis.

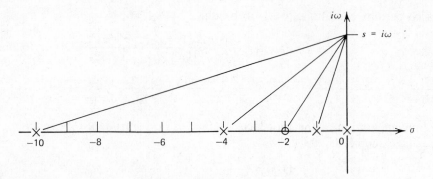

FIGURE 6.19 Testing the imaginary axis for loci crossings.

As ω goes from zero to ∞, $\underline{/G(s)}$ goes from $-90°$ to $-270°$. Therefore, ω must satisfy the angle condition at some point. By trial and error the value of ω which gives $\underline{/G(s)} = -180°$ is 5.2. By symmetry, $\underline{/G(s)} = +180°$ for $\omega = -5.2$.

For remote regions of the complex plane, all vectors from poles and zeros of $G(s)$ have approximately the same angle θ, as illustrated in Figure 6.20. Therefore, θ can be approximately by

$$\theta \times (\text{no. of zeros}) - \theta \times (\text{no. of poles}) \cong \pm 180°, \pm 540°, \ldots$$

For the example at hand

$$\theta = \frac{\pm 180°}{1 - 4} = \pm 60°$$

$$\theta = \frac{\pm 540°}{1 - 4} = \pm 180°$$

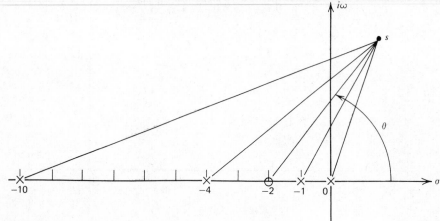

FIGURE 6.20 Asymptotes of loci in remote regions.

Beyond this, values of θ repeat. One can conclude that for remote points away from the real axis, the loci tend to follow asymptotes. In this case there are three such lines at $\theta = 60°$, $180°$, and $300°$. Each locus must connect a pole to a zero, and there are $n - m$ zeros at ∞. Therefore, there are $n - m$ asymptotes. In this example $n - m = 3$, and there are three asymptotes. The entire root locus plot for this case is sketched in Figure 6.21. The dotted lines and the negative real axis are the three asymptotes. Notice that one locus connects the pole at $s = -4$ with the zero at $s = -2$, and that two loci break away from the real axis in a symmetrical manner.

So far only angle condition (6.35b) has been employed to give the loci shapes, real axis segments, imaginary axis crossings, and asymptotes. In order to generate exact plots and evaluate values of K at points of interest, the other condition, equation (6.35a) must be introduced. Thus, roots of the characteristic equation, $N(s) + D(s) = 0$, lie on the root loci and satisfy

$$|G(s)| = 1 = \frac{K\,|s+2|}{|s|\,|s+1|\,|s+4|\,|s+10|}$$

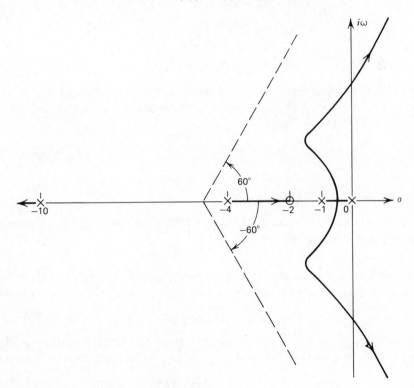

FIGURE 6.21 Complete root locus plot.

where K, the *loop gain*, determines the position of the roots of $W(s)$ on the loci. If $K = 0$, the poles of $W(s)$ are the poles of $G(s)$, because

$$W(s) = \frac{G(s)}{1 + G(s)} = \frac{K(s+2)}{K(s+2) + s(s+1)(s+4)(s+10)}$$

As K increases from zero, the poles of $W(s)$ move along the root loci. To determine K corresponding to a given pole of $W(s)$ on the root locus plot use

$$K = \frac{|s|\,|s+1|\,|s+4|\,|s+10|}{|s+2|}$$

Asymptotes of the root locus can be located by considering another form of equation (6.34),

$$G(s) = \frac{K(s^m + a_1 s^{m-1} + \cdots + a_m)}{(s^n + b_1 s^{n-1} + \cdots + b_n)} \tag{6.36}$$

where only the four coefficients,

$$a_1 = Z_1 + Z_2 + \cdots + Z_m$$
$$a_m = Z_1 Z_2 Z_3 \cdots Z_m$$
$$b_1 = P_1 + P_2 + \cdots + P_n$$
$$b_n = P_1 P_2 \cdots P_n$$

are of interest. Note that a_1 is the negative sum of zeros of $G(s)$, and b_1 is the negative sum of poles of $G(s)$. Dividing the numerator into the denominator changes expression (6.36) to

$$G(s) = \frac{K}{s^{n-m} + (b_1 - a_1)s^{n-m-1} + \cdots}$$

In order to satisfy condition (6.32), the denominator must be

$$s^{n-m} + (b_1 - a_1)s^{n-m-1} + \cdots = -K \tag{6.37}$$

Asymptotes are located and oriented by letting s get very large. As this happens expression (6.37) becomes

$$s^{n-m} = -K \tag{6.38}$$

This expression applies only if the asymptotes passed through the origin. If s_{asy} is the intersection point on the real axis, replace s with $s_1 + s_{asy}$ in form (6.37),

$$s_1^{n-m} + [(n-m)s_{asy} + (b_1 - a_1)]s_1^{n-m-1} + \cdots = -K \tag{6.39}$$

so that the origin is now at the intersection point. To satisfy equation (6.38) for large values of s, the coefficient in expression (6.39) must vanish. Therefore,

$$s_{asy} = -\frac{(b_1 - a_1)}{n - m}$$

or

$$S_{asy} = \frac{\Sigma[\text{Pole values of } G(s)] - \Sigma[\text{Zero values of } G(s)]}{n - m} \qquad (6.40)$$

For the example at hand, $s_{asy} = -13/3 = -4.33$.

The maximum value of K for stability corresponds to a pole of $W(s)$ on the $i\omega$ axis. Thus, at $s = \pm i5.2$

$$K = \frac{|5.2|\,|i5.2+1|\,|i5.2+4|\,|i5.2+10|}{|i5.2+2|} = 364.5$$

This accounts for the two poles at $\pm i5.2$. The other two poles corresponding to this value of K are found from

$$W(s) = \frac{K(s+2)}{s^4 + 15s^3 + 54s^2 + (40+K)s + 2K} = \frac{K(s+2)}{(s+Q_1)(s+Q_2)(s+Q_3)(s+Q_4)}$$

which implies

$$Q_1 + Q_2 + Q_3 + Q_4 = 15$$
$$Q_1 Q_2 Q_3 Q_4 = 2K$$

Since $Q_3 = i5.2$ and $Q_4 = -i5.2$, these become

$$Q_1 + Q_2 = 15$$
$$Q_1 Q_2 = 26.96$$

Solving simultaneously gives the values of Q_1 and Q_2 as 2.09 and 12.91, which correspond to the remaining pole locations of $W(s)$.

The procedure for constructing a root locus plot can now be presented in a logical manner for a unity feedback system. Assuming $G(s)$ is known and K is a variable:

(a) Factor $G(s)$ to determine its poles and zeros.
(b) Plot these poles and zeros to scale on the complex plane. Note that n should exceed m, and the poles of $W(s)$ are those of $G(s)$ if $K=0$, assuming form (6.31) for the transfer function.
(c) Sketch asymptotes using formula (6.40) for their intersection point and

$$\theta_{asy} = \frac{\pm 180°, \pm 540°, \ldots}{n - m} \qquad (6.41)$$

for their angles.
(d) The loci are drawn by showing each branch emanating from a pole of $G(s)$ as K varies from zero to infinity. Start by drawing segments on the real axis using the rule that loci lie to the left of an odd number of poles plus zeros.

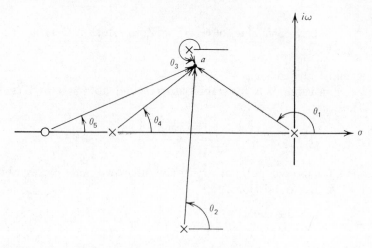

FIGURE 6.22 Departing angle from a complex pole.

(e)   If there are complex poles, find the angle at which the locus leaves these poles. Refer to Figure 6.22 and note that the sum of angles of vectors drawn to point a must satisfy

$$\sum_{i}^{n} \theta_i - \sum_{j}^{m} \theta_j = \pm 180°, \pm 540°, \ldots \qquad (6.42)$$
$$\text{poles} \qquad \text{zeros}$$

However, the asymptote associated with the locus from the pole considered in Figure 6.22 corresponds to $+180°$. Thus, for this complex pole

$$\theta_1 + \theta_2 + \theta_3 + \theta_4 - \theta_5 = 180°$$

is used. Since point a is assumed very close to the complex pole, θ_1, θ_2, θ_4, θ_5 may be taken as though their corresponding vectors are drawn to that pole. Then θ_3 is determined from

$$\theta_3 = 180° - \theta_1 - \theta_2 - \theta_4 + \theta_5$$

By symmetry the angle of departure from the conjugate pole is $-\theta_3$.

(f)   A branch of the loci lying between two poles on the real axis will *breakaway* from that axis and form two complex branches. This breakaway point is determined by the following technique. Referring to the example of Figure 6.23, the sum of angles of vectors drawn to a point just above the breakaway location must satisfy

$$\theta_1 + \theta_2 + \theta_3 + \theta_4 = 180°$$

The distance ε is small, such that ε^2 terms may be ignored. This

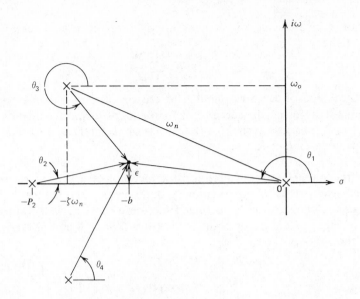

FIGURE 6.23 Locating the breakaway point.

condition becomes

$$\pi - \tan^{-1}\left(\frac{\varepsilon}{b}\right) + \tan^{-1}\left(\frac{\varepsilon}{P_2 - b}\right) + 2\pi - \tan^{-1}\left(\frac{\omega_o - \varepsilon}{\zeta\omega_n - b}\right)$$

$$+ \tan^{-1}\left(\frac{\omega_o + \varepsilon}{\zeta\omega_n - b}\right) = \pi$$

It is convenient to use the identity

$$\tan^{-1}\left(\frac{\omega_o - \varepsilon}{\zeta\omega_n - b}\right) - \tan^{-1}\left(\frac{\omega_o + \varepsilon}{\zeta\omega_n - b}\right) = \tan^{-1}\left(\frac{2\varepsilon(b - \zeta\omega_n)}{(b - \zeta\omega_n)^2 + \omega_o^2}\right) \quad (6.43)$$

Since all remaining angles are small the condition becomes

$$-\frac{\varepsilon}{b} + \frac{\varepsilon}{P_2 - b} - \frac{2\varepsilon(b - \zeta\omega_n)}{(b - \zeta\omega_n)^2 + \omega_o^2} = 0$$

and ε may be divided out leaving an expression for b only,

$$\frac{1}{P_2 - b} - \frac{1}{b} = \frac{2(b - \zeta\omega_n)}{(b - \zeta\omega_n)^2 + \omega_o^2}$$

The value of b is best found by trial and error methods. An alternate approach would be to plot K vs σ between two poles on the real axis. The value of σ which gives a maximum K is the breakaway point.

(g) For a given value of K poles of $W(s)$ on each locus are found by satisfying

$$K = \frac{\Pi\,|\mathbf{P}|}{\Pi\,|\mathbf{Z}|} \qquad (6.44)$$

The value of K corresponding to crossing the $i\omega$ axis may be determined by using the method of Figure 6.19 to first evaluate s at that point. Then use form (6.44). An alternate method is to use Routh's array to extract K.

So far the method of constructing root locus plots has been restricted to unity feedback systems. Many situations require elements in the feedback loop. Root locus diagrams can be constructed for these cases as well. Consider the diagram of Figure 6.15. Zeros occur at the roots $N_1(s) = 0$ and $D_2(s) = 0$. Poles occur at the roots of

$$N_1(s)N_2(s) + D_1(s)D_2(s) = 0$$

or

$$G(s)H(s) = -1 \qquad (6.45)$$

This implies that the root locus method can be applied by treating $G(s)H(s)$ as $G(s)$ was treated in the unity feedback case. The associated transfer function is

$$W(s) = \frac{G(s)}{G(s)H(s) + 1} \qquad (6.46)$$

Notice that *negative feedback*, which is the case when $R(s) - H(s)C(s)$ is the input to $G(s)$, results in poles of $W(s)$ occurring when $G(s)H(s) = -1$. This leads to the angle condition

$$\underline{/G(s)H(s)} = \pm 180°, \pm 540°, \dots$$

which helps stability because poles in the left-half plane tend to generate loci in that plane. Positive feedback would require this condition to be

$$\underline{/G(s)H(s)} = 0°, \pm 360°, \dots$$

which would tend to generate loci in the right-half plane. Therefore, negative feedback is generally recommended in single-loop control systems. There are situations in which positive feedback is essential, but these are beyond the intent of this text.

To illustrate the construction of a root locus plot for a nonunity feedback situation, consider an example with

$$G(s) = \frac{K}{s(s^2 + 4s + 8)}$$

$$H(s) = \frac{s+2}{s+1}$$

Substituting these into form (6.46) yields

$$W(s) = \frac{K(s+1)}{K(s+2) + s(s+1)(s^2 + 4s + 8)}$$

Following the root locus procedure with slight modifications for $H(s)$ gives:

(a) In factored form

$$G(s)H(s) = \frac{K(s+2)}{s(s+1)(s+2+i2)(s+2-i2)}$$

(b) The poles and zeros of $G(s)H(s)$ are plotted on Figure 6.24.

(c) Asymptotes are determined from

$$\theta_{asy} = \frac{\pm 180°, \pm 540°}{3} = \pm 60°, 180°$$

$$s_{asy} = -1$$

(d) On the real axis, loci exist between $s = 0$ and $s = -1$, and between $s = -2$ and $s = -\infty$. Remaining steps are identical to the unity feedback case. The complete root locus plot for this example is sketched in Figure 6.25. Notice that the loci leaving the complex poles of $G(s)H(s)$ do not approach the two asymptotes at $\theta_{asy} = \pm 60°$. Instead they go to a *break-in* point on the real axis. This is required in order to end the loci at zeros.

In summary, the treatment of linear control theory presented here is very brief and superficial. Development of the root locus concept was the major objective. This will next be applied to a realistic design problem in satellite attitude control. Of course, the topic of linear control theory is quite extensive and many texts are available for further study.

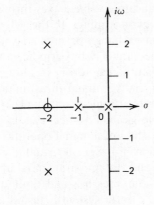

FIGURE 6.24 Poles and zeros of *G(s)H(s)*.

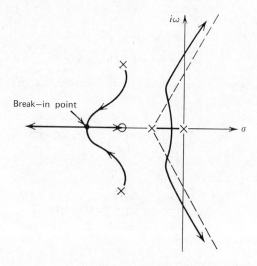

FIGURE 6.25 Complete root locus plot for nonunity feedback example.

6.2 DESIGN OF A BIAS MOMENTUM SYSTEM

The basic ideas of linear control theory will now be applied to satellite attitude control system design. Implementation of these concepts is best discussed by presenting the development of a preliminary system design for a realistic situation. An excellent example, and one which is of current interest, is the design procedure for a double-gimbaled momentum wheel attitude control system. Associated analyses and assumptions are discussed for a three-axis stabilized satellite configuration in geostationary orbit. This type of gyro device offers control torques about all three vehicle axes through wheel speed control and two-degree-of-freedom gyrotorquing. If the wheel size and speed are correctly selected, momentum exchange permits cancellation of cyclic torques without employing attitude jets, and only periodic momentum adjustment is necessary due to secular disturbance torques. This is in contrast to mass expulsion control systems, in that continual thrusting and roll, pitch, and yaw sensors are required for accurate attitude maintenance. Only a pitch/roll (earth) sensor is needed for the double-gimbaled wheel system described here. Considerations will be limited to on-station, nominal operations. Attitude acquisition requirements appear in Section 4.4. Specific satellite properties are assumed and appropriate control laws and system parameters are selected to counteract disturbance torques due to solar pressure and limited orbit control thruster misalignment. Stability of the control system is tested and responses obtained for impulsive, step, and cyclic disturbance torques.

6.2.1 Equations of Motion

A satellite configuration typical of body-stabilized vehicle technology was selected and is illustrated in Figure 6.26. Body axes and nominal wheel orientation are shown here. Specific mass and inertia properties used in this development are listed in Table 6.1. Also included are expected disturbance torque expressions and attitude pointing accuracy requirements. Development of the equations of motion follows the treatment in Chapter 5 for a rigid body with internal momentum. Linearization is permitted when angular deviations from nominal are small. The reference frame for attitude control of this type satellite is actually rotating with respect to an inertial frame at the orbit rate, ω_o. Nominal orientation of the double-gimbaled wheel is schematically illustrated in Figure 6.27a. Identifying the gravity gradient and external disturbance torques separately, Euler's equations for the rigid satellite and wheel are given by

$$\mathbf{T} + \mathbf{G} = \frac{d\mathbf{h}}{dt} = \left[\frac{d\mathbf{h}}{dt}\right]_b + \boldsymbol{\omega} \times \mathbf{h} \qquad (6.47)$$

where $\mathbf{T}$ is the disturbance torque due to solar pressure and thrust misalignment, $\mathbf{G}$ is the gravity gradient torque, and $\mathbf{h}$ is the total angular momentum, including the wheel. Unit vectors $\mathbf{i}$, $\mathbf{j}$, $\mathbf{k}$ correspond to body x, y, z principal axes, respectively. Thus

$$\mathbf{h} = \mathbf{h}_v + \mathbf{h}_w \qquad (6.48)$$

Angular momentum of the vehicle, $\mathbf{h}_v$ (excluding the wheel) is easily expressed in terms of components along the principal axes,

$$\mathbf{h}_v = I_x \omega_x \mathbf{i} + I_y \omega_y \mathbf{j} + I_z \omega_z \mathbf{k}$$

TABLE 6.1 Assumed Satellite Specifications and Disturbance Torques

Satellite mass	716 kg
Moments of inertia	$I_x = I_z = 2000 \text{ N} \cdot \text{m} \cdot \text{s}^2$
	$I_y = 400 \text{ N} \cdot \text{m} \cdot \text{s}^2$
Attitude accuracy requirements	pitch and roll = 0.05°
	yaw = 0.40°
Solar pressure torques	$\begin{cases} T_x = 2 \times 10^{-5}(1 - 2\sin \omega_o t) \text{ N} \cdot \text{m} \\ T_y = 10^{-4}(\cos \omega_o t) \text{ N} \cdot \text{m} \\ T_z = -5 \times 10^{-5}(\cos \omega_o t) \text{ N} \cdot \text{m} \end{cases}$
($t = 0$ at 6 A.M. or 6 P.M. orbital position)	
Thruster misalignment torque	$T_F = 8.5 \times 10^{-5} \text{ N} \cdot \text{m}$

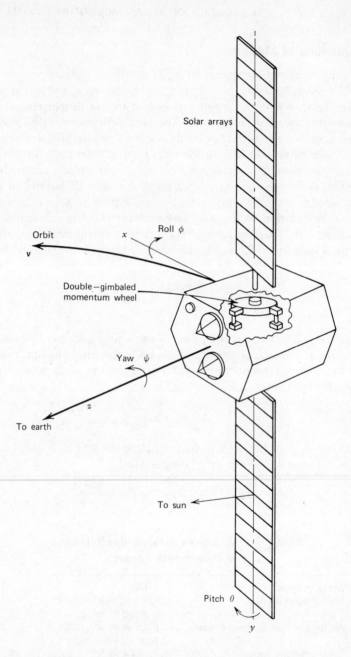

Solar arrays

Orbit

v

x

Roll ϕ

Double−gimbaled
momentum wheel

Yaw ψ

z

To earth

To sun

Pitch θ

y

FIGURE 6.26 Assumed body-stablized spacecraft configuration.

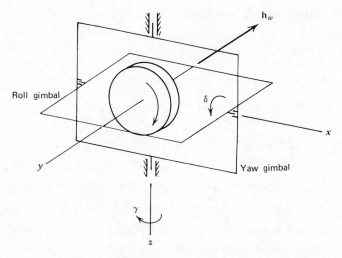

(a) Nominal orientation

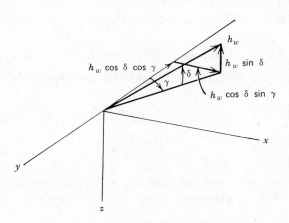

(b) Angular momentum components

FIGURE 6.27 Nomenclature for double-gimbaled momentum wheel.

Referring to Figure 6.27b, wheel momentum components are

$$h_{wx} = (\cos \delta \sin \gamma) h_w$$
$$h_{wy} = -(\cos \delta \cos \gamma) h_w \qquad (6.49)$$
$$h_{wz} = -(\sin \delta) h_w$$

where δ, γ are the roll and yaw gimbal angles, respectively. Combining these expressions leads to a new form of vector equation of motion,

$$
\begin{aligned}
\mathbf{T} + \mathbf{G} = &[I_x \dot{\omega}_x - \dot{\delta}(\sin \delta \sin \gamma) h_w + \dot{\gamma}(\cos \delta \cos \gamma) h_w \\
&+ (\cos \delta \sin \gamma) \dot{h}_w + \omega_y(\omega_z I_z - h_w \sin \delta) - \omega_z(\omega_y I_y - h_w \cos \delta \cos \gamma)]\mathbf{i} \\
&+ [I_y \dot{\omega}_y + \dot{\delta}(\sin \delta \cos \gamma) h_w + \dot{\gamma}(\cos \delta \sin \gamma) h_w - (\cos \delta \cos \gamma) \dot{h}_w \\
&+ \omega_z(\omega_x I_x + h_w \cos \delta \sin \gamma) - \omega_x(\omega_z I_z - h_w \sin \delta)]\mathbf{j} \\
&+ [I_z \dot{\omega}_z - \dot{\delta}(\cos \delta) h_w - (\sin \delta) \dot{h}_w + \omega_x(\omega_y I_y - h_w \cos \delta \cos \gamma) \\
&- \omega_y(\omega_x I_x + h_w \cos \delta \sin \gamma)]\mathbf{k} \qquad (6.50)
\end{aligned}
$$

The equations of motion can be linearized for small gimbal deflections. This permits replacing $\sin \delta$, $\sin \gamma$, $\cos \delta$, and $\cos \gamma$ with δ, γ, 1, and 1, respectively. Deviations from nominal wheel momentum h_n are also assumed small, allowing $h_w = h_n$. Control components of wheel momentum can now be defined using set (6.49),

$$h_{xc} = \gamma h_n, \; h_{zc} = -\delta h_n, \; \dot{h}_{yc} = -\dot{h}_w$$

Since attitude errors are also assumed small, the linearized gravity gradient torque components become

$$G_x = -3\omega_o^2(I_y - I_z)\phi$$
$$G_y = -3\omega_o^2(I_x - I_z)\theta \qquad (6.51)$$
$$G_z = 0$$

where $\omega_o = 7.28 \times 10^{-5}$ rad/s for geostationary orbit. The attitude rates and body rates are now simply related by

$$\omega_x = \dot{\phi} - \psi\omega_o$$
$$\omega_y = \dot{\theta} - \omega_o \qquad (6.52)$$
$$\omega_z = \dot{\psi} + \phi\omega_o$$

Equation (6.50) becomes three linearized expressions for ϕ, θ, and ψ,

$$T_x = I_x \ddot{\phi} + (a + \omega_o h_n)\phi + (b + h_n)\dot{\psi} + \dot{h}_{xc} - \omega_o h_{zc} \qquad (6.53a)$$
$$T_y = I_y \ddot{\theta} + e\theta + \dot{h}_{yc} \qquad (6.53b)$$
$$T_z = I_z \ddot{\psi} + (c + \omega_o h_n)\psi - (b + h_n)\dot{\phi} + \dot{h}_{zc} + \omega_o h_{xc} \qquad (6.53c)$$

where

$$a = 4\omega_o{}^2(I_y - I_z), \, b = -\omega_o(I_x - I_y + I_z)$$
$$c = \omega_o{}^2(I_y - I_x), \, e = 3\omega_o{}^2(I_x - I_z)$$

Set (6.53) represents the complete, linearized motion for roll, pitch, and yaw, respectively. Notice that pitch control is uncoupled from roll and yaw due to small angle assumptions. Thus, the problem reduces to pitch control and roll/yaw control. These are now treated separately.

6.2.2 The Pitch Loop

Design of an automatic pitch control system is straightforward because pitch motion is decoupled from the others and is, therefore, discussed before developing the more sophisticated roll/yaw autopilot. The linearized pitch equation, form (6.53b) can be written as

$$T_y = I_y\ddot{\theta} + 3\omega_o{}^2(I_x - I_z)\theta + \dot{h}_{yc}$$

This is further simplified, because the satellite of interest is symmetric such that $I_x = I_z$. Thus,

$$T_y \doteq I_y\ddot{\theta} + \dot{h}_{yc} \tag{6.54}$$

where $\dot{h}_{yc}$ represents rate of change of wheel speed to impose direct torque about the pitch axis. A typical pitch control loop is depicted in Figure 6.28. Equation (6.54) indicates that no natural damping is available and can be provided only through the control function $\dot{h}_{yc}$. A satisfactory form of control law is

$$\dot{h}_{yc} = K_p(\tau_p\dot{\theta} + \theta) \tag{6.55}$$

where the $\dot{\theta}$ term introduces damping, K_p and τ_p are the pitch autopilot gain and time constant, respectively. Pitch motion now becomes that of the classic, damped second order system,

$$T_y = I_y\ddot{\theta} + K_p\tau_p\dot{\theta} + K_p\theta \tag{6.56}$$

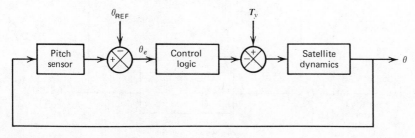

FIGURE 6.28 Pitch control loop.

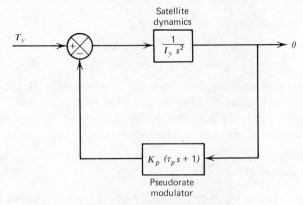

FIGURE 6.29 Implemented pitch control block diagram.

The associated transfer function is obtained directly as

$$\frac{\Theta(s)}{T_y(s)} = \frac{1}{I_y s^2 + K_p \tau_p s + K_p} \tag{6.57}$$

In the nominal operating mode, $\theta_{\text{ref}} = 0$, and sensor errors are ignored in preliminary considerations. The detailed pitch autopilot block diagram is relatively simple and is illustrated in Figure 6.29. Since available sensors can provide only direct measurement of angles and damping requires knowledge of angular rates, a *pseudorate modulator* is employed in this loop. Rate gyros are not appropriate here, because they cannot detect very low rates while satisfying long-life requirements. Basic conceptual components of pseudorate modulators are shown in Figure 6.30. A schmitt trigger with hysteresis is employed to

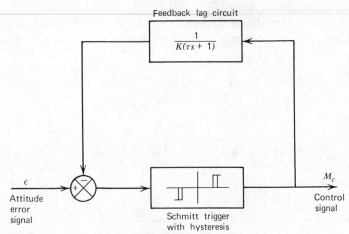

FIGURE 6.30 Schematic of pseudorate modulator.

produce a train of pulses whose average value has the form

$$M_c = K(\tau \dot{\varepsilon} + \varepsilon)$$

Such devices are bistable pulse generators with output pulses of constant amplitude which exist only as long as the input voltage exceeds a certain value. These devices are also used in the roll/yaw autopilot to generate roll/yaw rates. The loop shown in Figure 6.30 is sometimes called a *derived-rate increment* system, because it offers a method for synthesizing angular rates when direct measurement is not possible.

Equation (6.57) indicates that pitch motion has a natural frequency and damping factor given by

$$\omega_p = \sqrt{\frac{K_p}{I_y}}, \qquad \zeta_p = \frac{\tau_p}{2}\sqrt{\frac{K_p}{I_y}} \qquad (6.58)$$

Since overshoot is not desirable for such applications, parameter value selections are begun by setting $\zeta_p = 1$ (critical damping). A greater value would slow system response, while a smaller value permits overshoot. Thus, equation (6.46) becomes

$$\frac{\Theta(s)}{T_y(s)} = \frac{1}{I_y(s + \omega_p)^2} \qquad (6.59)$$

Selection of pitch autopilot gain K_p is based on two considerations, steady state error and response time. Since the limit allowed in pitch is specified, steady state errors must be well below this. Table 6.1 indicates the maximum magnitude of solar pressure torque is expected to be 10^{-4} N·m about the pitch axis. Assuming τ_p is much less than the orbital period, the maximum steady state error is estimated through the final value theorem as

$$\theta_{ss} = \frac{10^{-4}}{I_y \omega_p^2} \, \text{rad}$$

Selecting a gain value of $K_p = 0.275$ N·m/rad gives $\omega_p = 0.025$ rad/s and $\tau_p = 80$ s. This results in a maximum expected pitch error of $0.02°$ and validates the assumption, $\tau_p \ll$ orbital period. A root locus diagram for the pitch loop is shown in Figure 6.31 to check stability and evaluate gain value selections. This form is typical of second systems with two open loop poles at $s = 0$ and a single zero on the negative real axis. It is apparent that the system is stable for all values of K_p. The design value of 0.275 N·m/rad is depicted on the real axis, which is consistent with critical damping, $\zeta_p = 1$.

Responses produced by the pitch autopilot, with the gain and damping values selected above, are predictable analytically or numerically. Results for impulsive, step, and cyclic disturbance torques based on perturbations of Table 6.1 are plotted in Figures 6.32, 6.33, and 6.34, respectively. Since it is difficult

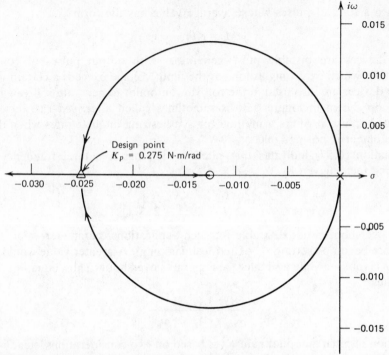

FIGURE 6.31 Root locus diagram for pitch loop.

to estimate the magnitude of impulsive torques which may act on the satellite, the response shown in Figure 6.32 is parameterized by the factor $T_y \tau_p / I_y$. A step input of magnitude $8.5 \times 10^{-5} \, \text{N} \cdot \text{m}$, representing an estimate of thruster misalignment torque about the pitch axis, produced the response of Figure 6.33. Both impulsive and step disturbances result in stable responses with the latter yielding a steady state error of 0.0177°, well below the allowable limit of

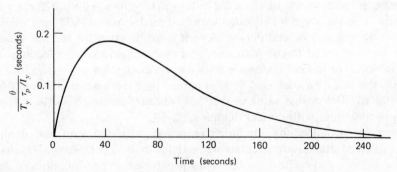

FIGURE 6.32 Pitch response to impulsive disturbance torque.

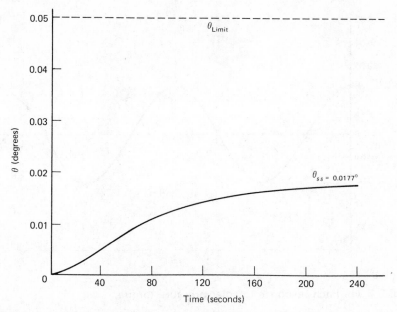

FIGURE 6.33 Pitch response to step disturbance torque.

0.05°. A cyclic disturbance torque due to solar pressure causes the periodic response of Figure 6.34 with amplitude of about 0.02°. Noting that the control term, $\dot{h}_{yc}$ can be expressed in terms of wheel speed, equation (6.55) becomes

$$I_w \dot{\Omega} = K_p(\tau_p \dot{\theta} + \theta) \tag{6.60}$$

where Ω is the wheel speed and I_w is the wheel moment of inertia about its symmetry axis. Effects of disturbance torques on wheel speed can be determined by applying the final value theorem. Assuming the initial wheel speed is nominal Ω_n, the Laplace transform of expression (6.60) gives

$$\Omega(s) = \frac{K_p}{I_w s}(\tau_p s + 1)\Theta(s) + \frac{\Omega_n}{s}$$

Apply equation (6.59) to eliminate $\Theta(s)$,

$$\Omega(s) = \left(\frac{K_p}{I_w I_y}\right)\frac{(\tau_p s + 1)}{s(s + \omega_p)^2}T_y(s) + \frac{\Omega_n}{s}$$

An impulsive disturbance torque of magnitude T_y leads to

$$\Omega_{ss} = \frac{T_y}{I_w} + \Omega_n \tag{6.61}$$

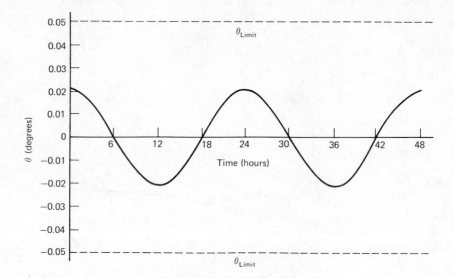

FIGURE 6.34 Pitch response to solar pressure torque.

indicating that a net change in wheel speed occurs even though $\theta_{ss} = 0$. A step disturbance torque leads to infinite wheel speeds if constant counter torquing is not provided through reaction jets or other devices. Thus, step disturbances lead to continual momentum dumping or wheel saturation. Cyclic disturbance torques result in cyclic wheel speeds centered about Ω_n. For solar torques considered here the amplitude of wheel momentum oscillation with respect to h_n is $1.37 \, \mathrm{N \cdot m \cdot s}$. This knowledge will aid in sizing h_n and the wheel torque motor as well as influencing saturation levels.

6.2.3 The Roll/Yaw Loop

Now that the pitch loop has been designed, the more sophisticated roll/yaw system can be attempted. Coupling between roll and yaw axes is a result of the momentum wheel orientation, which is nominally along the orbit normal. An attractive feature of bias momentum is that it does provide roll/yaw coupling while permitting accurate yaw control without a direct yaw sensor. Control torques are produced through gimbal deflections. Since direct roll sensing is available through the horizon sensor, this axis can be controlled directly in response to roll errors. Yaw deviations are sensed indirectly through coupling with the roll axis. This technique is sometimes referred to as *gyrocompassing*. Bias momentum magnitude h_n and system gains are selected to make the roll controller sensitive to yaw errors (i.e., artificially increasing the coupling effect). However, this must be done so that roll errors are not transformed into yaw errors. The general linearized equations for roll and yaw were given as expressions (6.53a) and (6.53c), respectively. Significant coupling effects are achieved through large values of bias momentum. This implies that a condition may be imposed on the nominal momentum,

$$h_n \gg \max\left[I_x\omega_o, I_y\omega_o, I_z\omega_o\right] \tag{6.62}$$

The linearized roll/yaw equations become

$$T_x = I_x\ddot{\phi} + \omega_o h_n\phi + h_n\dot{\psi} + \dot{h}_x - \omega_o h_z$$
$$T_z = I_z\ddot{\psi} + \omega_o h_n\psi - h_n\dot{\phi} + \dot{h}_z + \omega_o h_x \tag{6.63}$$

where the subscript c has been dropped, since h_x and h_z are understood to be products of gimbal deflections only.

It is interesting to briefly consider uncontrolled (gimbals fixed) yaw response, because this would be the situation if gimbal failure occured, or a fixed-gimbal, bias momentum system were being designed. Set (6.63) is simplified and expressed in Laplace variable form as

$$\begin{bmatrix} T_x(s) \\ T_z(s) \end{bmatrix} = \begin{bmatrix} I_x s^2 + \omega_o h_n & h_n s \\ -h_n s & I_z s^2 + \omega_o h_n \end{bmatrix}\begin{bmatrix} \Phi(s) \\ \Psi(s) \end{bmatrix}$$

Of primary concern here is yaw response to a yaw disturbance torque. Therefore, this expression is solved for the appropriate transfer function,

$$\frac{\Psi(s)}{T_z(s)} = \frac{I_x s^2 + \omega_o h_n}{(I_x s^2 + \omega_o h_n)(I_z s^2 + \omega_o h_n) + h_n^2 s^2}$$

A step yaw disturbance torque yields the response

$$\psi(t) = \frac{T_z}{\omega_o h_n}(1 - \cos \omega_o t) + \frac{I_x T_z}{h_n^2}\left[1 - \cos\left(\frac{h_n}{I_x}\right)t\right] \tag{6.64}$$

noting that $I_x = I_z$. Two phenomena are represented here. The first periodic term is associated with roll coupling into yaw at orbital rate ω_o, while the second is related to precessional effects of the momentum wheel. The latter motion is a short period oscillation whose frequency and magnitude depend on the value of nominal momentum. For example, if $h_n = 200\,\text{N}\cdot\text{m}\cdot\text{s}$ and $T_z = T_F$ (thrust misalignment torque in Table 6.1), then the yaw error given by form (6.64) becomes

$$\psi(t) = 0.334(1 - \cos \omega_o t) + 2.44 \times 10^{-4}(1 - \cos 0.10t) \tag{6.65}$$

where $\psi(t)$ is in degrees and t in seconds. Thus, the short period oscillation produces a very small yaw error, and orbit rate coupling results in a maximum allowable yaw error (0.40°) in 6 hr 46 min. To passively control this to within 0.40° the value of h_n must be at least $333 \, \text{N} \cdot \text{m} \cdot \text{s}$. This result implies that a fixed-gimbal wheel can be used as a bias momentum device. However, the required value of h_n turns out to be larger than for a double-gimbaled wheel device for the same mission and satellite.

With the gimbaled system in operation yaw error amplitude will differ from that obtained in equation (6.65), because artificial damping can be provided through selection of an actuator control law. This will also yield a steady state yaw error whose magnitude is a function of h_n. To investigate active roll/yaw control rewrite set (6.63) with the control terms in brackets

$$T_x = I_x \ddot{\phi} + \omega_o h_n \phi + h_n \dot{\psi} + [\dot{h}_x - \omega_o h_z] \tag{6.66a}$$

$$T_z = I_z \ddot{\psi} + \omega_o h_n \psi + [\dot{h}_z + \omega_o h_x - h_n \dot{\phi}] \tag{6.66b}$$

The roll control law is considered first because direct measurement of ϕ is available from the horizon sensor. Control terms in equation (6.66a) must be related to ϕ such that roll responses to disturbance torques are fast and well damped. This will also minimize coupling of roll errors into yaw. A law which satisfies these criteria is

$$M_{xc} = \dot{h}_x - \omega_o h_z = K\tau\dot{\phi} + K\phi - \omega_o h_n \phi \tag{6.67}$$

where M_{xc} is the roll control torque, K and τ are the roll autopilot gain and time constant, respectively. Notice that the first term on the right side introduces damping while the second and third transform the coefficient of ϕ in equation (6.66a) from $\omega_o h_n$ to the roll autopilot gain, resulting in

$$T_x = I_x \ddot{\phi} + K\tau\dot{\phi} + K\phi + h_n \dot{\psi} \tag{6.68}$$

This represents a classic second-order system with driving functions T_x and $-h_n \dot{\psi}$, the gyroscopic coupling term which permits yaw errors to appear in roll sensor outputs.

The ideal yaw response would be well damped and decoupled from roll. However, the yaw control law cannot provide direct damping, because yaw angles are not measurable by this system. Therefore, the control law uses roll angle and a pseudorate modulator to form the complement of equation (6.67),

$$M_{zc} = \dot{h}_z + \omega_o h_x - h_n \dot{\phi} = -kK(\tau\dot{\phi} + \phi) \tag{6.69}$$

where M_{zc} is the yaw control torque and k is the *yaw-to-roll gain ratio*. Equation (6.68) can be written as

$$-kK(\tau\dot{\phi} + \phi) = kh_n \dot{\psi} + kI_x \ddot{\phi} - kT_x \tag{6.70}$$

Then expressions (6.66b), (6.69), and (6.70) give a new yaw equation,

$$T_z = I_z \ddot{\psi} + kh_n \dot{\psi} + \omega_o h_n \psi + kI_x \ddot{\phi} - kT_x \tag{6.71}$$

This is interpreted as a damped second-order system with forcing functions T_z, kT_x, and $-kI_x\ddot{\phi}$. This last term is roll coupling and is not desirable because it represents yaw errors resulting from roll transients. Direct compensation for this in the control law is not practical, because it requires the second derivative of ϕ, which would introduce excessive noise into the system. Alternatively, it is convenient to define *orbit rate decoupled* momentum commands,

$$\left.\begin{array}{l} h_{xd} = h_{xc} \\ h_{zd} = h_{zc} - h_n\phi \end{array}\right\} \tag{6.72}$$

where h_{xd}, h_{zd} are the roll and yaw commands, respectively. The control laws, now written as

$$\left.\begin{array}{l} M_{xc} = \dot{h}_{xd} - \omega_o h_{zd} = K(\tau\dot{\phi} + \phi) \\ M_{zc} = \dot{h}_{zd} + \omega_o h_{xd} = -kK(\tau\dot{\phi} + \phi) \end{array}\right\} \tag{6.73}$$

represent an undamped oscillator at orbital frequency ω_o. Thus, oscillation of momentum commands at this rate has the effect of decoupling roll and yaw dynamics which arise from the orbital frequency. In other words, commands on gimbal motion are referred to an inertial frame as is the angular momentum vector, even though the satellite is pitching at the rate ω_o. In terms of gimbal angle deflections, set (6.72) becomes

$$h_{xd} = \gamma h_n$$
$$h_{zd} = -\delta h_n - h_n\phi$$

Using these forms with equations (6.73) permits the formulation of a control system block diagram. Figure 6.35 illustrates the roll/yaw autopilot. Roll sensor input is processed through a pseudorate modulator, and deflection commands γ and δ are produced according to the selected control laws.

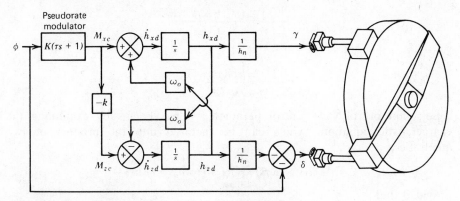

FIGURE 6.35 Schematic of roll/yaw control system.

To investigate system stability and responses roll/yaw equations are rewritten in Laplace form as

$$\begin{bmatrix} T_x(s) \\ T_z(s) \end{bmatrix} = \begin{bmatrix} I_x s^2 + K(\tau s + 1) & h_n s \\ -kK(\tau s + 1) & I_z s^2 + \omega_o h_n \end{bmatrix} \begin{bmatrix} \Phi(s) \\ \Psi(s) \end{bmatrix}$$

which yield a fourth-order characteristic equation,

$$\begin{aligned} \text{C.E.} = I_x I_z s^4 + K I_z \tau s^3 + (K I_z + I_x \omega_o h_n + kK\tau h_n)s^2 \\ + (\omega_o h_n K\tau + kKh_n)s + \omega_o h_n K \end{aligned} \tag{6.74}$$

The transfer functions for roll and yaw angle outputs associated with roll and yaw axis disturbance torques are now easily derived. Roll response is obtained from

$$\Phi(s) = \frac{T_x(s)\,(I_z s^2 + \omega_o h_n) - T_z(s)\,h_n s}{\text{C.E.}} \tag{6.75}$$

When $T_z = 0$,

$$\frac{\Phi(s)}{T_x(s)} = \frac{I_z s^2 + \omega_o h_n}{\text{C.E.}}$$

When $T_x = 0$,

$$\frac{\Phi(s)}{T_z(s)} = \frac{h_n s}{\text{C.E.}}$$

Yaw response is obtained from

$$\Psi(s) = \frac{T_z(s)\,(I_x s^2 + K\tau s + K) + kK(\tau s + 1)T_x(s)}{\text{C.E.}} \tag{6.76}$$

When $T_x = 0$,

$$\frac{\Psi(s)}{T_z(s)} = \frac{I_x s^2 + K\tau s + K}{\text{C.E.}}$$

When $T_z = 0$,

$$\frac{\Psi(s)}{T_x(s)} = \frac{kK(\tau s + 1)}{\text{C.E.}}$$

Selection of roll/yaw control parameter values begins by examining the characteristic equation, which can be factored into the product of two quadratics,

$$\text{C.E.} \cong (I_x s^2 + K\tau s + K)(I_z s^2 + kh_n s + \omega_o h_n) = 0$$

provided that

$$K\tau I_z \gg kh_n I_x \tag{6.77}$$

The roots of this equation yield natural frequencies and damping rates of each mode,

$$\omega_1 = \sqrt{\frac{K}{I_x}} \qquad \zeta_1 = \frac{\tau}{2}\sqrt{\frac{K}{I_x}} \qquad (6.78)$$

$$\omega_2 = \sqrt{\frac{\omega_o h_n}{I_z}} \qquad \zeta_2 = \frac{k}{2}\sqrt{\frac{h_n}{\omega_o I_z}}$$

The high frequency roots (ω_1, ζ_1) are characteristic of roll dynamics while the low frequency roots (ω_2, ζ_2) dominate yaw motion during yaw error correction. A large value of h_n will result in fast yaw corrections with respect to orbit period.

Roll autopilot gain K is selected to limit steady-state roll error caused by a constant roll torque and to provide fast response. From equation (6.75) the steady-state roll error produced by a constant roll torque is

$$\phi_{ss} = \frac{T_x}{K}$$

Based on the maximum roll torque resulting from solar pressure and the allowable roll error of 0.05°, this yields a minimum required value of $K = 0.07 \, \text{N} \cdot \text{m/rad}$ with a corresponding natural frequency of $\omega_1 = 0.006 \, \text{rad/s}$. Larger values of K will result in a small steady-state error and faster response time. Thus, a natural frequency of $\omega_1 = 0.025 \, \text{rad/s}$ is selected with a corresponding roll autopilot gain of $K = 1.25 \, \text{N} \cdot \text{m/rad}$. It should be noted that this choice is preliminary and must satisfy condition (6.77). Final selection of K will depend on a root-locus analysis of the system. Values of τ, roll time constant and k, yaw-to-roll gain ratio are selected to provide critical damping of roll and yaw dynamics. Thus, from set (6.78) for $\zeta_1 = \zeta_2 = 1$ and $h_n = 200 \, \text{N} \cdot \text{m} \cdot \text{s}$, $\tau = 80 \, \text{s}$ and $k = 0.054$. To justify this selection of h_n, the steady-state yaw offset expression is obtained directly from equation (6.76) for constant torques,

$$\psi_{ss} = \frac{T_z + kT_x}{\omega_o h_n} \qquad (6.79)$$

Since the contribution of roll torque is small (because $k = 0.054$), only yaw offset resulting from a yaw disturbance torque need be considered. The resulting condition on h_n is

$$h_n \geq \frac{|T_z|}{\omega_o |\psi_{ss}|_{max}} \qquad (6.80)$$

Taking the maximum amplitude of T_z, $5 \times 10^{-5} \, \text{N} \cdot \text{m}$, and the specified yaw error limit of 0.40° leads to a minimum value for nominal momentum

$$h_n \geq 99 \, \text{N} \cdot \text{m} \cdot \text{s}$$

Therefore, a value of $h_n = 200$ N · m · s satisfies this and insures that ψ_{ss} remains within accuracy limits until corrections are made. Furthermore, parameter selections satisfy conditions (6.62) and (6.77). Pitch control torques, which change the wheel speed, can be applied without exceeding a 1% variation in h_n. Therefore, 200 N · m · s appears to satisfy all criteria for this system. Final adjustment of parameter values now can be made with the aid of the system root locus diagram shown in Figure 6.36. This plot is based on the accurate characteristic equation given as form (6.74). The preliminary value of $K = 1.25$ N · m/rad does not quite correspond to $\zeta_1 = 1.0$. A final value of K is selected to insure near-critical damping (i.e., $K = 1.56$ N · m/rad). Table 6.2 gives the resulting roll/yaw autopilot parameter values. The root locus diagram depicts two second order systems. Yaw motion is represented by the inner loop and roll by the outer loop. The system appears stable for all values of $K > 0.1$ N · m/rad. Table 6.3 offers a summary of roll/yaw responses for impulsive and step inputs, where the step magnitude is 8.5×10^{-5} N · m, corresponding to thrust misalignment torque. This system is designed primarily to

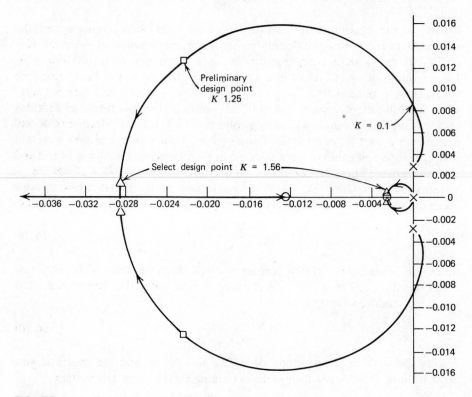

FIGURE 6.36 Root locus diagram for roll/yaw system.

TABLE 6.2 Summary of Roll/Yaw Control
System Parameters

Parameter	Value
h_n, nominal wheel momentum magnitude	200 N · m · s
K, roll autopilot gain	1.56 N · m/rad
τ, roll time constant	80 s
k, yaw-to-roll gain ratio	0.054
ω_1, roll dynamics natural frequency	0.0286 rad/s
ζ_1, roll dynamics damping ratio	0.998
ω_2, yaw dynamics natural frequency	0.00264 rad/s
ζ_2, yaw dynamics damping ratio	0.985

control solar pressure torques which are essentially cyclic. Roll and yaw responses are also cyclic but remain within specified limits. Thus, the resulting system is capable of maintaining accurate attitude orientation through the use of only a pitch-roll horizon sensor. An overall basic command flow chart is presented in Figure 6.37.

6.2.4 Torque Compensation

A further reduction in bias momentum magnitude is possible if solar pressure torques can be anticipated with confidence. For the double-gimbaled system just described preprogrammed compensation torques can be generated by the gimbal deflection method. This technique is also appropriate for fixed-gimbal, bias momentum systems. However, compensating torques must be applied with magnetic torquers or reaction thrusters. Magnetic devices would be preferred, but their effectiveness is lower than with gimbaled systems due to

TABLE 6.3 Summary of Roll/Yaw Responses

Case	Input Axis	Input Form	Steady-State Value (deg)
Roll	x-axis	Impulse	0
		Step	0.00312
	z-axis	Impulse	0
		Step	0
Yaw	z-axis	Impulse	0
		Step	0.33
	x-axis	Impulse	0
		Step	0.018

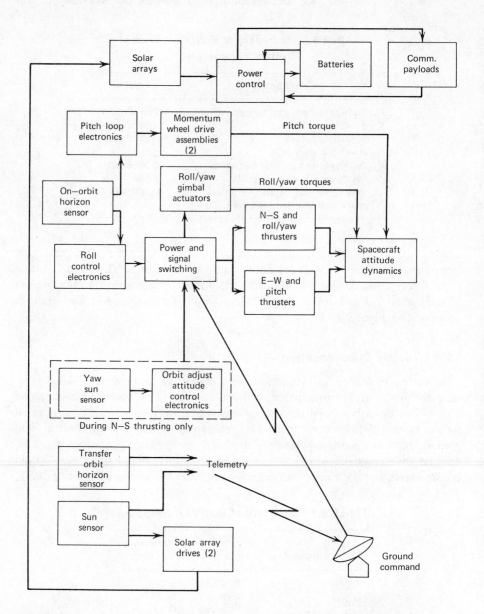

FIGURE 6.37 Command flow chart for double-gimbaled wheel system.

the varying nature of the earth's magnetic field at synchronous altitude. It is likely that only part of the perturbing torque profile can be anticipated. However, this approach should still work, at least to partially reduce the required size of h_n. Unpredictable perturbations can be expected because of internal mass shifts, structural bending, and so on. It would appear that on-orbit reprogramming of the control computer is highly desirable to maximize compensation and improve performance for given h_n. Figure 6.38 depicts

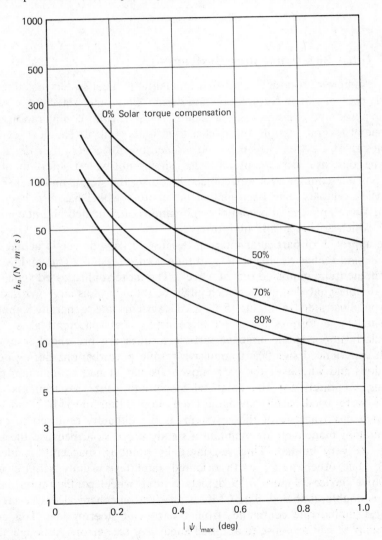

FIGURE 6.38 Effect of compensation on required bias momentum.

the effect of compensation on required bias momentum as a function of yaw error limit, $|\psi|_{max}$ and amount of compensation assuming $|T_z| = 5 \times 10^{-5} \, \text{N} \cdot \text{m}$. A double-gimbaled wheel system should be capable of 50–70% compensation, whereas magnetic torquers would be somewhat less effective. Bias momentum magnitude is related to the unpredictable part of yaw perturbing torque, $|\Delta T_z|$, by a condition based on form (6.80),

$$h_n \geq \frac{|\Delta T_z|}{|\psi|_{max}\omega_o} \tag{6.81}$$

6.2.5 Other Bias Momentum Systems

The example of a double-gimbaled momentum wheel system has illustrated attitude control system formulation and design. Parameter selection arguments were exposed and decision criteria stated. The particular configuration used is only one of several possible approaches to attitude control. The next section, in fact, describes a system which has no momentum for control purposes. However, the topic of bias momentum devices should not be left before mentioning other possible configuration selections. Of course, the simple and dual spinners are basic examples. The fixed-gimbal momentum wheel has also been mentioned. There are several potential combinations of gimbaled and nongimbaled wheels which appear in the literature.

One approach of particular interest, because of its wide use in actual flights, is the *control moment gyro* (CMG) attitude control system. This makes use of fixed-momentum, gimbaled sets of wheels. The more sophisticated versions use three, double-gimbaled gyros, which might be thought of as large gyroscopes of the type illustrated in Figure 5.1. CMG systems are generally capable of producing large torques about all three orthogonal spacecraft axes. This is particularly important if large perturbing torques are present. For example, Skylab experienced significant impulsive torque perturbations during docking operations and whenever the crew moved around. It had to maintain precise pointing accurately for the solar telescope and other experiments. Three CMG's were used, each having a fixed momentum of $3116 \, \text{N} \cdot \text{m} \cdot \text{s}$. An important disadvantage of these systems is the difficulty required to execute reorientation maneuvers or maintain a steady precession, because the gimbal angles are very limited. Thus, an inertially pointing spacecraft is ideal for CMG's, but others (e.g., earth oriented satellites) usually require use of alternative devices. Figure 6.39 depicts a three wheel configuration, each of which is double gimbaled. The CMG is an electromechanical device receiving electrical gimbal rate commands from a particular steering law. This control law usually acts in response to a sensed angular or rate error. Although all axes are highly coupled, this controller does not generally use all system degrees of

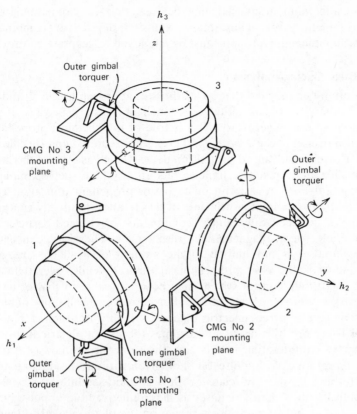

FIGURE 6.39 Double gimbal CMG configuration.

freedom for attitude corrections. Such laws are usually formulated to optimize relative momentum distribution within the spacecraft. Appropriate equations of motion can be thought of as the basic set for the double gimbaled wheel extended to represent a wheel nominally about each axis. Of course, the study of responses is more complicated because all axes are coupled and the control laws are more sophisticated.

6.3 DESIGN OF AN ALL-THRUSTER SYSTEM

The preceding section presented a very sophisticated and effective attitude control system which utilizes only an earth sensor for roll and pitch measurement. It was pointed out than an important advantage of this system is that no direct yaw sensing is required. One might then ask about implementation of a *zero-momentum* attitude control system. This term is used to describe the technique of actively controlling the orientation of a nonspinning satellite without using momentum exchange devices, that is, momentum wheels or control moment gyros. Thus, mass expulsion devices or combinations of thrusters and magnetic torquers must be employed to generate control torques.

6.3.1 Duty Cycle Analysis

It has often been argued that an all-thruster approach to the attitude control problem is expensive in terms of propellant mass, especially if accurate orientation control is necessary. This is true of the chemical propellant thrusters in current use, because *hard-limit* cycles are required and specific impulses are low. However, advanced propulsion devices, such as electric thrusters, offer much smaller and repeatable impulse bits at very high specific impulse, thus, permitting *soft-limit* cycles with much better propellant utilization efficiency. *Electric thrusters* employ the principle of accelerating positively charged particles (ions) or an electrically neutral combination of charged particles (plasma) through an electric or magnetic field, respectively. Therefore, exhaust velocity (specific impulse) is not limited to the amount of available energy in the propellant, as is the case with chemical devices. Thus, overall mass of an all-thruster attitude control system may be comparable to that of momentum exchange systems for the same vehicle. Furthermore, the absence of mechanical linkages and valves from electric devices is highly desirable for a long-life satellite. Therefore, this discussion assumes the use of electric thrusters.

A primary consideration in selecting thrusters for attitude control is their minimum reproducible impulse bit. For given attitude holding requirements and perturbing torques associated with a specific satellite, the type of limit cycle control (soft or hard) required is determined by the impulse bit sizes of the thrusters. A *hard-limit* cycle is shown on the phase plane of Figure 6.40a. Thrust is applied at both sides of the prescribed angular limit because the perturbing torque is zero or much smaller than the reaction torque. Figure 6.40b shows a *soft-limit* cycle which requires thrust at only one extremity. This cycle is permissible when the disturbing torques are large enough to decelerate the vehicle and reverse its direction before reaching the other limit. If the minimum impulse bit of a system can be controlled so that a soft-limit cycle is the normal mode of operation, orientation is controlled at a minimum propellant cost. Coupled with high specific impulse this efficient propellant utilization makes electric thrusters primary candidates for zero-momentum attitude control actuators.

Coupling among the axes can be ignored for small angular motions. Thus, the component equations for principal axes are simply

$$T_x = I_x \ddot{\phi}$$
$$T_y = I_y \ddot{\theta}$$
$$T_z = I_z \ddot{\psi}$$

$$(6.82)$$

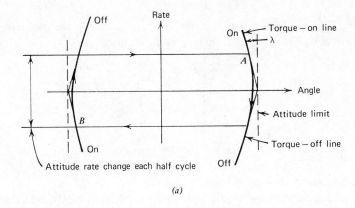

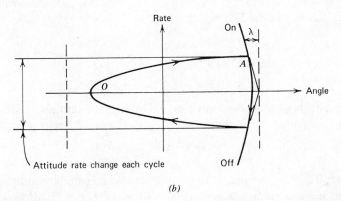

FIGURE 6.40 Phase plane diagram of limit cycles. (*a*) Hard-limit cycle. (*b*) Soft-limit cycle.

where T_x, T_y, and T_z are applied torque components, and ϕ, θ, and ψ are roll, pitch, and yaw, respectively. These equations are easily integrated and solved for rates at the attitude error limits. The required corrective *torque impulses* are taken as those which reverse the sign of the rate for each axis,

$$I_{M_\phi} = \int F_\phi l_\phi \, dt = -2 I_x \dot{\phi}_A$$

$$I_{M_\theta} = \int F_\theta l_\theta \, dt = -2 I_y \dot{\theta}_A$$

$$I_{M_\psi} = \int F_\psi l_\psi \, dt = -2 I_z \dot{\psi}_A$$

where F_ϕ, F_θ, and F_ψ are thrust levels, and l_ϕ, l_θ, and l_ψ are moment arms for these thrusters. *Force impulses* may now be defined as

$$I_{F_\phi} = \int F_\phi \, dt = -\frac{2}{l_\phi} I_x \dot{\phi}_A$$

$$I_{F_\theta} = \int F_\theta \, dt = -\frac{2}{l_\theta} I_y \dot{\theta}_A \qquad (6.83)$$

$$I_{F_\psi} = \int F_\psi \, dt = -\frac{2}{l_\psi} I_z \dot{\psi}_A$$

where integration is over the thrust interval.

The value of λ, or *torque-on line slope*, is a function of attitude thrust level. For cases of interest here, λ is extremely small and can be taken as zero. To ensure that pointing accuracy is maintained, the limits of attitude error for the soft-limit cycle are assumed to be

$$\phi(t_A) = 0.9\phi_{max}, \qquad \phi_o = -0.75\phi_{max}$$

$$\theta(t_A) = 0.9\theta_{max}, \qquad \theta_o = -0.75\theta_{max}$$

$$\psi(t_A) = 0.9\psi_{max}, \qquad \psi_o = -0.75\psi_{max}$$

to avoid the effects of sensor sensitivities and other uncertainties. Attitude limit components are sometimes determined from the over-all beam pointing error limit,

$$\varepsilon_{max} = \sqrt{\phi_{max}^2 + \theta_{max}^2 + (E\psi_{max})^2} \qquad (6.84)$$

For example, a 60° arc on the earth away from the subsatellite point results in an off-nadir angle, assuming a synchronous orbit, of $E = 8.27°$. If an accuracy of $\varepsilon_{max} = 0.1°$ is required, an acceptable combination of attitude angles is

$$\phi_{max} = \theta_{max} = 0.07°$$

$$\psi_{max} = 0.2°$$

As an example, consider the satellite specifications and disturbance torques listed in Table 6.1. Combining set (6.82) with the solar pressure torque functions in this table permits immediate integration. If all initial rates are taken as zero, the resulting equations are

$$\dot{\phi}(t) = \frac{2 \times 10^{-5}}{I_x} \left[t + \frac{2}{\omega_o} \cos \omega_o t - \frac{2}{\omega_o} \right] \qquad (6.85a)$$

$$\phi(t) = \phi_o + \frac{2 \times 10^{-5}}{I_x} \left[\frac{t^2}{2} + \frac{2}{\omega_o^2} \sin \omega_o t - \frac{2t}{\omega_o} \right] \qquad (6.85b)$$

$$\dot{\theta}(t) = \frac{10^{-4}}{\omega_o I_y} \sin \omega_o t \qquad (6.85c)$$

$$\theta(t) = \theta_o + \frac{10^{-4}}{\omega_o^2 I_y}(1 - \cos \omega_o t) \qquad (6.85d)$$

$$\dot{\psi}(t) = \frac{-5 \times 10^{-5}}{\omega_o I_z} \sin \omega_o t \qquad (6.85e)$$

$$\psi(t) = \psi_o - \frac{5 \times 10^{-5}}{\omega_o^2 I_z}(1 - \cos \omega_o t) \qquad (6.85f)$$

where the units are consistent with those of Table 6.1 and $\omega_o = 7.28 \times 10^{-5}$ rad/s. Equations (6.85b), (6.85d), and (6.85f) are used with

$$\phi(t_A) - \phi_o = 1.65\phi_{max}$$

$$\theta(t_A) - \theta_o = 1.65\theta_{max}$$

$$\psi(t_A) - \psi_o = 1.65\psi_{max}$$

to determine values of t_A once ϕ_{max}, θ_{max}, and ψ_{max} are specified. Since $\dot{\phi}$, $\dot{\theta}$, and $\dot{\psi}$ vary about the orbit, there is a range of values of t_A corresponding to each axis. The time between impulses is $2t_A$ which can easily be estimated and a range of values established. These calculations indicate that the time between impulses varies from several minutes to about an hour with corresponding impulses of 10^{-3} to 10^{-2} N · s when $l_\phi = l_\theta = l_\psi = 0.9$ m.

6.3.2 Automatic Control Requirements

It is a necessity to provide automatic attitude control during nominal modes of satellite operation. Assuming a synchronous orbit, the linearized equations of motion, including the gravity gradient contribution are quite similar to those of Section 6.2. Thus, set (6.53) excluding a momentum wheel, become

$$T_x = I_x \ddot{\phi} + a\phi + b\psi \qquad (6.86a)$$

$$T_y = I_y \ddot{\theta} + e\theta \qquad (6.86b)$$

$$T_z = I_z \ddot{\psi} + c\psi - b\dot{\phi} \qquad (6.86c)$$

where a, b, c, and e have the same definitions as before. The equations of motion now represent those of second-order systems without damping. Accurate pointing requirements cannot be maintained practically without fast damping of perturbations. Thus, control laws must again provide artificial damping. This is achieved by using forms similar to equation (6.55),

$$M_{xc} = K_x(\tau_x \dot{\phi} + \phi) - a\phi$$

$$M_{yc} = K_y(\tau_y \dot{\theta} + \theta) - e\theta$$

$$M_{zc} = K_z(\tau_z \dot{\psi} + \psi) - c\psi$$

Notice that M_{zc} requires that yaw be measured directly, implying the use of a yaw sensor. These expressions are added to the right-hand side of equations (6.86) to give

$$T_x = I_x\ddot{\phi} + K_x\tau_x\dot{\phi} + K_x\phi + b\dot{\psi} \tag{6.87a}$$

$$T_y = I_y\ddot{\theta} + K_y\tau_y\dot{\theta} + K_y\theta \tag{6.87b}$$

$$T_z = I_z\ddot{\psi} + K_z\tau_z\dot{\psi} + K_z\psi - b\dot{\phi} \tag{6.87c}$$

Note that roll and yaw are coupled through a factor of b.

Some response information can be obtained quickly through Laplace transform methods. The pitch equation is solved immediately as

$$\Theta(s) = \frac{T_y(s)}{I_ys^2 + K_y\tau_ys + K_y} \tag{6.88}$$

while roll and yaw must be handled simultaneously,

$$\begin{bmatrix} T_x(s) \\ T_z(s) \end{bmatrix} = \begin{bmatrix} I_xs^2 + K_x\tau_xs + K_x & bs \\ -bs & I_zs^2 + K_z\tau_zs + K_z \end{bmatrix} \begin{bmatrix} \Phi(s) \\ \Psi(s) \end{bmatrix} \tag{6.89}$$

The characteristic fourth-order equation (C.E.) takes the familiar form

$$\text{C.E.} = I_xI_zs^4 + (I_xK_z\tau_z + I_zK_x\tau_x)s^3 + (I_xK_z + I_zK_x + K_xK_z\tau_x\tau_z + b^2)s^2$$
$$+ K_xK_z(\tau_x + \tau_z)s + K_xK_z$$

If the condition

$$b^2 \ll \min[I_xK_z, I_zK_x, K_xK_z\tau_x\tau_z]$$

is satisfied, then the control response time is much less than the orbital period, and roll and yaw are effectively damped. Thus,

$$\text{C.E.} \cong (I_xs^2 + K_x\tau_xs + K_x)(I_zs^2 + K_z\tau_zs + K_z) \tag{6.90}$$

which represents two second order systems with the properties in roll,

$$\omega_1 = \sqrt{\frac{K_x}{I_x}} \qquad \zeta_1 = \frac{\tau_x}{2}\sqrt{\frac{K_x}{I_x}}$$

and in yaw,

$$\omega_2 = \sqrt{\frac{K_z}{I_z}} \qquad \zeta_2 = \frac{\tau_z}{2}\sqrt{\frac{K_z}{I_z}}$$

Since b is of order ω_o, roll/yaw coupling can be ignored here, leading quickly to

$$\Phi(s) = \frac{T_x(s)}{I_xs^2 + K_x\tau_xs + K_x} \tag{6.91a}$$

$$\Psi(s) = \frac{T_z(s)}{I_zs^2 + K_z\tau_zs + K_z} \tag{6.91b}$$

For impulsive disturbance torques,

$$\phi_{ss} = 0, \qquad \theta_{ss} = 0, \qquad \psi_{ss} = 0$$

For step disturbance torques,

$$\phi_{ss} = \frac{T_x}{K_x}, \qquad \theta_{ss} = \frac{T_y}{K_y}, \qquad \psi_{ss} = \frac{T_z}{K_z}$$

Thus, the system appears stable for the assumed control laws. Values of gains should be selected to insure that these steady state errors are less than specified limits. Control torque magnitudes should be sized to satisfy the soft limit cycle requirements of the preceding section.

Operational versions of some electric thrusters will have thrust vectoring capability such that a single unit, designed to make orbit adjustments, could also provide attitude control torques about two independent axes. However, only two independent control laws can be selected. To test this single-thruster control concept, consider the satellite depicted in Figure 6.41. If M_T is the torque component normal to roll torque and is produced by vectoring, then a possible set of control moment laws is

$$M_{xc} = K_x(\tau_x \dot{\phi} + \phi) - a\phi$$

$$M_{yc} = M_T \sin \delta - e\theta \qquad (6.92)$$

$$M_{zc} = M_T \cos \delta - c\psi$$

where δ is the thruster cant angle measured from the pitch axis and

$$M_T = K_T(\tau_T \dot{\theta} + \theta)$$

Notice again that M_{zc} requires that yaw be measured directly. The equations of motion become

$$T_x = I_x \ddot{\phi} + K_x \tau_x \dot{\phi} + K_x \phi + b\dot{\psi} \qquad (6.93a)$$

$$T_y = I_y \ddot{\theta} + K_T \tau_T \dot{\theta} \sin \delta + K_T \theta \sin \delta \qquad (6.93b)$$

$$T_z = I_z \ddot{\psi} + K_T \tau_T \, \dot{\theta} \cos \delta + K_T \theta \cos \delta - b\dot{\phi} \qquad (6.93c)$$

Again ignoring roll/yaw coupling through b, $\Psi(s)$ is obtained as

$$\Psi(s) = \frac{T_z}{I_z s^2} - \frac{T_y}{I_y s^2} \left[\frac{K_T \tau_T s \cos \delta + K_T \cos \delta}{I_y s^2 + K_T \tau_T s \sin \delta + K_T \sin \delta} \right] \qquad (6.94)$$

which is unstable for all input disturbance torques of interest. This is a result of limited vectoring capability. Therefore, three components of control torque and continuous yaw sensing are required to ensure a stable and practical system.

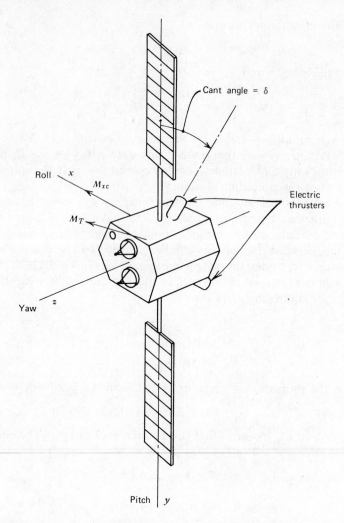

FIGURE 6.41 Assumed satellite configuration for an all-thruster control system.

EXERCISES

6.1 Show the steps in the derivation of equation (6.5) for the electronic integrator circuit of Figure 6.6.

6.2 Refer to the figure in order to derive the following transfer functions: (a) $B(s)/I(s)$, (b) $E(s)/I(s)$.

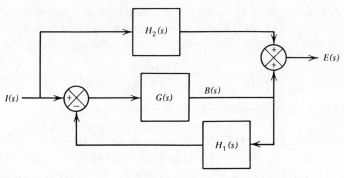

EXERCISE 6.2

6.3 In the figure, x_2 is the displacement of mass M_2 from its equilibrium position and f is a force applied to mass M_1. The massless cable moves the pulley without slipping. Derive a transfer function relating x_2 to f. Ignore gravity effects.

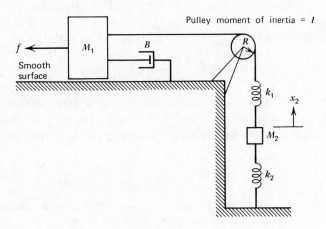

EXERCISE 6.3

6.4 Consider the following transfer function forms:

(a) $G(s) = \dfrac{4s^2 + 15s + 13}{s^3 + 6s^2 + 11s + 6}$

(b) $G(s) = \dfrac{8(s+1)}{s^4 + 4s^3 + 8s^2 + 8s}$

Expand each into partial fraction form.

6.5 A device for measuring thrust produced by an experimental reaction jet for satellite attitude control records the current going into a force-null voice coil. The transfer function relating current and force is

$$\frac{I(s)}{F(s)} = \frac{2 \times 10^{14}}{(s+1200)(s^2+400s+10^6)} \text{ (amp/N)}$$

(a) Plot the pole-zero diagram for this transfer function.
(b) Sketch the time response to a thrust step of 10 N.
(c) Estimate the maximum percentage of overshoot.
(d) What is the value of steady state current for a thrust of 10 N?

6.6 Prove the identity given by expression (6.43) which is helpful in determining the breakaway point.

6.7 Construct a root locus diagram for a unity, negative feedback loop with

$$G(s) = \frac{K(s+1)}{s(s+2)(s+4)(s^2+2s+9)}$$

Determine the range of positive K values for which the system is stable. Specify the value of ω at which the loci cross the $i\omega$ axis. Show asymptotes and breakaway points.

6.8 Consider a nonunity, positive feedback system with

$$G(s) = \frac{K(s+1)}{s^2+4s+9}, \qquad H(s) = \frac{3}{s+3}$$

where $K \geq 0$ sketch a root locus plot for this system and find the number of poles of $W(s)$ which occur in the right-half plane for $K = 10$.

6.9 Construct an approximate root locus diagram for a unity feedback loop with

$$G(s) = \frac{K}{(s+18)(s^2+12s+100)}$$

for both positive and negative values of K. Show asymptotes and breakaway points, if appropriate.

6.10 Derive expressions (6.51) for the gravity gradient torque components acting on the satellite of Figure 6.26. If a bias momentum wheel were not used with this configuration, would gravity gradients stabilize it in this orientation?

6.11 Show that the amplitude of wheel momentum oscillation with respect to its nominal value is $1.37 \text{ N} \cdot \text{m} \cdot \text{s}$, if the disturbance torque is

$$T_y = 10^{-4} \cos \omega_o t \text{ N} \cdot \text{m}$$

Assume the satellite properties of Section 6.2.2 with the momentum axis along the y-axis.

6.12 Derive a yaw offset expression for a gimbal-fixed bias momentum system which is analogous to expression (6.79) for the double-gimbaled system. Assume constant roll and yaw torque components.

6.13 The satellite shown has a single, fixed-gimbal momentum wheel control system with thrusters for roll/yaw torquing. These are offset from the yaw axis by an angle α. The control laws are

$$M_{xc} = -K(\tau\dot{\phi} + \phi)\cos\alpha$$
$$M_{zc} = K(\tau\dot{\phi} + \phi)\sin\alpha$$

(a) Derive the roll/yaw equations of motion in Laplace variable form.
(b) Determine the associated high and low system frequencies and corresponding damping factors.

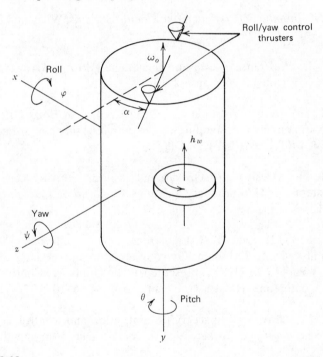

EXERCISE 6.13

6.14 Derive expression (6.64) for yaw response to a step disturbance torque about the yaw axis. Assume a gimbal-fixed bias momentum wheel and start by setting $\dot{h}_x = \dot{h}_z = 0$ in set (6.63). Using the satellite properties of Table 6.1, if $T_z = 8.5 \times 10^{-5}\,\text{N}\cdot\text{m}$ and $h_n = 100\,\text{N}\cdot\text{m}\cdot\text{s}$, how long would it take to exceed the specified yaw error limit of $0.4°$? Assume there is no initial error.

6.15 A double-gimbaled momentum wheel system has a bias momentum of $50\,\text{N}\cdot\text{m}\cdot\text{s}$ and is designed to hold yaw to within $0.2°$ in a synchronous orbit. How much unpredictable torque about the yaw axis can this system handle, while maintaining yaw to within limits.

6.16 Carry out the steps in deriving equations (6.85). Calculate the time between impulses for the synchronous satellite configuration assumed in Section 6.3.1. Consider each axis separately.

6.17 Derive expression (6.94) from set (6.93) and show that the response is unstable for a step torque, T_z.

REFERENCES

Clark, R. N., *Introduction to Automatic Control Systems*, Wiley and Sons, 1962, Chapters 1–6.

Clifford, F. J., "Lawrence Sperry and the Magic Pilot," *FAA Aviation News*, Vol. 8, *No. 5*, October 1969, pp. 12–13.

Kaplan, M. H., "Active Attitude and Orbit Control of Body-Oriented Geostationary Communications Satellites," *Progress in Astronautics and Aeronautics*, Vol. 33, MIT Press, 1974, pp. 29–56.

Kaplan, M. H., "Design and Operational Aspects of All-Electric Thruster Control Systems for Geostationary Satellites," *Journal of Spacecraft and Rockets*, Vol. 12, *No. 11*, Nov. 1975, pp. 682–688.

Powell, B. K., G. E. Lang, S. I. Lieberman, and S. C. Rybak, "Synthesis of Double Gimbal Control Moment Gyro Systems for Spacecraft Attitude Control," AIAA Paper No. 71–937, presented at the Guidance, Control and Flight Mechanics Conference, Hofstra University, August 16–18, 1971.

Sabroff, A. E., "Advanced Spacecraft Stabilization and Control Techniques," *Journal of Spacecraft and Rockets*, Vol. 5, *No. 12*, Dec. 1968, pp. 1377–1393.

Fundamentals and Methods of Astrodynamics

The study of controlled flight paths of man-made spacecraft is *astrodynamics*. This discipline is founded upon the principles of celestial mechanics in that the natural motion of a celestial body is equivalent to the trajectory of a probe during uncontrolled or *coasting* intervals. This chapter begins by presenting those areas of celestial mechanics which are basic to the development of astrodynamics. Then a treatment of the classical problem to describe the motion of a small spacecraft in the presence of two large attracting bodies is presented. This is called the *restricted three-body* problem. Techniques are developed for describing position and velocity in various types of conic paths, which lead to a universal set of formulas. Possible trajectories between two specified points in space are also investigated.

7.1 CELESTIAL MECHANICS

7.1.1 Potential of a Distributed Mass

Newton's law of gravitational attraction was introduced in its fundamental form in Section

1.2.1. Thus, any two mass particles m_1 and m_2 attract each other with a force of magnitude

$$F = G\frac{m_1 m_2}{r^2}$$ (7.1)

repeated here for convenience. Orbits of satellites about spherically symmetric, larger masses are Keplerian, and simply described. The real situation is, however, one that usually requires consideration of motion about a nonspherical mass. The force of attraction must then be calculated by considering contributions from each element of mass in the body. Thus, if m_1 in equation (7.1) is assumed to be dm_1, a typical element of a large body B, and m_2 a particle of mass whose motion is of interest, then the force due to dm_1 on m_2 becomes

$$d\mathbf{F} = G\frac{m_2\mathbf{r}_{12}}{r_{12}{}^3}\,dm_1$$

referring to Figure 7.1. Integrating over the entire body B gives the force of attraction on m_2,

$$\mathbf{F} = \int_B G\frac{m_2\mathbf{r}_{12}}{r_{12}{}^3}\,dm_1$$ (7.2)

To illustrate application of this form, consider the attraction of a homogeneous spherical shell on a unit mass at an outside point. This situation is depicted in Figure 7.2. The unit mass is at a distance D from the shell center of mass,

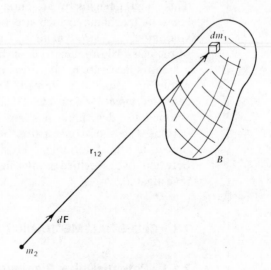

FIGURE 7.1 Attraction of a distributed mass.

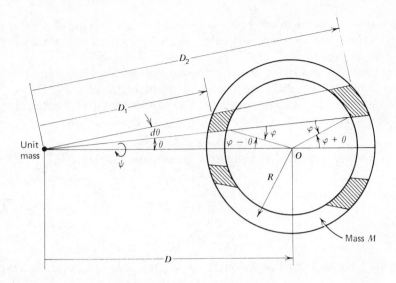

FIGURE 7.2 Attraction of a spherical homogenous shell.

point O. It is attracted toward the spherical shape with a force

$$\mathbf{F} = G \int \frac{\mathbf{r}}{r^3} \, dm$$

However, some simplification is possible by noting properties of symmetry. It is obvious that all components of $\mathbf{F}$ normal to the line between the unit mass and point O will cancel each other. This permits a quick calculation of F whose direction is toward O,

$$F = G \int \frac{\cos \theta}{r^2} \, dm$$

Performance of this integration can be accomplished by setting up the limits in terms of θ and then changing the variable to ϕ. The result is

$$F = \frac{GM}{D^2} \tag{7.3}$$

which should not be surprising after noting the symmetry properties. Thus, the attraction of a homogeneous spherical shell of mass M on an outside particle at distance D is the same as that of a particle of equal mass at the same distance. It is easy to show further that the attraction of any spherically symmetric mass distribution on an outside point is the same as if all the mass were concentrated at the center. It can be shown further that the attraction of a homogeneous spherical shell on an inside point is zero. Thus, the weight of a mass at the earth's center is zero. One might then ask why material below the surface is under great pressure.

It is clear that attraction of a distributed mass requires integration over that mass. Consider now the attraction of a set of n particles $P_1, P_2, \ldots, P_n$ of masses $m_1, m_2, \ldots, m_n$, respectively. The attraction on P_i by the remaining particles is determined as follows. Referring to Figure 7.3, let

$$\mathbf{r}_i = X_i \mathbf{I} + Y_i \mathbf{J} + Z_i \mathbf{K}$$

be the position vector of P_i in inertial coordinates. Define $\mathbf{r}_{ij}$ as

$$\mathbf{r}_{ij} = \mathbf{r}_j - \mathbf{r}_i$$

so that its magnitude is

$$r_{ij} = \sqrt{r_i^2 + r_j^2 - 2\mathbf{r}_i \cdot \mathbf{r}_j}$$

or

$$r_{ij} = \sqrt{r_i^2 + r_j^2 - 2 r_i r_j \cos \gamma_{ij}} \tag{7.4}$$

Now the force of attraction $\mathbf{F}_i$ on P_i by $m_1, m_2, \ldots, m_{i-1}, m_{i+1}, \ldots, m_n$ is

$$\mathbf{F}_i = G m_i \sum_{\substack{j=1 \\ j \neq i}}^{n} \frac{m_j}{r_{ij}^3} \mathbf{r}_{ij} \tag{7.5}$$

This force of attraction can be expressed as a potential function if it is conservative, i.e., if

$$\oint_C \mathbf{F}_i \cdot d\mathbf{r}_i = 0$$

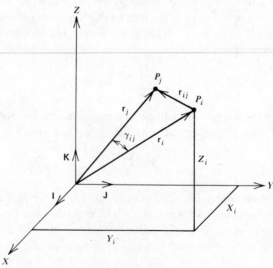

FIGURE 7.3 Attraction among a set of particles.

where $\mathbf{r}_i$ is the inertial position of P_i and C is any closed path. Since it is a function of distance only, $\mathbf{F}_i$ is, in fact, conservative and can be represented as

$$\mathbf{F}_i = m_i \, \nabla_i U \tag{7.6}$$

where U is the gravitational potential function (i.e., the negative of potential energy, V of Chapter 1), and

$$\nabla_i = \mathbf{I} \frac{\partial}{\partial X_i} + \mathbf{J} \frac{\partial}{\partial Y_i} + \mathbf{K} \frac{\partial}{\partial Z_i}$$

in Cartesian form. Thus, the potential at point P_i is

$$U_i = G \sum_{\substack{j=1 \\ j \neq i}}^{n} \frac{m_j}{r_{ij}} \tag{7.7}$$

noting that

$$\nabla_i \frac{1}{r_{ij}} = \frac{1}{r_{ij}^{3}} (\mathbf{r}_j - \mathbf{r}_i)$$

The concept of gravitational potential permits some convenience, because it is a scalar function of position. Such a potential can be used to describe the attraction properties of a distributed mass. Consider the situation in Figure 7.4.

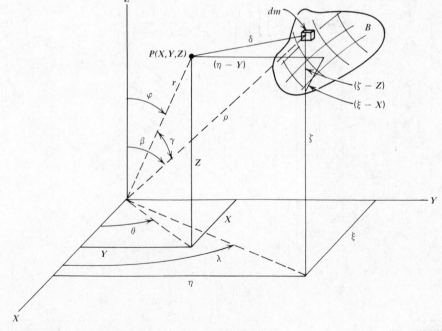

FIGURE 7.4 Potential of a distributed mass.

The potential of point $P(X,Y,Z)$ due to the attraction of dm is

$$dU(X,Y,Z) = G\frac{dm}{\delta}$$

The potential at $P(X,Y,Z)$ due to the entire body B, is

$$U(X,Y,Z) = G\int_B \frac{dm}{\delta}$$

If $\Gamma(\xi,\eta,\zeta)$ is the density at point ξ,η,ζ, then this integral becomes

$$U(X,Y,Z) = G\iiint_B \frac{\Gamma(\xi,\eta,\zeta)\,d\xi\,d\eta\,d\zeta}{[(\xi-X)^2+(\eta-Y)^2+(\zeta-Z)^2]^{1/2}} \qquad (7.8)$$

Since Cartesian coordinates are somewhat cumbersome, spherical coordinates are introduced. Referring to Figure 7.5, and noting that X, Y, Z become r, ϕ, θ, and ξ, η, ζ, become ρ, β, λ,

$$dm = (\rho^2 \sin \beta\, d\rho\, d\beta\, d\lambda)\Gamma(\rho,\beta,\lambda)$$

Expression (7.8) is now rewritten as

$$U(r,\phi,\theta) = G\iiint_B \frac{\Gamma(\rho,\beta,\lambda)\rho^2 \sin \beta\, d\rho\, d\beta\, \lambda}{[r^2+\rho^2-2\,r\rho \cos \gamma]^{1/2}} \qquad (7.9)$$

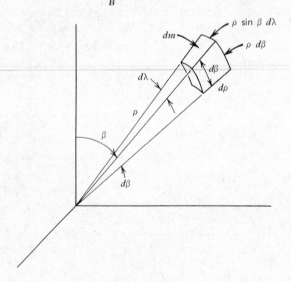

FIGURE 7.5 Spherical coordinate definitions.

If the point at which the potential is required, $P(r,\phi,\theta)$ is at a greater radial distance from the origin of coordinates than any part of B, then $r > \rho$ and

$$\left[1 - 2\frac{\rho}{r}\cos\gamma + \left(\frac{\rho}{r}\right)^2\right]^{-1/2}$$

is the *generating function* of a series of Legendre polynomials (spherical harmonics). In practice, the origin of coordinates can be selected to satisfy $r > \rho$ for all points in B if P is outside the body. Thus

$$[r^2 + \rho^2 - 2r\rho\cos\gamma]^{-1/2} = \frac{1}{r}\sum_{k=0}^{\infty}\left(\frac{\rho}{r}\right)^k P_k(\cos\gamma) \qquad (7.10)$$

The Legendre polynomials $P_k(\cos\gamma)$ can be determined from Rodriques' formula,

$$P_k(\nu) = \frac{1}{2^k k!}\frac{d^k(\nu^2-1)^k}{d\nu^k} \qquad (7.11)$$

where ν can be replaced by $\cos\gamma$ after differentiation. For example, the first four polynomials are

$$P_0(\cos\gamma) = 1, \qquad P_1(\cos\gamma) = \cos\gamma$$

$$P_2(\cos\gamma) = \frac{1}{4}(3\cos 2\gamma + 1), \qquad P_3(\cos\gamma) = \frac{1}{8}(5\cos 3\gamma + 3\cos\gamma)$$

Referring to Figure 7.6, the cosine law of spherical trigonometry gives

$$\cos\gamma = \cos\beta\cos\phi + \sin\beta\sin\phi\cos(\theta - \lambda)$$

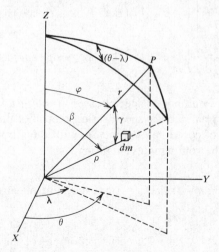

FIGURE 7.6 Spherical trigonometry nomenclature.

which is to be used to convert the variables from γ to the more convenient set ϕ, β, λ. The *addition theorem* of spherical harmonics permits

$$P_k(\cos \gamma) = P_k(\cos \phi) P_k(\cos \beta)$$
$$+2 \sum_{j=1}^{k} \frac{(k-j)!}{(k+j)!} \cos j(\theta - \lambda) P_k^{j}(\cos \phi) P_k^{j}(\cos \beta)$$

where $P_k^{j}(\nu)$ is the *associated Legendre function* of the first kind of degree k and order j, that is,

$$P_k^{j}(\nu) = (1 - \nu^2)^{j/2} \frac{d^j}{d\nu^j} P_k(\nu)$$

Inserting these results into expression (7.9) yields

$$U(r,\phi,\theta) = \frac{Gm}{r} + \sum_{k=1}^{\infty} \frac{A_k}{r^{k+1}} P_k(\cos \phi) + \sum_{k=1}^{\infty} \sum_{j=1}^{k} \frac{B_k^{j}}{r^{k+1}} P_k^{j}(\cos \phi) \cos j\theta$$
$$+ \sum_{k=1}^{\infty} \sum_{j=1}^{k} \frac{C_k^{j}}{r^{k+1}} P_k^{j}(\cos \phi) \sin j\theta \qquad (7.12)$$

where m, A_k, B_k^{j}, and C_k^{j} are defined as

$$m = \iiint_B (\rho,\beta,\lambda)\rho^2 \sin \beta \, d\rho \, d\beta \, d\lambda$$

$$A_k = G \iiint_B \rho^{k+2}\Gamma(\rho,\beta,\lambda) P_k(\cos \beta) \sin \beta \, d\rho \, d\beta \, d\lambda$$

$$B_k^{j} = 2G \frac{(k-j)!}{(k+j)!} \iiint_B \rho^{k+2}\Gamma(\rho,\beta,\lambda) P_k^{j}(\cos \beta) \cos j\lambda \sin \beta \, d\rho \, d\beta \, d\lambda$$

$$C_k^{j} = 2G \frac{(k-j)!}{(k+j)!} \iiint_B \rho^{k+2}\Gamma(\rho,\beta,\lambda) P_k^{j}(\cos \beta) \sin j\lambda \sin \beta \, d\rho \, d\beta \, d\lambda$$

There are a limited number of special cases of interest which allow $U(r,\phi,\theta)$ to take on simpler forms. A basic specification typical of this type of problem is to assume the origin of coordinates coincides with the attracting body center of mass. This permits the following evaluations

$$A_1 = G \iiint_B \rho \cos \beta \, dm = 0 \qquad \text{(first moment of mass about } XY \text{ plane)}$$

$$B_1^{1} = G \iiint_B \rho \sin \beta \cos \lambda \, dm = 0 \qquad \text{(first moment of mass about } YZ \text{ plane)}$$

$$C_1{}^1 = G \iiint_B \rho \sin \beta \sin \lambda \, dm = 0 \qquad \text{(first moment of mass about } XZ \text{ plane)}$$

This brings expression (7.12) to

$$U(r,\phi,\theta) = \frac{GM}{r} + \sum_{k=2}^{\infty} \left[\frac{A_k}{r^{k+1}} P_k(\cos \phi) + \sum_{j=1}^{k} \frac{B_k{}^{j'}}{r^{k+1}} P_k{}^j(\cos \phi) \cos j\theta \right.$$

$$\left. + \sum_{j=1}^{k} \frac{C_k{}^j}{r^{k+1}} P_k{}^j(\cos \phi) \sin j\theta \right] \qquad (7.13)$$

If, for example, the mass distribution is symmetrical about the Z-axis, then it is an *oblate* attracting body, which simplifies the density distribution, $\Gamma = \Gamma(\rho,\beta)$ and implies

$$B_k{}^j = C_k{}^j = 0$$

for all $j \geq 1$. The potential of an oblate body B is now extracted from form (7.13) as

$$U(r,\phi) = \frac{Gm}{r} \left[1 - \sum_{k=2}^{\infty} J_k \left(\frac{R_B}{r} \right)^k P_k(\cos \phi) \right] \qquad (7.14)$$

where R_B is the body radius at its equator (i.e., intersection with XY plane) and

$$-GmJ_kR_B{}^k = A_k = 2\pi G \iint_B \rho^{k+2} \Gamma(\rho,\beta) P_k(\cos \beta) \sin \beta \, d\rho \, d\beta$$

defines J_k, the *zonal harmonic coefficient* which is obtainable from satellite orbit observations about the body. Odd numbered terms ($k = 3, 5, 7, \ldots$) represent antisymmetries about the equatorial plane. This provides a way to describe a *pear-shaped* planet. Unfortunately, most planets are not simply oblate, but have asymmetries about the Z-axis as well. In many cases oblateness effects on orbits are predominant and *first order* perturbations may be analyzed by accounting for only low degree terms such as J_2 and J_3.

7.1.2 Potential of the Earth

In 1961 the International Astronomical Union (IAU) adopted a standard form for the earth's external gravitational potential. Equation (7.13) is essentially that form, written as

$$U_\oplus(r,\phi,\theta) = \frac{\mu_\oplus}{r} \left[1 + \sum_{k=1}^{\infty} \sum_{j=0}^{k} \left(\frac{R_\oplus}{r} \right)^k P_k{}^j(\cos \phi) \{ C_k{}^j \cos j\theta + S_k{}^j \sin j\theta \} \right] \qquad (7.15)$$

where the center of coordinates is taken as the earth center of mass. Zonal harmonics correspond to $j = 0$, thus, define J_k as

$$J_k = -C_k^0$$

This leads to an alternate form of expression (7.15),

$$U_\oplus(r,\phi,\theta) = \frac{\mu_\oplus}{r}\left[1 - \sum_{k=2}^{\infty}\left(\frac{R_\oplus}{r}\right)^k J_k P_k(\cos\phi)\right.$$

$$\left. + \sum_{k=2}^{\infty}\sum_{j=1}^{k}\left(\frac{R_\oplus}{r}\right)^k P_k^j(\cos\phi)\{C_k^j \cos j\theta + S_k^j \sin j\theta\}\right] \quad (7.16)$$

Values of the various coefficients can be obtained, in principle, from satellite observations. However, as the degree and order increase, the effect of each harmonic on satellite motion decreases, thus, increasing the difficulty of extracting the value of succeeding coefficients. As of 1968, the values of J_k up to degree 7 and C_k^j, S_k^j up to degree four were approximated by:

$J_2 = 1082.7 \times 10^{-6}$	$C_2^1 = 0$	$S_2^1 = 0$
$J_3 = -2.56 \times 10^{-6}$	$C_2^2 = 1.57 \times 10^{-6}$	$S_2^2 = -0.897 \times 10^{-6}$
$J_4 = -1.58 \times 10^{-6}$	$C_3^1 = 2.10 \times 10^{-6}$	$S_3^1 = 0.16 \times 10^{-6}$
$J_5 = -0.15 \times 10^{-6}$	$C_3^2 = 0.25 \times 10^{-6}$	$S_3^2 = -0.27 \times 10^{-6}$
$J_6 = 0.59 \times 10^{-6}$	$C_3^3 = 0.077 \times 10^{-6}$	$S_3^3 = 0.173 \times 10^{-6}$
$J_7 = -0.44 \times 10^{-6}$	$C_4^1 = -0.58 \times 10^{-6}$	$S_4^1 = -0.46 \times 10^{-6}$
	$C_4^2 = 0.074 \times 10^{-6}$	$S_4^2 = 0.16 \times 10^{-6}$
	$C_4^3 = 0.053 \times 10^{-6}$	$S_4^3 = 0.004 \times 10^{-6}$
	$C_4^4 = -0.0065 \times 10^{-6}$	$S_4^4 = 0.0023 \times 10^{-6}$

Unfortunately, the magnitude of each succeeding coefficient does not necessarily decrease, but the factor $(R_\oplus/r)^k$ tends to diminish each term. It should be noted that this representation is especially sensitive to near-surface mass and density anomalies in the attracting body. The earth does not have this type of large scale anomalies, but the moon does. Thus, it is difficult to represent the moon's gravitational field with form (7.16). Of course, if $(r/R_\oplus)$ is large, effects of asphericities are reduced. Only low orbits experience effects of the higher degree terms in a harmonic expansion.

7.1.3 The n-Body Problem

Returning to the problem of a set of n particles, $P_1, P_2 \ldots, P_n$, the equation of motion of the ith particle P_i is given by expression (7.5). Since

$$\mathbf{F}_i = m_i \frac{d^2\mathbf{r}_i}{dt^2}$$

from Newton's law for constant mass, expression (1.1), a potential function can be defined for n-bodies by

$$m_i \frac{d^2 \mathbf{r}_i}{dt^2} = \nabla_i U \tag{7.17}$$

One form of U which satisfies form (7.5) is

$$U = \frac{G}{2} \sum_{i=1}^{n} \sum_{\substack{j=1 \\ j \neq i}}^{n} \frac{m_i m_j}{r_{ij}} \tag{7.18}$$

Obviously, any constant may be added to U without changing the outcome of equation (7.17). The function U may be thought of as the total work done by gravitational forces in assembling the system of particles from a state of infinite separation. Thus, the system potential energy is simply $-U$.

As pointed out in Section 2.2 a complete description of motion of this system of particles requires $6n$ integrals, three components each of position and velocity for each particle referred to an inertial frame. It was also pointed out that the relative motion between two particles (two-body problem) requires only six integrals of motion. However, only 10 integrals can be obtained. These consist of:

1. *Conservation of total linear momentum.* There are no forces external to the system. Therefore, the vector sum of forces must be zero,

$$\sum_{i=1}^{n} m_i \frac{d^2 \mathbf{r}_i}{dt^2} = G \sum_{i=1}^{n} \sum_{\substack{j=1 \\ j \neq i}}^{n} \frac{m_i m_j}{r_{ij}{}^3} (\mathbf{r}_j - \mathbf{r}_i) = 0 \tag{7.19}$$

The first integral gives

$$\sum_{i=1}^{n} m_i \frac{d \mathbf{r}_i}{dt} = \mathbf{c}_1' = \text{constant vector} \tag{7.20}$$

and the second yields

$$\sum_{i=1}^{n} m_i \mathbf{r}_i = \mathbf{c}_1 t + \mathbf{c}_2 \tag{7.21}$$

Noting that the mass center of this set of particles is at

$$\mathbf{r}_{cm} = \frac{\sum\limits_{i=1}^{n} m_i \mathbf{r}_i}{\sum\limits_{i=1}^{n} m_i} = \frac{\mathbf{c}_1 t + \mathbf{c}_2}{M} \tag{7.22}$$

where $M = \sum\limits_{i=1}^{n} m_i$. Thus, conservation of linear momentum provides six constants of motion: $\mathbf{c}_2/M$ represents the three components of initial center of mass position, and $\mathbf{c}_1/M$ is the constant velocity of the mass center.

2. *Conservation of total angular momentum.* Since no net torque acts on this system,

$$\sum_{i=1}^{n} \mathbf{F}_i \times \mathbf{r}_i = 0$$

Replacing $\mathbf{F}_i$ by its equivalent inertia and attraction forms gives

$$\sum_{i=1}^{n} m_i \frac{d^2\mathbf{r}_i}{dt^2} \times \mathbf{r}_i = G \sum_{i=1}^{n} \sum_{\substack{j=1 \\ j \neq i}}^{n} \frac{m_i m_j}{r_{ij}^3} (\mathbf{r}_j - \mathbf{r}_i) \times \mathbf{r}_i = 0$$

which points out that since $\mathbf{r}_i \times \mathbf{r}_i = 0$ for all i,

$$\sum_{i=1}^{n} \sum_{\substack{j=1 \\ j \neq i}}^{n} (\mathbf{r}_j \times \mathbf{r}_i) = 0$$

Also, note that

$$\frac{d^2\mathbf{r}_i}{dt^2} \times \mathbf{r}_i = \frac{d}{dt}\left(\frac{d\mathbf{r}_i}{dt} \times \mathbf{r}_i\right)$$

Thus,

$$\sum_{i=1}^{n} m_i \left(\mathbf{r}_i \times \frac{d\mathbf{r}_i}{dt}\right) = \mathbf{c}_3 \qquad (= \text{total angular momentum}) \qquad (7.23)$$

3. *Conservation of total energy.* Using form (7.17), generate the expression

$$\sum_{i=1}^{n} m_i \frac{d^2\mathbf{r}_i}{dt^2} \cdot \frac{d\mathbf{r}_i}{dt} = \sum_{i=1}^{n} \nabla_i U \cdot \frac{d\mathbf{r}_i}{dt} = \frac{dU}{dt} \qquad (7.24)$$

This result is obtained by remembering that

$$\nabla_i U \cdot \frac{d\mathbf{r}_i}{dt} = \frac{\partial U}{\partial X_i}\frac{dX_i}{dt} + \frac{\partial U}{\partial Y_i}\frac{dY_i}{dt} + \frac{\partial U}{\partial Z_i}\frac{dZ_i}{dt}$$

A first integral of the left-hand side of equation (7.24) is obtained by using

$$\int \frac{d^2\mathbf{r}_i}{dt^2} \cdot \frac{d\mathbf{r}_i}{dt} \, dt = \frac{1}{2}\frac{d\mathbf{r}_i}{dt} \cdot \frac{d\mathbf{r}_i}{dt} = \frac{T_i}{m_i}$$

which leads immediately to

$$T - U = c \qquad (= \text{total energy of the system}) \qquad (7.25)$$

where

$$T = \frac{1}{2}\sum_{i=1}^{n} m_i \frac{d\mathbf{r}_i}{dt} \cdot \frac{d\mathbf{r}_i}{dt} \qquad (7.26)$$

No further integrals of general form are obtainable for the n-body problem. Summarizing, the 10 constants of integration are the components of $\mathbf{c}_1$, $\mathbf{c}_2$, and $\mathbf{c}_3$, plus scalar c.

7.1.4 Disturbed Two-Body Motion

There are many situations in which two-body motion is of primary interest, but other bodies influence or disturb this motion. Thus, the disturbed motion of two bodies can be treated by solving equation (7.19) for the $i = 1$ term. Repeat this for the $i = 2$ term and take the difference of resulting expressions to get the motion of P_2 with respect to P_1,

$$\frac{d^2\mathbf{r}}{dt^2} + \frac{\mu}{r^3}\mathbf{r} = -G\sum_{j=3}^{n}\left(\frac{m_j}{d_j^3}\mathbf{d}_j + \frac{m_j}{\rho_j^3}\boldsymbol{\rho}_j\right) \tag{7.27}$$

where nomenclature is defined in Figure 7.7, and related definitions are $\mathbf{r} = \mathbf{r}_2 - \mathbf{r}_1$, $\boldsymbol{\rho}_j = \mathbf{r}_j - \mathbf{r}_1$, $\mathbf{d}_j = \mathbf{r} - \boldsymbol{\rho}_j = \mathbf{r}_2 - \mathbf{r}_j$, and $\mu = G(m_1 + m_2)$. The right-hand side of equation (7.27) can be rewritten by using the identity

$$\frac{\mathbf{d}_j}{d_j^3} + \frac{\boldsymbol{\rho}_j}{\rho_j^3} = -\nabla\left(\frac{1}{d_j} - \frac{1}{\rho_j^3}\mathbf{r}\cdot\boldsymbol{\rho}_j\right) \tag{7.28}$$

where ∇ is the gradient operator which operates only on the components of the independent vector from P_1 to P_2, (i.e., $\mathbf{r}$) and on dependent vectors and

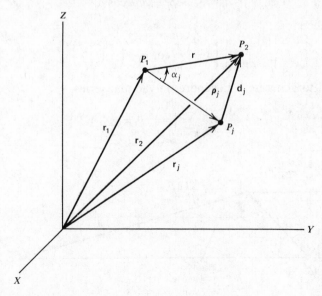

FIGURE 7.7 Nomenclature for disturbed two-body motion.

functions of **r**. Thus, equation (7.27) becomes

$$\frac{d^2\mathbf{r}}{dt^2}+\frac{\mu}{r^3}\mathbf{r}=G\sum_{j=3}^{n}m_j\nabla\left(\frac{1}{d_j}-\frac{1}{\rho^3}\mathbf{r}\cdot\boldsymbol{\rho}_j\right) \tag{7.29}$$

The form on the right-hand side has the appearance of a *disturbance potential*, which prompts the definition of a function D_j as

$$D_j=Gm_j\left(\frac{1}{d_j}-\frac{1}{\rho_j^3}\mathbf{r}\cdot\boldsymbol{\rho}_j\right) \tag{7.30}$$

This leads to further simplification of form (7.29)

$$\frac{d^2\mathbf{r}}{dt^2}+\frac{\mu}{r^3}\mathbf{r}=\nabla\sum_{j=3}^{n}D_j \tag{7.31}$$

This form represents the motion of P_2 with respect to P_1 in the presence of a disturbing potential due to $n-2$ other particles.

Consider a typical situation of interest in spaceflight. Assume $r\ll\rho_j$ and either form (7.27) or (7.31) will, in principle, give a solution of motion. However, in such cases, $\nabla\sum_{j=3}^{n}D_j$ is the difference of two large and almost equal vectors, as depicted in Figure 7.8. One way to avoid this difficulty is now offered. Since

$$d_j=\sqrt{r^2+\rho_j^2-2r\rho_j\cos\alpha_j}$$

and $\rho_j>r$, then

$$\frac{1}{d_j}=\frac{1}{\rho_j}\left[1-2\frac{r}{\rho_j}\cos\alpha_j+\left(\frac{r}{\rho_j}\right)^2\right]^{-1/2}$$

can be used to generate a Legendre polynomial series,

$$\frac{1}{d_j}=\frac{1}{\rho_j}\sum_{k=0}^{\infty}\left(\frac{r}{\rho_j}\right)^k P_k(\cos\alpha_j)$$

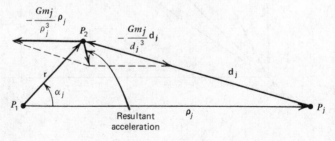

FIGURE 7.8 Typical situation of interest for the disturbed motion of two bodies.

This leads to a more appropriate form of D_j,

$$D_j = \frac{Gm_j}{\rho_j} \left[1 + \sum_{k=2}^{\infty} \left(\frac{r}{\rho_j} \right)^k P_k (\cos \alpha_j) \right] \tag{7.32}$$

7.1.5 Sphere of Influence

Section 3.4 introduced the idea of a *sphere of influence* in connection with the patched conic technique for interplanetary transfers. It identifies a region in which a body is the primary attracting mass. This concept also arises in the treatment of disturbed two-body motion, that is, to determine whether a spacecraft is under the influence of one large body while perturbed by another large body or is under the influence of the second large body and perturbed by the first. The method used in Chapter 3 bypassed this question by treating each trajectory leg as a two-body problem to obtain initial conditions for the following segment. The succeeding attracting body was not considered until these conditions could be calculated. Accurate interplanetary computations must account for secondary large bodies. Thus, the sphere of influence concept is important to distinguish mission phases. Application of this idea requires that two descriptions of motion be compared in order to select an origin about which motion takes place. This selection should accommodate numerical computations. Consider the situation in Figure 7.9 in which P_1 and P_3 are the primary attracting bodies. Motion of P_2 with respect to P_1 is obtained directly from result (7.27) by dropping subscript j,

$$\frac{d^2 \mathbf{r}}{dt^2} + \frac{G(m_1 + m_2)}{r^3} \mathbf{r} = -Gm_3 \left(\frac{\mathbf{d}}{d^3} + \frac{\boldsymbol{\rho}}{\rho^3} \right) \tag{7.33a}$$

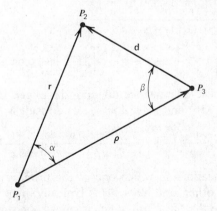

FIGURE 7.9 Definition of sphere of influence.

Similarly, the motion of P_2 with respect to P_3 is

$$\frac{d^2\mathbf{d}}{dt^2} + \frac{G(m_2+m_3)\mathbf{d}}{d^3} = -Gm_1\left(\frac{\mathbf{r}}{r^3} - \frac{\boldsymbol{\rho}}{\rho^3}\right) \tag{7.33b}$$

Laplace postulated that the advantage of either form depends on the magnitude ratio of disturbing force to corresponding central attraction. This ratio is determined for each of the two bodies, P_1 and P_3, and the ratio which gives the smaller value is associated with the primary body of attraction. For P_2 relative to P_1 this ratio is

$$\frac{\text{disturbing force}}{\text{central force}} = \frac{Gm_3\left[\left(\dfrac{\mathbf{d}}{d^3}+\dfrac{\boldsymbol{\rho}}{\rho^3}\right)\cdot\left(\dfrac{\mathbf{d}}{d^3}+\dfrac{\boldsymbol{\rho}}{\rho^3}\right)\right]^{1/2}}{G(m_1+m_2)/r^2} \tag{7.34}$$

To simplify this expression note that

$$\cos\beta = \frac{\rho}{d} - \frac{r}{d}\cos\alpha$$

and

$$\frac{d}{\rho} = \left[1 - 2\frac{r}{\rho}\cos\alpha + \left(\frac{r}{\rho}\right)^2\right]^{1/2}$$

Substituting these into expression (7.34) leads to

$$\frac{\text{disturbing force}}{\text{central force}} = \frac{m_3}{m_1+m_2}\left(\frac{r/\rho}{d/\rho}\right)^2\left[1 - 2\left(\frac{d}{\rho}\right)\left(1 - \frac{r}{\rho}\cos\alpha\right) + \left(\frac{d}{\rho}\right)^4\right]^{1/2} \tag{7.35a}$$

Analogous expressions for motion of P_2 relative to P_3 give

$$\frac{\text{disturbing force}}{\text{central force}} = \frac{Gm_1\left[\left(\dfrac{\mathbf{r}}{r^3}-\dfrac{\boldsymbol{\rho}}{\rho^3}\right)\cdot\left(\dfrac{\mathbf{r}}{r^3}-\dfrac{\boldsymbol{\rho}}{\rho^3}\right)\right]^{1/2}}{G(m_2+m_3)/d^2}$$

$$= \frac{m_1}{m_2+m_3}\left(\frac{r}{\rho}\right)^{-2}\left[1 - 2\frac{r}{\rho}\cos\alpha + \left(\frac{r}{\rho}\right)^2\right]\left[1 - 2\left(\frac{r}{\rho}\right)^2\cos\alpha + \left(\frac{r}{\rho}\right)^4\right]^{1/2} \tag{7.35b}$$

To determine the sphere of influence dividing the effects of P_1 and P_3 equate forms (7.35a) and (7.35b) and solve for r/ρ. The result is

$$\left(\frac{r}{\rho}\right)^4 = \frac{m_1(m_1+m_2)}{m_3(m_2+m_3)}\left(\frac{d}{\rho}\right)^4\left[\frac{1 - 2(r/\rho)^2\cos\alpha + (r/\rho)^4}{1 - 2(d/\rho)[1-(r/\rho)\cos\alpha] + (d/\rho)^4}\right]^{1/2} \tag{7.36}$$

There are many situations in which one of the primary attracting masses is much larger than the other, and mass m_2 is typically much smaller than either primary mass. Assume that $m_3 \gg m_1$, which implies $r \ll \rho$ since m_1 would have a small sphere of influence. If small terms are neglected in equation (7.36) the

TABLE 7.1 Planetary Spheres of Influence

Planet	Mass Ratio $\times 10^4$ (Planet/Sun)	Radius of Sphere of Influence (10^5 km)
Mercury	0.00164	1.12
Venus	0.0245	6.16
Earth	0.0304	9.29
Mars	0.00324	5.78
Jupiter	9.55	482
Saturn	2.86	545
Uranus	0.436	519
Neptune	0.518	868
Pluto	0.025	341

ratio (r/ρ) can be determined explicitly as

$$\frac{r}{\rho} = \left[\left(\frac{m_3}{m_1}\right)^{2/5} (1 + 3\cos^2\alpha)^{1/10} + \frac{2}{5}\cos\alpha\left(\frac{1 + 6\cos^2\alpha}{1 + 3\cos^2\alpha}\right)\right]^{-1} \qquad (7.37)$$

Further simplification can be accomplished by noting that the second term is much smaller than the first and that

$$1 < (1 + 3\cos^2\alpha)^{1/10} < 1.15$$

Replacing this by unity and ignoring the second term in expression (7.37) yields

$$\frac{r}{\rho} \cong \left(\frac{m_1}{m_3}\right)^{2/5} \qquad (7.38)$$

which is a sphere about P_1. Inside this sphere of influence of P_1 it is appropriate to consider the motion of a spacecraft to be dominated by m_1 and use P_1 as the origin of coordinates. Outside this sphere use P_3 as the origin. Each planet of the solar system has an associated sphere of influence relative to the sun. Equation (7.38) gives an estimate of the radius of each in Table 7.1. The moon has a sphere of influence relative to earth of radius 66,100 km.

7.2 RESTRICTED THREE-BODY PROBLEM

7.2.1 Lagrangian Points

It has been noted that the general three-body problem does not have an analytical solution. However, certain special situations have solutions. Such cases are typically referred to as *restricted three-body* problems and were originally studied by Lagrange. Fortunately, many realistic cases of interest permit treatment as restricted three-body situations. The most obvious example is a spacecraft moving in the earth-moon system. This is used to illustrate the method of analysis. Certain assumptions are required which permit a straightforward solution at a slight loss of accuracy. Assume earth and moon move in circles around their center of mass (barycenter) and the satellite at point P has negligible mass, as depicted in Figure 7.10. Absolute acceleration of P is obtained in terms of the moving coordinate system, x, y, z by application of equation (1.34),

$$\mathbf{a}_P = \mathbf{a}_O + \mathbf{n} \times (\mathbf{n} \times \mathbf{r}) + \dot{\mathbf{n}} \times \mathbf{r} + \ddot{\mathbf{r}}_b + 2\mathbf{n} \times \dot{\mathbf{r}}_b \qquad (7.39)$$

where $\mathbf{n}$ is the angular velocity of the earth-moon system, $\mathbf{n} = n\mathbf{i}_z$. This acceleration is balanced by the attraction forces. Thus,

$$\mathbf{a}_P = -\frac{\mu_\oplus}{r_\oplus^3}\mathbf{r}_\oplus - \frac{\mu_{\mathbb{C}}}{r_{\mathbb{C}}^3}\mathbf{r}_{\mathbb{C}} = \nabla\left(\frac{\mu_\oplus}{r_\oplus} + \frac{\mu_{\mathbb{C}}}{r_{\mathbb{C}}}\right) \qquad (7.40)$$

Equating forms (7.39) and (7.40), and taking components yields the three equations of motion for P in the rotating coordinate system:

$$\ddot{x} - 2n\dot{y} - n^2 x = \frac{\partial}{\partial x}\left(\frac{\mu_\oplus}{r_\oplus} + \frac{\mu_{\mathbb{C}}}{r_{\mathbb{C}}}\right) \qquad (7.41a)$$

$$\ddot{y} + 2n\dot{x} - n^2 y = \frac{\partial}{\partial y}\left(\frac{\mu_\oplus}{r_\oplus} + \frac{\mu_{\mathbb{C}}}{r_{\mathbb{C}}}\right) \qquad (7.41b)$$

$$\ddot{z} = \frac{\partial}{\partial z}\left(\frac{\mu_\oplus}{r_\oplus} + \frac{\mu_{\mathbb{C}}}{r_{\mathbb{C}}}\right) \qquad (7.41c)$$

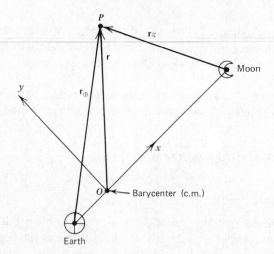

FIGURE 7.10 Restricted three-body problem nomenclature.

Typical of such problems is the characterization of motion rather than an exact solution of the equations. One important aspect of this problem is the question of existence of equilibrium points in the earth-moon system. Satellites placed at such points would not experience any relative acceleration, thus, permitting a stationary position in the rotating frame. Secondly, if these points exist, the question of stability must be answered. Existence of equilibrium points can be investigated by obtaining an integral of motion related to energy. This is done by multiplying equations (7.41) by $2\dot{x}$, $2\dot{y}$, $2\dot{z}$, respectively. Then add the resulting forms,

$$2\dot{x}\ddot{x}+2\dot{y}\ddot{y}+2\dot{z}\ddot{z}-2n^2x\dot{x}-2n^2y\dot{y}$$

$$=2\dot{x}\frac{\partial}{\partial x}\left(\frac{\mu_\oplus}{r_\oplus}+\frac{\mu_{\mathbb{C}}}{r_{\mathbb{C}}}\right)+2\dot{y}\frac{\partial}{\partial y}\left(\frac{\mu_\oplus}{r_\oplus}+\frac{\mu_{\mathbb{C}}}{r_{\mathbb{C}}}\right)+2\dot{z}\frac{\partial}{\partial z}\left(\frac{\mu_\oplus}{r_\oplus}+\frac{\mu_{\mathbb{C}}}{r_{\mathbb{C}}}\right)$$

This can be integrated over time to get

$$\dot{x}^2+\dot{y}^2+\dot{z}^2-n^2(x^2+y^2)=\frac{2\mu_\oplus}{r_\oplus}+\frac{2\mu_{\mathbb{C}}}{r_{\mathbb{C}}}-C \tag{7.42}$$

where C is known as *Jacobi's constant*. If an *effective potential*, $U(x,y,z)$ is defined as

$$U(x,y,z)=\frac{\mu_\oplus}{r_\oplus}+\frac{\mu_{\mathbb{C}}}{r_{\mathbb{C}}}+\frac{1}{2}n^2(x^2+y^2) \tag{7.43}$$

then the effective potential energy of a point in the rotating frame is $-U(x, y, z)$. Equation (7.42) can now be rewritten as

$$U(x,y,z)-\frac{1}{2}C=\frac{1}{2}(\dot{x}^2+\dot{y}^2+\dot{z}^2)$$

Note that the right-hand side must be greater than zero. Thus, a given initial position (x_0,y_0,z_0) and initial velocity in the rotating frame $(\dot{x}_0,\dot{y}_0,\dot{z}_0)$ must yield a value C_0 which satisfies

$$2U(x,y,z)\geq C_0 \tag{7.44}$$

The equality corresponds to no relative motion, that is, a stationary situation. Therefore, contours of constant U values correspond to loci of zero relative velocity positions. Figure 7.11 depicts a number of these with $C_1>C_2>C_3\cdots$. Out-of-plane motion is not consistent with equilibrium points. Thus, only x,y, plane motion is considered. Points L_1, L_2, L_3, L_4, and L_5 are the *Lagrangian points*. Potential energy *wells* are revealed in the neighborhoods of both earth and moon, but *peaks* occur at equilateral triangular points L_4 and L_5, and at saddle points L_1, L_2, and L_3. Notice that a particle may have relative

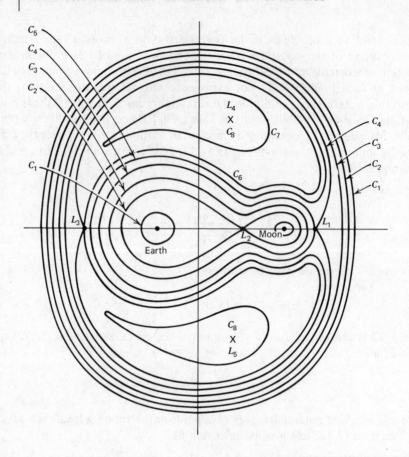

FIGURE 7.11 Contours of zero relative velocity in x,y plane.

motion only in regions corresponding to values of C higher than its own. For example, if C is very large, then condition (7.44) is satisfied only when x and y are very large or when $r_\oplus$ or r_C is very small. A particle with $C = C_1$ can move only in regions close to earth or moon, or at a great distance from both.

7.2.2 Stability of Equilateral Points

The three saddle points, L_1, L_2, and L_3 are unstable equilibrium positions, because any displacement away from such a point would result in a complete departure from that neighborhood. Equilateral points L_4 and L_5 corresponds to minimum values of C which implies that particles with $C = C_8$ are free to move anywhere in the x, y plane. Thus, the question of bounded motion near these two points is an important one. Particle motion at L_4 or L_5 is said to be *stable* if, when perturbed, it oscillates indefinitely around the point. To test the stability of the equilateral points consider the situation in Figure 7.12. Distances are normalized such that the moon moves around earth in a circle of radius 1. Equation (7.33a) can be applied to obtain the effective attraction of the moon, which is simply the moon's attraction at earth,

$$\frac{\mu_{\text{C}}}{1^3}(\mathbf{OM} - \mathbf{EM}) = -\mu_{\text{C}}\mathbf{i}$$

The attraction of earth is $-GM_{\oplus}\mathbf{i}$, and centrifugal force is $n^2\mathbf{EO} = n^2\mathbf{i}$, where

$$n^2 = \frac{G(M_{\oplus} + M_{\text{C}})}{1^3}$$

from the two-body problem. Since attraction forces are balanced by centrifugal force at L_4, the sum of forces must be zero. For a small, in-plane displacement from O, correction to the earth's attraction is

$$\frac{GM_{\oplus}}{1^3}(2x\mathbf{i} - y\mathbf{j})$$

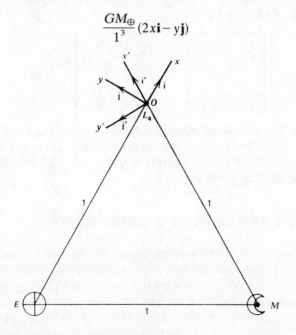

FIGURE 7.12 Stability of triangular points.

and the correction for lunar attraction is

$$\frac{GM_{\mathbb{C}}}{1^3}(2x'\mathbf{i}' - y\mathbf{j}') = GM_{\mathbb{C}}\nabla(x'^2 - \frac{1}{2}y'^2)$$

$$= GM_{\mathbb{C}}\nabla\left[\left(\frac{1}{2}x + \frac{\sqrt{3}}{2}y\right)^2 - \frac{1}{2}\left(-\frac{\sqrt{3}}{2}x + \frac{1}{2}y\right)^2\right]$$

$$= GM_{\mathbb{C}}\left[\left(-\frac{1}{4}x + \frac{3\sqrt{3}}{4}y\right)\mathbf{i} + \left(\frac{3\sqrt{3}}{4}x + \frac{5}{4}y\right)\mathbf{j}\right]$$

Equate the sum of the effective gravitational attractions to the acceleration and evaluate in the rotating x,y coordinates of Figure 7.12. This is accomplished by applying the relative motion equation and writing $M_{\mathbb{C}}/(M_{\oplus}+M_{\mathbb{C}}) = \rho$ and $M_{\oplus}/(M_{\oplus}+M_{\mathbb{C}}) = 1-\rho$,

$$\ddot{x} - 2n\dot{y} - n^2x = 2(1-\rho)n^2x + \rho n^2\left(-\frac{1}{4}x + \frac{3\sqrt{3}}{4}y\right) \qquad (7.45a)$$

$$\ddot{y} + 2n\dot{x} - n^2y = -(1-\rho)n^2y + \rho n^2\left(\frac{3\sqrt{3}}{4}x + \frac{5}{4}y\right) \qquad (7.45b)$$

Scaling the time so that $n = 1$ and trying a solution of form $x = Ae^{\omega t}$, $y = be^{\omega t}$ without initial conditions yields

$$\begin{bmatrix} s^2 - \left(3 - \frac{9}{4}\rho\right) & -\left(2s + \frac{3\sqrt{3}}{4}\rho\right) \\ \left(2s - \frac{3\sqrt{3}}{4}\rho\right) & \left(s^2 - \frac{9}{4}\rho\right) \end{bmatrix}\begin{bmatrix} x \\ y \end{bmatrix} = 0$$

where $s = \omega/n$. The associated characteristic equation is $s^4 + s^2 + 27\rho(1-\rho)/4 = 0$, which has pure imaginary roots provided that $\rho(1-\rho) \leq 1/27$. For the earth-moon system, $\rho = 1/82.6$. Therefore, $\rho(1-\rho) = 1/83.6$, indicating L_4 and L_5 have stable motion about them.

7.3 POSITION AND VELOCITY IN CONIC ORBITS

The concept of conic orbits was introduced in Chapter 2 in the discussion of two-body and central force motion. This section expands that basic material using geometrical and physical interpretations for determining spacecraft position and velocity about a central body of attraction. The time equation will also be treated in a general manner. Then universal formulas describing conic position and velocity without prior knowledge of orbital energy level are developed.

7.3.1 Geometric and Kinetic Properties of Conic Sections

There are three basic conic shapes of interest; ellipse, parabola, and hyperbola. Of course, the circle is a special case of an ellipse. Nomenclature to be used for these conics is presented in Figure 7.13. Geometric interpretations will require definition of important points and parameters:

$$F = \text{focus}$$

$$F^* = \text{vacant focus}$$

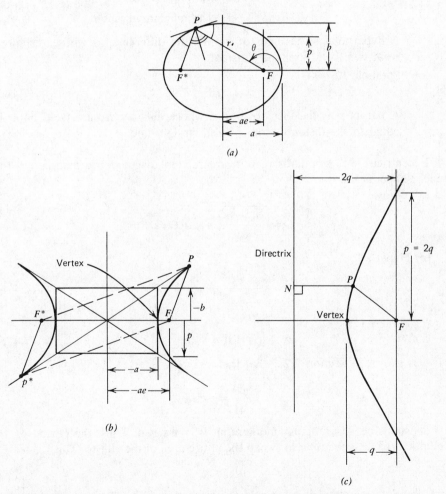

FIGURE 7.13 Geometric nomenclature for conics. (*a*) Ellipse. (*b*) Hyperbola. (*c*) Parabola.

Each shape has a fundamental geometric definition:

(a)   Ellipse $(a > 0)$

$$PF + PF^* = 2a \qquad (7.46a)$$

An ellipse is the locus of points, the sum of whose distances from two fixed points is constant.

(b)   Hyperbola $(a < 0)$

$$\left. \begin{array}{l} PF^* - PF = -2a \text{ (real branch)} \\ P^*F - P^*F^* = -2a \text{ (imaginary branch)} \end{array} \right\} \qquad (7.46b)$$

A hyperbola is the locus of points, the difference of whose distances from two fixed points is constant.

(c)   Parabola $(a = \infty)$

$$PF = PN \qquad (7.46c)$$

A parabola is the locus of points whose distance from a fixed point is equal to the distance from a fixed straight line.

Eccentricity and semilatus rectum are related to geometric parameters for each shape as follows:

(a) Ellipse

$$e = \frac{\sqrt{a^2 - b^2}}{a}, p = a(1 - e^2) \qquad (7.47a)$$

(b) Hyperbola

$$e = \frac{\sqrt{a^2 + b^2}}{-a}, p = a(1 - e^2) \qquad (7.47b)$$

(c) Parabola

$$e = 1, p = 2q \qquad (7.47c)$$

It was shown in Section 2.2.3 that the areal rate,

$$\frac{dA}{dt} = \frac{1}{2} r^2 \dot{\theta}$$

is constant, because angular momentum is conserved. Thus, the period of an elliptic orbit, τ is the time to sweep the entire area of the ellipse. This led to

$$\mu = n^2 a^3$$

which relates semimajor axis to the mean motion. The energy integral, sometimes called the *vis-viva integral*, is now constructed from the two-body equation of motion given by expression (2.5),

$$\frac{d^2\mathbf{r}}{dt^2} + \frac{\mu}{r^3}\mathbf{r} = 0$$

Dotting this with $d\mathbf{r}/dt$ and noting that

$$\frac{d\mathbf{r}}{dt} \cdot \frac{d^2\mathbf{r}}{dt^2} = \frac{1}{2}\frac{d}{dt}\left(\frac{d\mathbf{r}}{dt} \cdot \frac{d\mathbf{r}}{dt}\right)$$

and

$$\frac{d\mathbf{r}}{dt} \cdot \mathbf{r} = \frac{1}{2}\frac{d}{dt}(\mathbf{r} \cdot \mathbf{r}) = \frac{1}{2}\frac{d}{dt}(r^2)$$

gives the first integral as

$$\frac{1}{2}\left(\frac{d\mathbf{r}}{dt} \cdot \frac{d\mathbf{r}}{dt}\right) = \mu\frac{1}{r} + \mathscr{E}$$

or

$$\frac{1}{2}v^2 = \frac{\mu}{r} + \mathscr{E} \qquad \text{(vis-viva integral)}$$

where $\mathscr{E}$ is the energy per unit mass and is related to semimajor axis by

$$\mathscr{E} = -\frac{\mu}{2a}$$

which was explained in Section 2.2.3. Thus, velocity magnitude is given by

$$v^2 = \frac{2\mu}{r} - \frac{\mu}{a} \tag{7.48}$$

7.3.2 Position and Velocity Formulas

Position and velocity components are now developed for each conic. Consider first a parabolic orbit. Noting that $e = 1$,

$$r = \frac{p}{1 + \cos\theta}$$

and the identity

$$1 + \cos\theta = 2\cos^2\left(\frac{\theta}{2}\right)$$

radial distance r can be expressed as

$$r = \frac{p}{2\cos^2\left(\frac{\theta}{2}\right)} = \frac{p}{2}\left[\frac{\cos^2\left(\frac{\theta}{2}\right) + \sin^2\left(\frac{\theta}{2}\right)}{\cos^2\left(\frac{\theta}{2}\right)}\right] = \frac{p}{2}\left[1 + \tan^2\left(\frac{\theta}{2}\right)\right] \tag{7.49}$$

In order to relate this to time, note that angular momentum magnitude is simply

$$h = r^2 \frac{d\theta}{dt} = \sqrt{\mu p}$$

Combining equation (7.49) with this yields

$$4\sqrt{\frac{\mu}{p^3}}\, dt = \sec^4\left(\frac{\theta}{2}\right) d\theta$$

which is integrated to the form

$$4\sqrt{\frac{\mu}{p^3}}\, t\, \bigg|_{t_v}^{t} = 2\left[\tan\left(\frac{\theta}{2}\right) + \frac{1}{3}\tan^3\left(\frac{\theta}{2}\right)\right]$$

where t_v is the time of vertex (or periapsis) passage. This is finally expressed as

$$2\sqrt{\frac{\mu}{p^3}}\, (t - t_v) = \tan\left(\frac{\theta}{2}\right) + \frac{1}{3}\tan^3\left(\frac{\theta}{2}\right) \qquad (7.50)$$

If t is given, it appears difficult to extract θ from this cubic equation. However, use of the identity

$$\frac{1}{3}\left(\lambda^3 - \frac{1}{\lambda^3}\right) = \left(\lambda - \frac{1}{\lambda}\right) + \frac{1}{3}\left(\lambda - \frac{1}{\lambda}\right)^3$$

can ease this situation somewhat. Let

$$\tan\left(\frac{\theta}{2}\right) = \lambda - \frac{1}{\lambda}$$

which permits form (7.50) to be written as

$$2\sqrt{\frac{\mu}{p^3}}\, (t - t_v) = \frac{1}{3}\left(\lambda^3 - \frac{1}{\lambda^3}\right) \qquad (7.51)$$

A second and third substitution of

$$\lambda = -\tan w$$
$$\lambda^3 = -\tan s$$

allows

$$\lambda^3 - \frac{1}{\lambda^3} = 2\cot 2s \qquad (7.52)$$

Now a three-step sequence can be outlined to calculate θ for given t in a parabolic orbit:

1. Use expressions (7.51) and (7.52) in the combined form

$$\cot 2s = 3\sqrt{\frac{\mu}{p^3}}(t - t_v)$$

to calculate s.

2. Then use

$$\tan^3 w = \tan s$$

to obtain w.

3. Finally, apply

$$\tan\left(\frac{\theta}{2}\right) = -\tan w + \cot w = 2\cot 2w$$

to complete the sequence.

It remains to determine position $\mathbf{r}$ and velocity $\mathbf{v}$ at the corresponding value of true anomaly θ. Set up an in-plane coordinate system with unit vectors $\mathbf{i}_\xi$ and $\mathbf{i}_\eta$ as defined in Figure 7.14. Then the position vector can be expressed as

$$\mathbf{r} = r\cos\theta\,\mathbf{i}_\xi + r\sin\theta\,\mathbf{i}_\eta \qquad (7.53)$$

Using form (7.49),

$$r\cos\theta = \frac{p}{2}\left[1 - \tan^2\left(\frac{\theta}{2}\right)\right]$$

$$r\sin\theta = p\tan\left(\frac{\theta}{2}\right) \qquad (7.54)$$

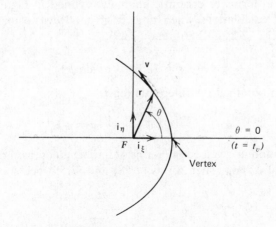

FIGURE 7.14 Position and velocity in parabolic orbit.

which permits equation (7.53) to become

$$\mathbf{r} = \frac{p}{2}\left[1 - \tan^2\left(\frac{\theta}{2}\right)\right]\mathbf{i}_\xi + p\tan\left(\frac{\theta}{2}\right)\mathbf{i}_\eta$$

Differentiating this directly leads to $\mathbf{v}$,

$$\mathbf{v} = \sqrt{\frac{\mu}{p}}\left[-\sin\theta\,\mathbf{i}_\xi + (1 + \cos\theta)\mathbf{i}_\eta\right] \tag{7.55}$$

Noting that $1 + \cos\theta = p/r$ and using expression (7.54) permits rewriting form (7.55) as

$$\mathbf{v} = -\frac{\sqrt{\mu p}}{r}\tan\left(\frac{\theta}{2}\right)\mathbf{i}_\xi + \frac{\sqrt{\mu p}}{r}\mathbf{i}_\eta \tag{7.56}$$

Thus, position and velocity in parabolic orbits can be determined for given time past periapsis by the preceding analysis.

An analogous treatment is used to derive corresponding expressions for an elliptic orbit. Remembering that

$$r = \frac{p}{1 + e\cos\theta}$$

and noting that the position vector in terms if $\mathbf{i}_\xi$, $\mathbf{i}_\eta$, defined in Figure 7.14, is simply that given by form (7.53), and the velocity is obtained by differentiating $\mathbf{r}$,

$$\mathbf{v} = \sqrt{\frac{\mu}{p}}\left[-\sin\theta\mathbf{i}_\xi + (e + \cos\theta)\mathbf{i}_\eta\right] \tag{7.57}$$

which differs from expression (7.55) only by the eccentricity. It turns out that writing $\mathbf{r}$ and $\mathbf{v}$ in terms of eccentric anomaly, defined in Figure 2.5, permits simplification of calculations when time is given. Rewrite expression (7.53) using forms (2.23), (2.25), (2.26), and (2.29),

$$\mathbf{r} = a(\cos\psi - e)\mathbf{i}_\xi + \sqrt{ap}\sin\psi\,\mathbf{i}_\eta \tag{7.58}$$

Velocity is again obtained by differentiating $\mathbf{r}$,

$$\mathbf{v} = -\frac{\sqrt{\mu a}}{r}\sin\psi\,\mathbf{i}_\xi + \frac{\sqrt{\mu p}}{r}\cos\psi\,\mathbf{i}_\eta \tag{7.59}$$

To describe position and velocity in terms of initial values, assume that at $t = t_0$, $\mathbf{r} = \mathbf{r}_0$, $\mathbf{v} = \mathbf{v}_0$, and $\psi = \psi_0$. Then forms (7.58) and (7.59) become

$$\mathbf{r}_0 = a(\cos\psi_0 - e)\mathbf{i}_\xi + \sqrt{ap}\sin\psi_0\mathbf{i}_\eta$$

$$\mathbf{v}_0 = -\frac{\sqrt{\mu a}}{r_0}\sin\psi_0\mathbf{i}_\xi + \frac{\sqrt{\mu p}}{r_0}\cos\psi_0\mathbf{i}_\eta \tag{7.60}$$

Now unit vectors $\mathbf{i}_\xi$, $\mathbf{i}_\eta$ can be eliminated by solving for these and substituting the results into equations (7.58) and (7.59). Thus,

$$\mathbf{i}_\xi = \frac{\cos \psi_0}{r_0}\mathbf{r}_0 - \sqrt{\frac{a}{\mu}} \sin \psi_0 \mathbf{v}_0$$

$$\mathbf{i}_\eta = \sqrt{\frac{a}{p}}\frac{\sin \psi_0}{r_0}\mathbf{r}_0 + \frac{a}{\sqrt{\mu p}}(\cos \psi_0 - e)\mathbf{v}_0$$

(7.61)

and the new expressions for $\mathbf{r}$ and $\mathbf{v}$ are

$$\mathbf{r}(t) = \frac{a}{r_0}[\cos (\psi - \psi_0) - e\cos \psi_0]\mathbf{r}_0$$

$$+ \sqrt{\frac{a^3}{\mu}}[\sin (\psi - \psi_0) - e(\sin \psi - \sin \psi_0)]\mathbf{v}_0 \qquad (7.62a)$$

$$\mathbf{v}(t) = -\frac{\sqrt{\mu a}}{r r_0} \sin (\psi - \psi_0)\mathbf{r}_0 + \frac{a}{r}[\cos (\psi - \psi_0) - e \cos \psi]\mathbf{v}_0 \quad (7.62b)$$

Of course, this form requires knowledge of ψ in order to calculate $\mathbf{r}$ and $\mathbf{v}$. At a given time, ψ is obtained through Kepler's equation given as form (2.32) and expressed here as

$$\sqrt{\frac{\mu}{a^3}}(t - t_0) = (\psi - \psi_0) - e(\sin \psi - \sin \psi_0) \qquad (7.63)$$

Use of expression (2.25) permits rewriting this as

$$\sqrt{\frac{\mu}{a^3}}(t - t_0) = (\psi - \psi_0) + \frac{\mathbf{r}_0 \cdot \mathbf{v}_0}{\sqrt{\mu a}}[1 - \cos (\psi - \psi_0)]$$

$$- \left(1 - \frac{r_0}{a}\right) \sin (\psi - \psi_0) \qquad (7.64)$$

It is implied that $\mathbf{r}_0$, $\mathbf{v}_0$, and t_0 are also given. Thus, a and e are calculated by the methods of Section 3.1. Then ψ_0 is uniquely obtained from

$$e \cos \psi_0 = \left(1 - \frac{r_0}{a}\right)$$

$$e \sin \psi_0 = \frac{\mathbf{r}_0 \cdot \mathbf{v}_0}{\sqrt{\mu a}}$$

In principle, equation (7.64) is solved for ψ at time t. Several methods are available in the literature for extracting ψ from Kepler's equation. However, only the question of existence and uniqueness is addressed here. The establishment of universal position and velocity formulae will preclude the need for these methods. To test uniqueness, define *mean anomaly* as

$$M = n(t - t_v) \qquad (7.65)$$

and a function $F(\psi)$ as

$$F(\psi) = \psi - e \sin \psi - M$$

Note that Kepler's equation becomes

$$M = \psi - e \sin \psi \qquad (7.66)$$

Let M vary over half an orbit which insures generality because of symmetry, $k\pi \leq M \leq (k+1)\pi$, where $k = 0, 1, 2, 3, \ldots$ Then ψ varies over the same range and $F(\psi)$ has limits obtained by considering $F(k\pi) = k\pi - M \leq 0$, because $M \geq k\pi$. Then $F[(k+1)\pi] = (k+1)\pi - M > 0$, because $M \leq (k+1)\pi$. Therefore, $F(\psi)$ vanishes at least once in the interval and there is at least one solution. Uniqueness is tested by checking the slope of $F(\psi)$, that is,

$$\frac{dF}{d\psi} = 1 - e \cos \psi$$

This must be positive because $e < 1$ and $\cos \psi < 1$. Thus, $F(\psi)$ can be zero only once in the interval implying only one solution.

Now consider determination of position and velocity in hyperbolic orbits. The procedure used is analogous to that of elliptic orbits. However, instead of using the concept of eccentric anomaly, which is defined only for ellipses, an area is employed as the position variable. This area is based on a reference geometric shape (as is eccentric anomaly), called the *equilateral hyperbola*, depicted in Figure 7.15 with labeled points defined as A, the vertex or pericenter and C, the center of the hyperbola. The Cartesian formula for a hyperbola in terms of this nomenclature is

$$\frac{(CR)^2}{a^2} - \frac{(PR)^2}{b^2} = 1$$

Remembering the hyperbolic identity, $\cosh^2 H - \sinh^2 H = 1$, define H such that

$$(CR) = |a| \cosh H$$
$$(PR) = |b| \sinh H$$

where H turns out to be related to area CAQ by

$$\text{Area } CAQ = \frac{a^2}{2} H \qquad (7.67)$$

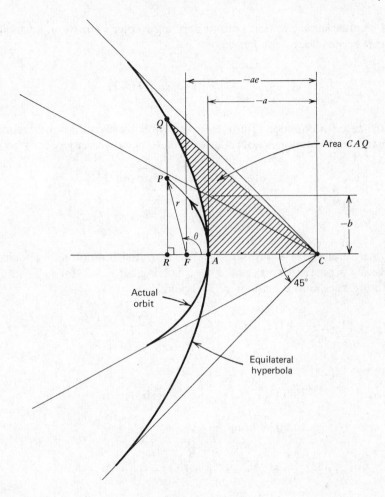

FIGURE 7.15 Definition of equilateral hyperbola.

This permits a simple expression for the hyperbola,

$$r = a(1 - e \cosh H) \qquad (7.68)$$

Several identities can be generated by equating this with the conic equation,

$$\cos \theta = \frac{e - \cosh H}{e \cosh H - 1}, \quad \cosh H = \frac{e + \cos \theta}{1 + e \cos \theta} \qquad (7.69)$$

$$\sin \theta = \frac{\sqrt{e^2 - 1} \sinh H}{e \cosh H - 1}, \quad \sinh H = \frac{\sqrt{e^2 - 1} \sin \theta}{1 + e \cos \theta}$$

In a manner similar to that used in deriving Kepler's equation, an analogous form can be developed for hyperbolic paths,

$$\sqrt{\frac{\mu}{(-a)^3}}\,(t-t_v)=e\sinh H-H \tag{7.70}$$

As is the case for equation (7.64), this is solvable by several known techniques, but again such a solution is not required. Position and velocity vectors become

$$\mathbf{r}=-a(e-\cosh H)\mathbf{i}_\xi+\sqrt{-ap}\sinh H\,\mathbf{i}_\eta$$

$$\mathbf{v}=-\frac{\sqrt{-\mu a}}{r}\sinh H\,\mathbf{i}_\xi+\frac{\sqrt{\mu p}}{r}\cosh H\,\mathbf{i}_\eta \tag{7.71}$$

These can similarly be expressed in terms of initial position and velocity. Let $H=H_0$ at $t=t_0$. Then $\mathbf{r}=\mathbf{r}_0$ and $\mathbf{v}=\mathbf{v}_0$. Solving set (7.71) for $\mathbf{i}_\xi$, $\mathbf{i}_\eta$ at t_0 and substituting back into equations (7.71) yields

$$\mathbf{r}(t)=\left\{1+\frac{a}{r_0}[\cosh(H-H_0)-1]\right\}\mathbf{r}_0+\left\{t-\frac{\sinh(H-H_0)-(H-H_0)}{\sqrt{\mu/(-a)^3}}\right\}\mathbf{v}_0$$

$$\tag{7.72}$$

$$\mathbf{v}(t)=-\frac{\sqrt{-\mu a}}{rr_0}\sinh(H-H_0)\,\mathbf{r}_0+\left\{1+\frac{a}{r}[\cosh(H-H_0)-1]\right\}\mathbf{v}_0$$

where $(H-H_0)$ is obtained from

$$\sqrt{\frac{\mu}{(-a)^3}}\,(t-t_0)=-(H-H_0)+\frac{\mathbf{r}_0\cdot\mathbf{v}_0}{\sqrt{-\mu a}}[\cosh(H-H_0)-1]$$

$$+\left(1-\frac{r_0}{a}\right)\sinh(H-H_0) \tag{7.73}$$

7.3.3 Battin's Universal Formulas

The three conic shapes have been treated individually without attempting to solve for position and velocity at a specified time for elliptic and hyperbolic orbits. R. H. Battin is credited with the development of a useful set of universal formulae for obtaining $\mathbf{r}$ and $\mathbf{v}$ which apply to all three conics without a priori knowledge of the specific conic shape. Typically, the problem is one in which initial position and velocity are given, with a determination of $\mathbf{r}$ and $\mathbf{v}$ to be made for some later time. It turns out that two transcendental functions must be introduced to make the solution work. Define these as

$$S(x) = \frac{1}{3!} - \frac{x}{5!} + \frac{x^2}{7!} - \cdots = \begin{cases} \dfrac{\sqrt{x} - \sin\sqrt{x}}{(\sqrt{x})^3} & \text{(for } x > 0\text{)} \\[2ex] \dfrac{\sinh\sqrt{-x} - \sqrt{-x}}{(\sqrt{-x})^3} & \text{(for } x < 0\text{)} \end{cases} \qquad (7.74)$$

$$C(x) = \frac{1}{2!} - \frac{x}{4!} + \frac{x^2}{6!} - \cdots = \begin{cases} \dfrac{1 - \cos\sqrt{x}}{x} & \text{(for } x > 0\text{)} \\[2ex] \dfrac{\cosh\sqrt{-x} - 1}{-x} & \text{(for } x < 0\text{)} \end{cases} \qquad (7.75)$$

These are graphically represented in Figure 7.16. Derivatives of $S(x)$ and $C(x)$

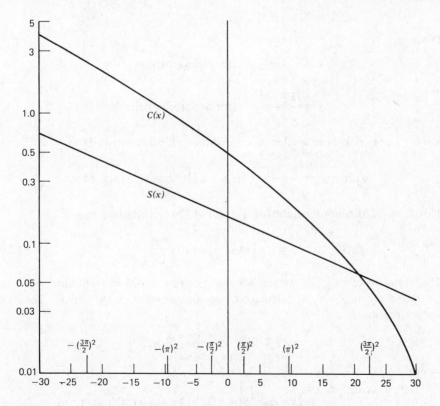

FIGURE 7.16 Transcendental functions for universal formulas.

take on the simple forms

$$\frac{dS}{dx} = \frac{1}{2x}[C(x) - 3S(x)]$$

$$\frac{dC}{dx} = \frac{1}{2x}[1 - xS(x) - 2C(x)]$$

and $S(x)$ is related to $C(x)$ by the expression

$$[1 - xS(x)]^2 = C(x)[2 - xC(x)] = 2C(4x)$$

Note that the semimajor axis can be obtained from r_0, v_0 by

$$a = \left(\frac{2}{r_0} - \frac{v_0^2}{\mu}\right)^{-1} \tag{7.76}$$

which gives the proper sign according to the conic shape. For convenience define α_0 as

$$\alpha_0 = \frac{1}{a}$$

Then set

$$x = \frac{\psi - \psi_0}{\sqrt{\alpha_0}} \qquad \text{(for elliptic orbits)}$$

$$x = \frac{H - H_0}{\sqrt{-\alpha_0}} \qquad \text{(for hyperbolic orbits)} \tag{7.77}$$

Kepler's equation is now identical for either ellipse or hyperbola,

$$\sqrt{\mu}(t - t_0) = \frac{\mathbf{r}_0 \cdot \mathbf{v}_0}{\sqrt{\mu}} x^2 C(\alpha_0 x^2) + (1 - r_0 \alpha_0) x^3 S(\alpha_0 x^2) + r_0 x \tag{7.78}$$

This form is also appropriate for a parabola by setting $\alpha_0 = 0$ and

$$x = \sqrt{p}\left[\tan\left(\frac{\theta}{2}\right) - \tan\left(\frac{\theta_0}{2}\right)\right] \tag{7.79}$$

The value of x can be extracted by a Newton iteration technique. Assume $t_0 = 0$ and try an $x = x_1$ to calculate t_1 using expression (7.78) and x_1. The next approximation is

$$x_2 = x_1 - \frac{\sqrt{\mu}\, t_1 - \sqrt{\mu}\, t}{(\sqrt{\mu}\, dt/dx)_{x = x_1}}$$

where

$$\sqrt{\mu}\,\frac{dt}{dx} = \frac{\mathbf{r}_0 \cdot \mathbf{v}_0}{\sqrt{\mu}}[x - \alpha_0 x^3 S(\alpha_0 x^2)] + (1 - r_0 \alpha_0) x^2 C(\alpha_0 x^2) + r_0$$

In general

$$x_{n+1} = x_n - \frac{\sqrt{\mu}\, t_n - \sqrt{\mu}\, t}{(\sqrt{\mu}\, dt/dx)_{x=x_n}} \qquad (7.80)$$

Once x is determined to desired accuracy, then position and velocity at time $t(t_0 = 0)$ are obtained universally from

$$\mathbf{r}(t) = \left[1 - \frac{x^2}{r_0} C(\alpha_0 x^2)\right]\mathbf{r}_0 + \left[t - \frac{x^3}{\sqrt{\mu}} S(\alpha_0 x^2)\right]\mathbf{v}_0$$

$$\mathbf{v}(t) = \frac{\sqrt{\mu}}{rr_0}[\alpha_0 x^3 S(\alpha_0 x^2) - x]\mathbf{r}_0 + \left[1 - \frac{x^2}{r} C(\alpha_0 x^2)\right]\mathbf{v}_0 \qquad (7.81)$$

The preceding development can be formulated as a computer program. Given the initial position and velocity at $t = 0$ as $\mathbf{r}_0$, $\mathbf{v}_0$ and some desired time t, the following sequence will yield $\mathbf{r}(t)$ and $\mathbf{v}(t)$:

1. Compute α_0 using equation (7.76).
2. Determine x by the iterative technique given by expression (7.80).
3. Compute $S(\alpha_0 x^2)$ and $C(\alpha_0 x^2)$ using forms (7.74) and (7.75), respectively.
4. Compute $\mathbf{r}(t)$, $\mathbf{v}(t)$ from equations (7.81).

In principle, the conic shape does not have to be known, but α_0 gives this in any case. Thus, if a computer is used to obtain $\mathbf{r}(t)$, $\mathbf{v}(t)$, it should also present α_0 as an information item.

To illustrate application of the universal formulas, consider the example depicted in Figure 7.17. An earth satellite is observed to have position and velocity at some

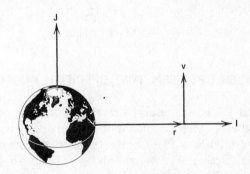

FIGURE 7.17 Initial situation for universal formulas example.

TABLE 7.2   Iteration to Determine x

Iteration i	x_i	$S(\alpha_0 x_i^2)$	$C(\alpha_0 x_i^2)$	$\sqrt{\mu}\, t_i$	$\left(\sqrt{\mu}\,\dfrac{dt}{dx}\right)_{x=x_i}$
1	1500.0	0.628	3.46	2.40×10^9	8.77×10^6
2	1227.4	0.411	1.95	8.67×10^8	3.32×10^6
3	968.0	0.294	1.21	3.10×10^8	1.29×10^6
4	732.4	0.231	0.845	1.09×10^8	5.19×10^5
5	534.6	0.199	0.665	3.95×10^7	2.24×10^5
6	388.7	0.183	0.583	1.60×10^7	1.09×10^5
7	304.8	0.177	0.550	8.66×10^6	6.74×10^4
8	277.4	0.175	0.541	6.97×10^6	5.68×10^4
9	274.8	0.175	0.540	6.82×10^6	5.58×10^4
10	274.8	0.175	0.540	6.82×10^6	5.58×10^4

reference point of

$$\mathbf{r} = 10^4\, \mathbf{I}\ \text{km}$$

$$\mathbf{v} = 9.2\, \mathbf{J}\ \text{km/s}$$

where $\mathbf{I}$ and $\mathbf{J}$ are the unit vectors of an inertial frame. The mission controllers want to determine the position and velocity three hours later. Assuming $t_0 = 0$, then $t - t_0 = 10{,}800$ s. Following the preceding sequence gives:

1. $\alpha_0 = -1.234 \times 10^{-5}\ \text{km}^{-1}$ (hyperbolic)
2. Guessing an initial value of $x_1 = 1500$ produced the iterative sequence listed in Table 7.2. By the tenth step it is apparent that $x = 274.8$.
3. As a by-product of the iteration for x, the values of $S(\alpha_0 x^2)$ and $C(\alpha_0 x^2)$ are 0.175 and 0.540, respectively.
4. Position and velocity at $t = 10{,}800$ s are

$$\mathbf{r} = -30{,}778\, \mathbf{I} + 46{,}442\, \mathbf{J}\ (\text{km})$$

$$\mathbf{v} = -3.62\, \mathbf{I} + 2.47\, \mathbf{J}\ (\text{km/s})$$

7.4   TRAJECTORIES BETWEEN TWO SPECIFIED POINTS

Sections 3.1 and 7.3 have presented methods for determining position and velocity of a satellite at some specified time for a given set of initial conditions. The problem of establishing flight paths between two given points about a central body of attraction is now considered. Flights to the outer planets present situations in which the heliocentric end points are restricted by any of several physical or economic constraints. Earth launch date is always subject to

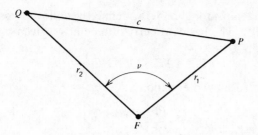

FIGURE 7.18 Relationship between end points and attracting center.

funding and equipment pressures. Planetary alignment for scientific data collection' and multiply flybys is another factor. Therefore, the solution of this problem has practical implications for planetary mission planners.

7.4.1 Survey of Possible Flight Paths

Begin the analysis by considering the geometric relationship between end points and the central body. Referring to Figure 7.18, points P and Q are the initial and final locations of interest about attractive center F. To illustrate the possible ways to connect P and Q, let $e = 1$ for a parabolic orbit. Then there are two possibilities as depicted in Figure 7.19. Both trajectories have the same energy ($\mathscr{E} = 0$), but each represents a different transfer time.

Consider all possible paths from P to Q, where $r_1 < r_2$. No loss of generality is involved due to this restriction, because transfer times and other critical parameters are the same in both directions for a given flight path. Elliptic transfers are reviewed first. Note that only one focus and two points are specified. If the vacant focus F^* is given, then only one path is possible. Otherwise, there are an infinite number of vacant foci which are permissible. It is the locus of vacant foci that establishes satisfactory elliptic orbits between P

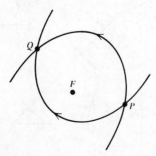

FIGURE 7.19 Two parabolic paths between P and Q.

and Q. Referring again to Figure 7.18, define the following constants: $r_1 = FP$, $r_2 = FQ$, and $c = PQ$. The basic property of an ellipse, given by form (7.46a), is rewritten here as

$$PF^* + PF = QF^* + QF = 2a$$

This is reconstructed as two equations

$$\begin{aligned} PF^* &= 2a - r_1 \\ QF^* &= 2a - r_2 \end{aligned} \tag{7.82}$$

As illustrated in Figure 7.20, F^* is the intersection of two circles of radius $2a - r_1$ and $2a - r_2$. By varying a the locus of F^* can be generated, and several observations can be made:

(a) There is a minimum value of a, a_m, for which the corresponding circles just touch. If $a < a_m$, no elliptic path is possible, that is, the energy is too low for an elliptic orbit to reach both P and Q. Thus, a_m corresponds to the lowest possible energy path, and this energy is

$$\mathscr{E}_m = -\frac{\mu}{2a_m} \tag{7.83}$$

and the value of a_m is obtained by observing that

$$(2a_m - r_2) + (2a_m - r_1) = c$$

which yields

$$2a_m = \frac{1}{2}(r_1 + r_2 + c) \tag{7.84}$$

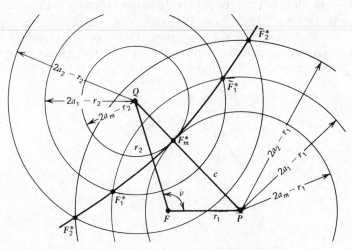

FIGURE 7.20 Locus of vacant foci for elliptic transfers.

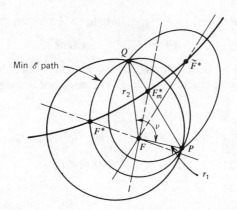

FIGURE 7.21 Minimum and nonminimum energy elliptic paths.

Introduce a new constant s as

$$2s = r_1 + r_2 + c \tag{7.85}$$

which permits form (7.84) to become

$$2a_m = s \tag{7.86}$$

The vacant focus corresponding to a_m is F_m^* and is located on c such that

$$PF_m^* = s - r_1$$
$$QF_m^* = s - r_2$$

(b) If $a > a_m$ there are two possible vacant foci, referred to as F^*, $\tilde{F}^*$. Each of these corresponds to a different elliptic path connecting P and Q, but both paths have the same a. The vacant foci are equidistant from line PQ. Refer to Figure 7.21 which shows a minimum energy path and a pair of equal energy ellipses for $a > a_m$. Since $FF^* = 2ae$ and $F\tilde{F}^* = 2ae$, the fact that a is fixed for this pair leads to $e < \tilde{e}$, because $FF^* < F\tilde{F}^*$ by the selection of labels. Therefore, the ellipse corresponding to $\tilde{F}^*$ has a greater eccentricity and smaller p [i.e., $p = a(1 - e^2)$] than its associated ellipse defined by F^*.

(c) The shape of the locus of vacant foci can be determined by observing that set (7.82) can be rewritten as

$$-r_1 = PF^* - 2a$$
$$r_2 = -(QF^* - 2a)$$

Then add to get

$$r_2 - r_1 = PF^* - QF^*$$

Noting that r_1 and r_2 are constant for a given situation leads to

$$PF^* - QF^* = \text{constant} \tag{7.87}$$

comparison with definition (7.46b) indicates that the locus of F^* is a hyperbola with P and Q as foci. The constant is the length of the major axis, i.e.,

$$r_2 - r_1 = -2a_H \tag{7.88}$$

Eccentricity of this locus is obtained by noting that

$$c = -2a_H e_H$$

Thus

$$e_H = \frac{c}{r_2 - r_1} \tag{7.89}$$

and the asymptotes of this hyperbola have slopes of $\pm (b/a)$, where

$$\frac{b}{a} = \sqrt{e^2 - 1} = \left[\frac{c^2}{(r_2 - r_1)^2} - 1 \right]^{1/2}$$

application of the cosine law permits this to be rewritten as

$$\text{slopes} = \pm \frac{2\sqrt{r_1 r_2}}{r_2 - r_1} \sin \left(\frac{\nu}{2} \right) \tag{7.90}$$

Thus, asymptotes are effected by the angle between FP and FQ.

Since the minimum energy elliptic transfer is of particular interest, it will now be reviewed in detail. Begin by noticing that potential energy depends only on the distance from F. Thus, the minimum energy transfer corresponds to having minimum kinetic energy at P and Q. Since P could be any point along this path, kinetic energy is minimum all along the path. To obtain parameter p_m and eccentricity e_m for the minimum energy ellipse, use

$$FF_m^* = 2a_m e_m$$

Referring to Figure 7.22,

$$(FF_m^*)^2 = [(s - r_2) \sin \alpha]^2 + [r_2 - (s - r_2) \cos \alpha]^2$$

From the cosine law and the definition of s, obtain

$$\cos \alpha = \frac{2s(s - r_1)}{r_2 c} - 1$$

Finally, combining results leads to

$$(2e_m a_m)^2 = s^2 - \frac{4s}{c}(s - r_1)(s - r_2) \tag{7.91}$$

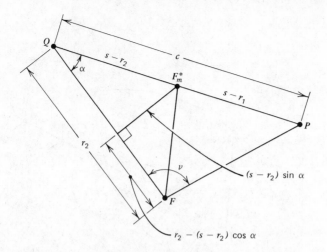

FIGURE 7.22 Geometry of minimum energy ellipse.

Further reduction is possible by using $p = a(1 - e^2)$, written as

$$(2ea)^2 = 4a(a - p)$$

Using appropriate subscripts this becomes

$$(2e_m a_m)^2 = 4\frac{s}{2}\left(\frac{s}{2} - p_m\right)$$

where form (7.86) is applied. Thus,

$$(2e_m a_m)^2 = s^2 - 2sp_m$$

Using this with result (7.91) yields

$$p_m = \frac{2}{c}(s - r_1)(s - r_2) = \frac{r_1 r_2}{c}(1 - \cos \nu) \qquad (7.92)$$

It is easy to obtain e_m from p_m,

$$e_m{}^2 = 1 - \frac{2p_m}{s} \qquad (7.93)$$

Next consider possible hyperbolic paths between P and Q. Since F is the focus of such hyperbolas, P and Q must lie on the branch which is concave relative to F. Nomenclature for this situation is illustrated in Figure 7.23. The path must satisfy geometric definition (7.46b), expressed here as

$$PF^* - r_1 = QF^* - r_2 = -2a$$

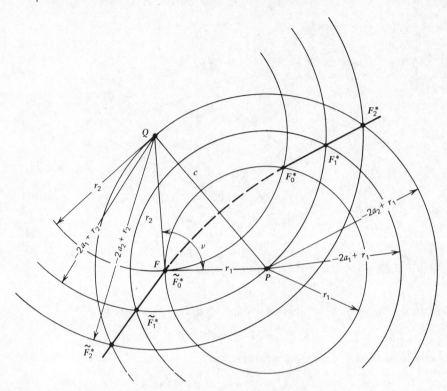

FIGURE 7.23 Locus of vacant foci for hyperbolic transfers.

Possible locations of F^* are again the intersections of two circles for each value of a. When the vacant focus is at F_0^*, P and Q are joined by a straight line path, corresponding to $a = 0$, $\mathscr{E} = \infty$. Important observations can be made from this figure:

(a) All vacant foci F^* for concave paths fall outside the circles of radius r_1 about P and radius r_2 about Q. For each value of $-a > 0$ there are two possible solutions as illustrated in Figure 7.24. Note that $e > \tilde{e}$, $p > \tilde{p}$.

(b) The locus of vacant foci for hyperbolic paths is the conjugate branch of the locus for elliptic paths, because

$$-(PF^* - QF^*) = r_2 - r_1 \qquad (7.94)$$

which is the negative of form (7.87).

Finally consider parabolic paths between P and Q. As noted previously, only two possibilities exist since parabolas have only one focus and one value of a.

Since this geometric shape is defined as the locus of points equidistant from F and a straight line, construct two circles of radius r_1 and r_2 centered at P and Q, respectively. The two possible directrices must be tangent to both circles simultaneously, as shown in Figure 7.25. The axes of possible parabolas are normal to these directrices. It is not difficult to see that these axes are parallel to the asymptotes of the loci of the vacant foci for elliptic and hyperbolic transfers from P to Q. The slopes were already obtained as expression (7.90) and it turns out that

$$p_1 = \frac{4(s-r_1)(s-r_2)}{c^2}\left(\sqrt{\frac{s}{2}} + \sqrt{\frac{s-c}{2}}\right)^2 \tag{7.95}$$

$$p_2 = \frac{4(s-r_1)(s-r_2)}{c^2}\left(\sqrt{\frac{s}{2}} - \sqrt{\frac{s-c}{2}}\right)^2 \tag{7.96}$$

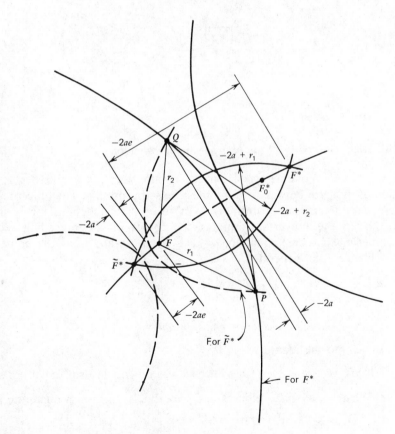

FIGURE 7.24 Two hyperbolic paths for one value of a.

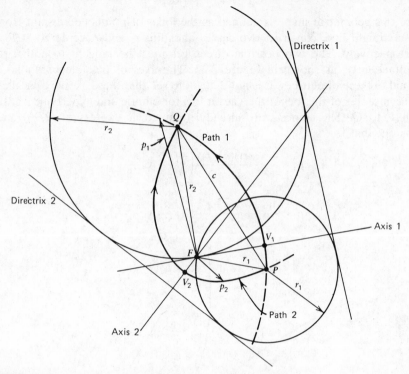

FIGURE 7.25 Geometry of the two parabolic transfers.

Now that each possible type of transfer path has been analyzed a general solution of the path between two points seems appropriate. The objective is to express parameter and eccentricity in a form independent of orbit shape, analogous to the universal formulas for conics of the preceding section. Remembering the conic equation, $r = p/(1 + e \cos \theta)$, and using the nomenclature of Figure 7.26, generate

$$e \cos \theta_1 = \frac{p}{r_1} - 1$$

and

$$e \cos \theta_2 = e \cos (\theta_1 + \nu) = \frac{p}{r_2} - 1$$

Insert these into the identity

$$\cos^2 (\theta_1 + \nu) - 2 \cos (\theta_1 + \nu) \cos \theta_1 \cos \nu + \cos^2 \theta_1 - \sin^2 \nu = 0$$

and employ $ae^2 = a - p$ for ellipses and hyperbolas, to form a quadratic in p,

$$ac^2 p^2 - r_1 r_2 (1 - \cos \nu)[2a(r_1 + r_2) - r_1 r_2 (1 + \cos \nu)]p$$

$$+ a r_1^2 r_2^2 (1 - \cos \nu)^2 = 0 \qquad (7.97)$$

where the first coefficient was obtained by applying the cosine law. Remembering that $2s = r_1 + r_2 + c$, the term in brackets becomes

$$[2a(r_1+r_2)-r_1r_2(1+\cos v)] = -2s(s-c-2a)-2ac \qquad (7.98)$$

At this point the two types of paths must be treated separately to arrive at an expression for p. Start with the elliptic transfer by defining α and β such that

$$\sin\left(\frac{\alpha}{2}\right) = \sqrt{\frac{r_1+r_2+c}{4a}} = \sqrt{\frac{s}{2a}}$$

$$\sin\left(\frac{\beta}{2}\right) = \sqrt{\frac{r_1+r_2-c}{4a}} = \sqrt{\frac{s-c}{2a}} \qquad (7.99)$$

These lead to the following identities

$$s-c-2a = -2a\cos^2\left(\frac{\beta}{2}\right)$$

$$s = 2a\sin^2\left(\frac{\alpha}{2}\right) \qquad (7.100)$$

$$c = 2a\left[\sin^2\left(\frac{\alpha}{2}\right)-\sin^2\left(\frac{\beta}{2}\right)\right]$$

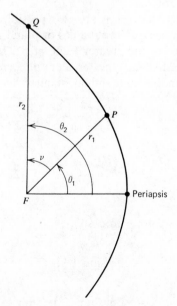

FIGURE 7.26 Nomenclature for true anomaly of P and Q.

which restructure expression (7.98) as

$$-2s(s-c-2a)-2ac=2a^2\left[\sin^2\left(\frac{\alpha+\beta}{2}\right)+\sin^2\left(\frac{\alpha-\beta}{2}\right)\right]$$

Before placing this into equation (7.97), note that

$$r_1 r_2(1-\cos \nu)=2(s-r_1)(s-r_2) \qquad (7.101)$$

Then the quadratic in p becomes

$$ac^2 p^2 - 4a^2(s-r_1)(s-r_2)\left[\sin^2\left(\frac{\alpha+\beta}{2}\right)+\sin^2\left(\frac{\alpha-\beta}{2}\right)\right]p$$
$$+4a(s-r_1)^2(s-r_2)^2=0 \qquad (7.102)$$

Using the last identity in set (7.100) and multiplying form (7.102) by c^2/a gives

$$c^4 p^2 - 4a(s-r_1)(s-r_2)\left[\sin^2\left(\frac{\alpha+\beta}{2}\right)+\sin^2\left(\frac{\alpha-\beta}{2}\right)\right]c^2 p$$
$$+16a^2(s-r_1)^2(s-r_2)^2 \sin^2\left(\frac{\alpha+\beta}{2}\right)\sin^2\left(\frac{\alpha-\beta}{2}\right)=0$$

with roots

$$p=\frac{4a(s-r_1)(s-r_2)}{c^2}\sin^2\left(\frac{\alpha\pm\beta}{2}\right) \qquad (7.103)$$

which reconfirms the fact that there are two possible elliptic paths for each value of $a>a_m$, but each has different values of p and e. The upper sign (+) in result (7.103) corresponds to the greater value of p. The two possibilities were illustrated in Figure 7.21, with the greater p corresponding to F^*. The hyperbolic transfer is handled in a similar manner. Define γ and δ such that

$$\sinh\left(\frac{\gamma}{2}\right)=\sqrt{\frac{r_1+r_2+c}{-4a}}=\sqrt{\frac{s}{-2a}}$$

$$\sinh\left(\frac{\delta}{2}\right)=\sqrt{\frac{r_1+r_2-c}{-4a}}=\sqrt{\frac{s-c}{-2a}} \qquad (7.104)$$

yielding the identities

$$s-c-2a=-2a\cosh^2\left(\frac{\delta}{2}\right)$$

$$s=-2a\sinh^2\left(\frac{\gamma}{2}\right) \qquad (7.105)$$

$$c=-2a\left[\sinh^2\left(\frac{\gamma}{2}\right)-\sinh^2\left(\frac{\delta}{2}\right)\right]$$

Thus, for hyperbolic transfer expression (7.98) becomes

$$-2s(s-c-2a)-2ac = -2a^2\left[\sinh^2\left(\frac{\gamma+\delta}{2}\right)+\sinh^2\left(\frac{\gamma-\delta}{2}\right)\right]$$

Using equation (7.101) and the last identity in set (7.105) leads to the roots of quadratic (7.97)

$$p = \frac{-4a(s-r_1)(s-r_2)}{c^2}\sinh^2\left(\frac{\gamma\pm\delta}{2}\right) \tag{7.106}$$

which also confirms the two possible hyperbolic paths for given a. The (+) sign corresponds to F^* in Figure 7.24.

7.4.2 Lambert's Time-of-Flight Theorem

A very important property of the flight from P to Q is the time. In fact, the flight time must be related to the type of path taken. Lambert discovered that this time depends only on the value of a for a given situation. He postulated that traverse time is a function of major axis, the sum of distances between center of force and end points, and the length of the chord between these end points, which in functional form is

$$t = t(a, r_1 + r_2, c)$$

Thus, time is independent of p and e. Lagrange later proved this theorem analytically, and its validity is assumed here. Analytical formulation of flight time is possible by noting that since P and Q are fixed, the time of flight is also fixed as long as $r_1 + r_2 = $ constant and a is constant. This implies that F and F^* may be moved around as long as $(r_1 + r_2)$ and a remain unchanged. Thus, the locus of F is $r_1 + r_2 = $ constant. (i.e., an ellipse with foci at P and Q). This is illustrated in Figure 7.27. The major axis of this ellipse is $(r_1 + r_2)$. Now the locus of F^* is $PF^* + QF^* = 4a - (r_1 + r_2)$ which is a constant. This locus is also an ellipse with foci at P and Q and major axis $[4a - (r_1 + r_2)]$. Figure 7.27 shows this locus and possible corresponding pairs of F and F^*. A limiting case occurs when F and F^* are at extremities of their respective ellipses as represented by F_2, F_2^*. This corresponds to rectilinear motion between P and Q. In general, the relative positions of F and F^* on their respective ellipses must satisfy

$$QF^* + QF = PF^* + PF = 2a$$

Referring back to Figure 7.20, if F^* is on the lower branch of its locus when F, P, and Q are all fixed, then the arched triangle PQF, shown in Figure 7.28 does not contain F^* when $\nu < 180°$. If F^* were on the upper branch of its locus, then the corresponding triangle would contain F^* when $\nu < 180°$.

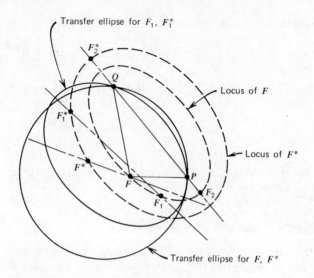

FIGURE 7.27   Geometry related to Lambert's theorem.

The preceding geometrical innovation permits a straightforward calculation of flight time. For rectilinear motion velocity magnitude is simply

$$v = \frac{dr}{dt}$$

and energy gives

$$v^2 = \mu\left(\frac{2}{r} - \frac{1}{a}\right)$$

Thus, time is extracted from

$$dt = \frac{1}{\sqrt{\mu}} \frac{r\,dr}{\sqrt{2r - r^2/a}} \tag{7.107}$$

The geometry of this situation dictates

$$QF_2 + PF_2 = r_1 + r_2$$
$$QF_2 - PF_2 = c$$

Take the difference and the sum of these equations to extract

$$PF_2 = s - c$$
$$QF_2 = s$$

which are the end points of the time integration of expression (7.107),

$$t = \frac{1}{\sqrt{\mu}} \int_{s-c}^{s} \frac{r\,dr}{\sqrt{2r - r^2/a}} \tag{7.108}$$

This integral is valid for any type of conic transfer between P and Q, even though the argument of rectilinear motion was developed for an elliptic path. F_2 and F_2^* of Figure 7.27 can similarly be associated with parabolic and hyperbolic transfers.

In order to perform integration (7.108) each type of conic path must be considered separately. For an ellipse use α and β as defined in set (7.99). Then the limits of integration become

$$s - c = a(1 - \cos \beta)$$

$$s = a(1 - \cos \alpha)$$

Since r varies between PF_2 and QF_2 which are both less than $2a$, a more convenient variable of integration is introduced by the definition

$$r = a(1 - \cos \sigma)$$

with differential form

$$dr = a \sin \sigma \, d\sigma$$

The limits of integration take on simple forms,

$$\text{at} \quad r = s - c, \sigma = \beta$$

$$\text{at} \quad r = s, \qquad \sigma = \alpha$$

which brings expression (7.108) to

$$t = \frac{1}{\sqrt{\mu}} \int_\beta^\alpha \frac{a(1 - \cos \sigma)(a \sin \sigma \, d\sigma)}{\left[2a(1 - \cos \sigma) - \dfrac{a^2(1 - \cos \sigma)^2}{a}\right]^{1/2}}$$

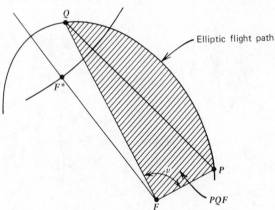

FIGURE 7.28 Vacant focus outside arched triangle.

The denominator reduces to $\sqrt{a}\sin\sigma$, and flight time is simply

$$t = \sqrt{\frac{a^3}{\mu}} \int_\beta^\alpha (1-\cos\sigma)\,d\sigma \tag{7.109}$$

which yields

$$t = \frac{\tau}{2\pi}[(\alpha-\sin\alpha)-(\beta-\sin\beta)] \tag{7.110}$$

Referring again to Figure 7.27 notice that F_2^* came from the lower branch of the locus of F^*. Thus, the traverse from P to Q is direct and the apoapsis is not encountered enroute. This implies that the time given by expression (7.110) is less than that associated with F^* from the upper branch. Consider the situation in Figure 7.29. Here F^* is moved from the upper branch of its locus to become $\tilde{F}_2^*$ with F_2 at the same position as before. The associated rectilinear motion is from P to $\tilde{F}_2^*$ to Q. Therefore, the corresponding traverse time from P to Q is $t+2\times$(time from Q to $\tilde{F}_2^*$), or

$$\tilde{t} = t + \frac{2}{\sqrt{\mu}} \int_s^{2a} \frac{r\,dr}{\sqrt{2r-r^2/a}} = t + 2\sqrt{\frac{a^3}{\mu}} \int_\alpha^\pi (1-\cos\sigma)\,d\sigma$$

which integrates to

$$\tilde{t} = \tau - \frac{\tau}{2\pi}[(\alpha-\sin\alpha)+(\beta-\sin\beta)] \tag{7.111}$$

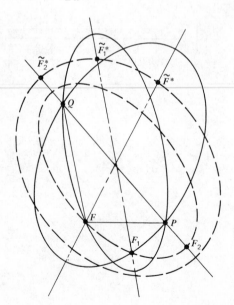

FIGURE 7.29 Geometry for traverse time when apoapsis is encountered.

In summary, for elliptic transfers between P and Q, the time required is given by equation (7.110) or (7.111), depending on which vacant focus is selected. Note that the minimum energy transfer path corresponds to $s = 2a_m$, $\alpha_m = \pi$, and $\tau_m = \pi(s^3/2\mu)^{1/2}$. Since there is only one vacant focus for this special case,

$$t_m = \tilde{t}_m = \frac{\tau_m}{2\pi}[\pi - (\beta_m - \sin \beta_m)]$$

where $\sin (\beta_m/2) = [(s-c)/s]^{1/2}$.

Traverse time for hyperbolic paths may be determined in an analogous manner. Equation (7.108) is integrated using γ, δ as defined in set (7.104). The variable of integration is here defined by

$$r = a(1 - \cosh \Sigma)$$

and the short traverse time becomes

$$t = \sqrt{\frac{-a^3}{\mu}}[(\sinh \gamma - \gamma) - (\sinh \delta - \delta)] \tag{7.112}$$

which corresponds to selecting F^* from the upper branch. If F^* were selected from the lower branch, then the path would have a traverse time of

$$\tilde{t} = t + \frac{2}{\sqrt{\mu}} \int_o^{s-c} \frac{r\,dr}{\sqrt{2r - r^2/a}}$$

or

$$\tilde{t} = \sqrt{\frac{-a^3}{\mu}}[(\sinh \gamma - \gamma) + (\sinh \delta - \delta)] \tag{7.113}$$

Equations (7.112) and (7.113) give the possible hyperbolic transfer times from P to Q for a given value of a. These expressions can be used to obtain the traverse time for parabolic passage, since the parabolic case can be treated as the result of $|a| \to \infty$. Simply take the limit of equations (7.112) and (7.113) to obtain the two possible passage times corresponding to two paths in Figure 7.25,

$$t_1 = \lim_{-a \to \infty} t = \frac{1}{3}\sqrt{\frac{2}{\mu}}[s^{3/2} - (s - c)^{3/2}] \tag{7.114}$$

$$t_2 = \lim_{-a \to \infty} \tilde{t} = \frac{1}{3}\sqrt{\frac{2}{\mu}}[s^{3/2} + (s - c)^{3/2}] \tag{7.115}$$

These expressions are similarly obtainable from equations (7.110) and (7.111) by taking the limit as $a \to \infty$.

Developments of preceding paragraphs permit a summarization of all possible paths between P and Q through classification according to shape and value of ν. Figure 7.30 illustrates these eight cases and labels them as $E1$, $E2$, $E3$, $E4$, $P1$, $P2$, $H1$, and $H2$. These are all distinguishable and four represent elliptic transfers. The four cases $E1$, $E2$, $P1$, and $H1$, correspond to values of $\nu < 180°$, with $E1$ and $E2$ separated because the arched triangle PQF may or may not include F^*. Similarly, $E3$ and $E4$ both correspond to $\nu > 180°$, and they differ by the location of F^*. Flight time and parameter for each case are listed as follows,

Case $E1$: (7.110) and (7.103)

$$t = \frac{\tau}{2\pi}[(\alpha - \sin \alpha) - (\beta - \sin \beta)], \qquad p = \frac{4a(s - r_1)(s - r_2)}{c^2} \sin^2\left(\frac{\alpha + \beta}{2}\right)$$

Case $E2$: (7.111) and (7.103)

$$t = \tau - \frac{\tau}{2\pi}[(\alpha - \sin \alpha) + (\beta - \sin \beta)], \qquad p = \frac{4a(s - r_1)(s - r_2)}{c^2} \sin^2\left(\frac{\alpha - \beta}{2}\right)$$

Case $P1$: (7.114) and (7.95)

$$t = \frac{1}{3}\sqrt{\frac{2}{\mu}}[s^{3/2} - (s - c)^{3/2}], \qquad p = \frac{4(s - r_1)(s - r_2)}{c^2}\left[\sqrt{\frac{s}{2}} + \sqrt{\frac{s - c}{2}}\right]^2$$

Case $H1$: (7.112) and (7.106)

$$t = \sqrt{\frac{-a^3}{\mu}}[(\sinh \gamma - \gamma) - (\sinh \delta - \delta)], \qquad p = \frac{-4a(s - r_1)(s - r_2)}{c^2} \sinh^2\left(\frac{\gamma + \delta}{2}\right)$$

Case $E3$: $\{\tau - (7.110)\}$ and (7.103)

$$t = \tau - \frac{\tau}{2\pi}[(\alpha - \sin \alpha) - (\beta - \sin \beta)], \qquad p = \frac{4a(s - r_1)(s - r_2)}{c^2} \sin^2\left(\frac{\alpha + \beta}{2}\right)$$

Case $E4$: $\{\tau - (7.111)\}$ and (7.103)

$$t = \frac{\tau}{2\pi}[(\alpha - \sin \alpha) + (\beta - \sin \beta)], \qquad p = \frac{4a(s - r_1)(s - r_2)}{c^2} \sin^2\left(\frac{\alpha - \beta}{2}\right)$$

Case $P2$: (7.115) and (7.96)

$$t = \frac{1}{3}\sqrt{\frac{2}{\mu}}[s^{3/2} + (s - c)^{3/2}], \qquad p = \frac{4(s - r_1)(s - r_2)}{c^2}\left[\sqrt{\frac{s}{2}} - \sqrt{\frac{s - c}{2}}\right]^2$$

Case $H2$: (7.113) and (7.106)

$$t = \sqrt{\frac{-a^3}{\mu}}[(\sinh \gamma - \gamma) + (\sinh \delta - \delta)], \qquad p = \frac{-4a(s - r_1)(s - r_2)}{c^2} \sinh^2\left(\frac{\gamma - \delta}{2}\right)$$

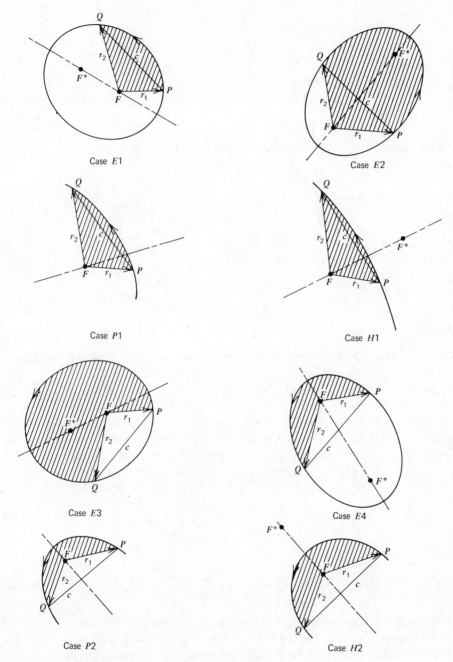

FIGURE 7.30 Classification of all conic transfers.

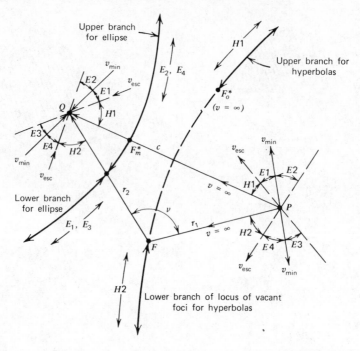

FIGURE 7.31 Departure and arrival directions for transfers between P and Q.

Figure 7.31 depicts all possible departure directions from P and arrival directions at Q for the cases shown in Figure 7.30. Notice that there is a zone in which no path may enter. This is the area bounded by triangle FPQ, the sides of which correspond to infinite velocity. All loci of vacant foci are also shown, and portions labeled according to the associated cases. The velocity vector at P is $\mathbf{v}_1$ and can be obtained by writing position vector $\mathbf{r}_2$ in terms of initial position and velocity. Referring to Section 7.3.2, this becomes

$$\mathbf{r}_2 = \left[1 - \frac{r_2}{p} (1 - \cos \nu) \right] \mathbf{r}_1 + \frac{r_1 r_2}{\sqrt{\mu p}} \sin \nu \, \mathbf{v}_1$$

and is easily solved for $\mathbf{v}_1$,

$$\mathbf{v}_1 = \frac{\sqrt{\mu p}}{r_1 r_2 \sin \nu} \left\{ \mathbf{r}_2 - \left[1 - \frac{r_2}{p} (1 - \cos \nu) \right] \mathbf{r}_1 \right\} \tag{7.116}$$

Typically, the flight time is specified because of mission constraints. Solution of the time equations for appropriate values of a is difficult. Although there are techniques for extracting these values, it is sufficient to present performance in a graphical form. Figure 7.32 does this in a nondimensionalized manner.

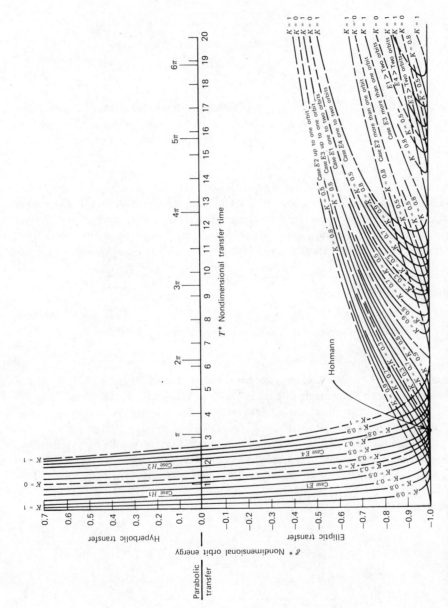

FIGURE 7.32 Summary of possible transfers between two points.

Parameters are defined as follows:

$$\mathscr{E}^* = \frac{\mathscr{E}}{|\mathscr{E}_{min}|} = -\frac{a_{min}}{a}$$

$$T^* = n_{max}t$$

$$K = 1 - \frac{c}{s}$$

where $n_{max} = [\mu/a_{min}^3]^{1/2}$ and $a_{min} = s/2$. For a given situation and value of flight time, T^* is calculated. Then a number of transfers may be possible as is true for a value of $T^* = 3\pi$. The utility of this plot can be demonstrated by illustration.

As an example, consider a flight from earth to Venus. The spacecraft is to arrive 280 days after leaving earth, with a heliocentric longitude change of 48° in the direction of earth motion. Ignoring effects of planetary attraction on flight time, transfer path options may be determined by using Figure 7.32. The situation is depicted in Figure 7.33 with the transfer taken from Venus to Earth because $r_{\venus} < r_{\oplus}$. Numerical values are

$$r_1 = r_{\venus} = 1.083 \times 10^8 \text{ km}, \ r_2 = r_{\oplus} = 1.496 \times 10^8 \text{ km}, \ \nu = 48°$$

Calculated quantities become

$$c = 1.115 \times 10^8 \text{ km}, \ s = 1.847 \times 10^8 \text{ km}, \ K = 0.396$$

$$a_m = 9.234 \times 10^7 \cdot \text{km}, \ n_{max} = 4.105 \times 10^{-7} \text{ rad/s}$$

$$T^* = 9.93$$

These values of T^* and K lead to six possible transfers from Figure 7.32. Traverses of less than one orbit correspond to case $E2$ with $a = 1.42 \times 10^8$ km and case $E3$ with $a = 1.37 \times 10^8$ km. The other possibilities correspond to traverses of one to two orbits and have values of $a = 1.11 \times 10^8$ km for case $E1$, $a = 1.01 \times 10^8$ km for E_4, $a = 9.42 \times 10^7$ km for $E2$, and $a = 9.25 \times 10^7$ km for $E3$. Each possibility represents a different launch time and velocity increment. Consider the case in which $a = 9.42 \times 10^7$ km. The heliocentric velocity leaving earth is obtained from equation (7.116)

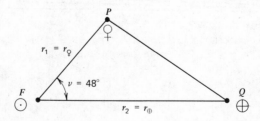

FIGURE 7.33 Example of Earth–Venus transfer.

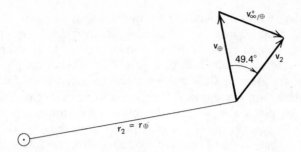

FIGURE 7.34 Earth departure for Venus transfer.

by interchanging subscripts 1 and 2. Values obtained are $\alpha = 163.9°$, $\beta = 77.1°$, $p = 3.832 \times 10^7$ km, and $\mathbf{v}_2 = 1.87 \times 10^{-7} [\mathbf{r}_1 - (0.0649)\mathbf{r}_2]$. Thus, the magnitude is $v_2 = 21.54$ km/s at an angle of 49.4° away from $\mathbf{v}_\oplus$, illustrated in Figure 7.34. Finally, the value for velocity $v_{\infty/\oplus}$ is 22.70 km/s, which requires $\Delta v \cong 17.4$ km/s to escape from a low parking orbit. Refer to Section 3.4 for escape trajectory relationships.

7.5 OBSERVATIONAL PROBLEMS OF ORBIT DETERMINATION

The process of orbit determination has been limited here to obtaining first approximations of the fundamental orbital elements and satellite position and velocity at a given time in an inertial coordinate frame. Many other methods of solving the two-body equation appear in the literature, and the interested reader should refer to those. Nevertheless, some of the observational problems associated with orbit determination are offered as an introduction to operational aspects of space vehicle tracking.

7.5.1 Time Measurement

The measurement of time in determining orbital elements accurately is critical. Consider an observer on the earth at a given *meridian* (a north-south line normal to the equator). He looks at a clock and notes a specific number. This can be interpreted as the distance between the observer and the reference meridan at the Greenwich Observatory. The earth is divided into 24 time zones, each equivalent to one hour of rotation past an inertial reference direction. In actuality, the observer measures local *mean solar time*: that is, mean time at a particular meridan is defined as 12 hours plus the hour angle. This allows measurement from midnight. Thus, the hour angle is measured to the west of a given meridian and gives the time since noon passage of a fictitious sun that has the same period as the true sun, but moves at a constant speed along the equatorial plane. If an observer reads h hours on an accurate clock and he is n time zones west of Greenwich, then the *Greenwich mean time* is $h + n$, which is also referred to as *universal time*, U.T. Astronomers use another time reference called, *Julian date*, J.D. This is simply a continuing count of each day elapsed since the arbitrarily selected date, January 1, 4713 B.C. Each Julian date is measured from noon to noon so that astronomers do not have to change dates during the night observations. Julian dates for corresponding Gregorian dates are obtainable from the *American Ephemeris and Nautical Almanac*. For example, the Julian date for July 4, 1976 at 8:00 A.M. is 2442963.83333.

A satellite observer must relate his position to an inertial direction. Thus, his meridian is only related to the *prime* or Greenwich meridian. An inertial system of the type shown in Figure 7.35 locates object B by using right ascension α and declination δ. Now a relationship is needed between the prime meridian and this inertial frame. If the earth's axis of rotation coincides with the Z axis, then a unique angle, θ_g locates the Greenwich meridian as illustrated in Figure 7.36. This angle is known as the *mean sidereal time* of the prime meridian, since X is in the direction of the *mean vernal equinox*. The observer has a *local sidereal time*, θ and is at an *east longitude*, λ_E, the angle between prime and observer meridians measured eastward in the equatorial plane. Thus,

$$\theta = \theta_g + \lambda_E \qquad (7.117)$$

where $0 \le \theta \le 2\pi$. Usually, λ_E is known and knowledge of θ_g quickly yields θ. Practical calculation of θ_g at 12 midnight or 0^{hr} U.T. is accomplished by an

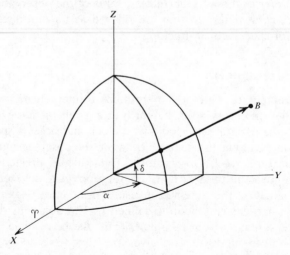

FIGURE 7.35 Inertial frame using right ascension and declination.

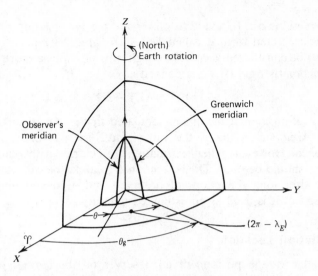

FIGURE 7.36 Rotating meridians referred to the inertial frame.

empirical formula,

$$\theta_{g_0} = 99.6909833° + 3600.7689°\, T_u + 0.00038708°\, T_u^2 \qquad (7.118)$$

where time T_u is measured in centuries as

$$T_u = \frac{\text{J.D.} - 2415020.0}{36525}$$

and θ_{g_0} is in degrees. Thus, Greenwich sidereal time at 0^{hr} U.T., θ_{g_0}, is obtainable directly as a function of Julian date. Sidereal time at any other time of day can be obtained if it is known that there is one extra sidereal day for every *tropical year*, the time required by the sun (365.24219879 mean solar days) to make an apparent revolution of the ecliptic from vernal equinox to vernal equinox. This is slightly shorter than the sidereal year owing to precession of the equinoxes. Thus, the rate

$$\frac{d\theta}{dt} = \left(1 + \frac{1}{365.24219879}\right) \text{REV/yr}$$

permits a fairly accurate approximation of Greenwich sidereal time,

$$\theta_g = \theta_{g_0} + (t - t_0)\frac{d\theta}{dt} \qquad (7.119)$$

where $t - t_0$ is time since 0^{hr} U.T. The rotation of the earth is not exactly constant in that it experiences periodic and secular variations in the rotation rate. Corrections can be made to universal time for accurate time representation. This corrected form is called *Ephemeris time*. Whenever the mean solar second is not considered accurate enough, this provides a more constant time. Thus, Ephemeris time (E.T.) is defined as

$$E.T. = U.T. + \Delta T \tag{7.120}$$

where ΔT is an annual increment tabulated in the *American Ephemeris and Nautical Almanac*. Actually, the value of ΔT is small and typically of little consequence. However, when calculating trajectories with planetary destinations, E.T. should be used. The correction ΔT cannot be calculated in advance, but only after long reductions of observed and predicted longitudes of the Moon. Hence, it is used as an estimated quantity.

7.5.2 Station Location

Knowledge of the position of an observer on the surface of the earth is critical for orbital observations. This position is defined by the *station coordinates*. Since earth is not a perfect sphere, a model of its shape must be adopted. For preliminary orbit determination, an oblate spheroid model is quite satisfactory. Cross sections parallel to the equator are perfect circles, and meridians are ellipses whose semimajor and semiminor axes are denoted by a and b, respectively. This type of model is analogous to that used for the gravitational field of form (7.16) with only the J_2 term taken as nonzero. The semimajor axis is just equal to the equatorial radius of earth, $R_\oplus$. *Flattening* is defined as the difference of the largest and smallest semiaxes of the ellipse of revolution, divided by the semimajor axis. Thus,

$$f = \frac{a-b}{a} \tag{7.121}$$

Since

$$\frac{b}{a} = \sqrt{1-e^2}$$

flattening is quickly related to eccentricity by

$$e^2 = 2f - f^2 \tag{7.122}$$

The value of f has been measured for earth and is approximately 1/298.3.

Station coordinates associated with a flattened earth must be specifically defined such that the elliptic meridians are taken into account. Referring to Figure 7.37, *geocentric latitude* ϕ' is defined as the acute angle between the equator and a line connecting the geometric center of the coordinate system with a point on the surface of the reference ellipsoid. It is convenient to introduce the *reduced latitude*, ψ_c defined as the acute angle between the

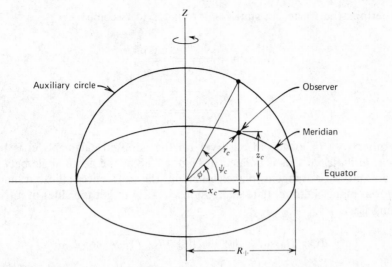

FIGURE 7.37 Geocentric latitude.

equator and a line from the origin of the coordinates to a point defined by the vertical projection of the station onto an auxiliary circle of radius $R_\oplus$. These two angles are important in defining the position of an observer. In terms of geocentric latitude, the station coordinates are

$$x_c = r_c \cos \phi', \qquad z_c = r_c \sin \phi'$$

Using reduced latitude yields

$$x_c = R_\oplus \cos \psi_c, \qquad z_c = R_\oplus \sqrt{1 - e^2} \sin \psi_c \qquad (7.123)$$

and

$$r_c = \sqrt{x_c^2 + z_c^2} = R_\oplus \sqrt{1 - e^2 \sin^2 \psi_c} \qquad (7.124)$$

Geocentric latitude can be expressed in terms of reduced latitude by

$$\sin \phi' = \frac{z_c}{r_c} = \frac{\sqrt{1 - e^2} \sin \psi_c}{\sqrt{1 - e^2 \sin^2 \psi_c}} \qquad (7.125)$$

$$\cos \phi' = \frac{x_c}{r_c} = \frac{\cos \psi_c}{\sqrt{1 - e^2 \sin^2 \psi_c}} \qquad (7.126)$$

Expression (7.125) yields

$$\sin \psi_c = \frac{\sin \phi'}{\sqrt{1 - e^2 \cos^2 \phi'}}$$

This permits the Cartesian station coordinates to become

$$x_c = \frac{R_\oplus \sqrt{1-e^2} \cos \phi'}{\sqrt{1-e^2 \cos^2 \phi'}} \tag{7.127}$$

$$z_c = \frac{R_\oplus \sqrt{1-e^2} \sin \phi'}{\sqrt{1-e^2 \cos^2 \phi'}} \tag{7.128}$$

Another way to locate an observer on the adopted earth model is by the *geodetic latitude, ϕ.* This is defined in Figure 7.38 as the acute angle between a line normal to a tangent plane touching the reference ellipsoid and the equatorial plane. Differentiate expressions (7.123) to get the sides of a triangle defining ϕ,

$$-dx_c = R_\oplus \sin \psi_c \, d\psi_c, \qquad dz_c = R_\oplus \sqrt{1-e^2} \cos \psi_c \, d\psi_c$$

then

$$ds = \sqrt{(-dx_c)^2 + (dz_c)^2} = R_\oplus \sqrt{1-e^2 \cos^2 \psi_c} \, d\psi_c$$

and

$$\sin \phi = -\frac{dx_c}{ds} = \frac{\sin \psi_c}{\sqrt{1-e^2 \cos^2 \psi_c}} \tag{7.129}$$

$$\cos \phi = \frac{dz_c}{ds} = \frac{\sqrt{1-e^2} \cos \psi_c}{\sqrt{1-e^2 \cos^2 \psi_c}}$$

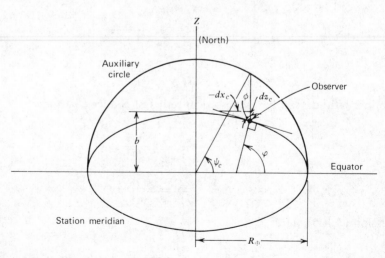

FIGURE 7.38 Geodetic latitude.

Combine these to obtain

$$\sqrt{1-e^2\cos^2\psi_c}=\frac{\sqrt{1-e^2}}{\sqrt{1-e^2\sin^2\phi}}$$

This is then used with set (7.129) to get

$$\sin\psi_c=\frac{\sqrt{1-e^2}\sin\phi}{\sqrt{1-e^2\sin^2\phi}}$$

$$\cos\psi_c=\frac{\cos\phi}{\sqrt{1-e^2\sin^2\phi}}$$

Finally, the station coordinates become

$$x_c=\frac{R_\oplus\cos\phi}{\sqrt{1-e^2\sin^2\phi}} \tag{7.130}$$

$$z_c=\frac{R_\oplus(1-e^2)\sin\phi}{\sqrt{1-e^2\sin^2\phi}} \tag{7.131}$$

So far, only the rectangular coordinates of an observer on the adopted ellipsoid are known. However, there are local variations above and below this reference surface which are quite common. For example, Telescope Peak is 3369 m above sea level while Death Valley is 86 m below sea level, both of which are within 32 km of each other. Figure 7.39 illustrates the components of *elevation deviation H*, which is normal to the adopted ellipsoid. These are expressed as

$$\Delta x_c=H\cos\phi \tag{7.132}$$

$$\Delta z_c=H\sin\phi$$

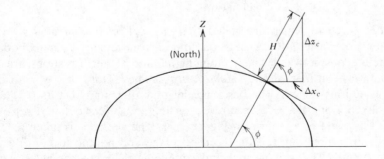

FIGURE 7.39 Elevation deviation.

Now add these quantities to the relationships for x_c and z_c to get more precise expressions in terms of flattening, geodetic latitude, and elevation. The resulting expressions are

$$x_c = \left[\frac{R_\oplus}{\sqrt{1 - (2f - f^2) \sin^2 \phi}} + H \right] \cos \phi \qquad (7.133)$$

$$z_c = \left[\frac{R_\oplus (1 - f)^2}{\sqrt{1 - (2f - f^2) \sin^2 \phi}} + H \right] \sin \phi \qquad (7.134)$$

Another type of latitude is *astronomical latitude*, ϕ_a, which is the acute angle formed by the intersection of the local gravity direction and the equatorial plane. This type of measurement is a function of the local gravitational field which is affected by anomalies such as mountains, bodies of water, etc. Differences between geodetic and astronomical latitude are termed *station errors*. In practical situations it is easiest to measure astronomical latitude at a given site, because a plumb bob can be used to locate the direction of local gravity, while the normal to the equator can be determined celestially. When station errors cannot be removed from this measurement, astronomical latitude can be used as a first approximation to geodetic latitude, because differences are usually quite small.

7.5.3 Basic Elements and Transformations

To define a Keplerian orbit uniquely, five parameters are required,

e, eccentricity
p, parameter
Ω, position of ascending node
ω, argument of periapsis
i, inclination

the latter three of which are called the *classical orientation angles* and are illustrated in Figure 7.40. A sixth parameter, true anomaly, uniquely defines the satellite position at a given moment in time. Transformations are often required between inertial and satellite orbit frames. Thus, it is wise to define appropriate unit vectors. The basic inertial frame has unit vectors **I**, **J**, and **K**, with **I** taken along the reference direction (e.g., ♈). Next define a set of unit vectors $\mathbf{i}_p$, $\mathbf{i}_q$, and $\mathbf{i}_z$, as shown in Figure 7.41. The vector $\mathbf{i}_p$ is taken along the periapsis or **e** direction, $\mathbf{i}_q$ is in the orbit plane and normal to $\mathbf{i}_p$ in the flight direction, and $\mathbf{i}_z$ is orbit normal and completes the right-handed system. A vector expressed in **I**, **J**, **K** components may be written in terms of $\mathbf{i}_p$, $\mathbf{i}_q$, $\mathbf{i}_z$ by a

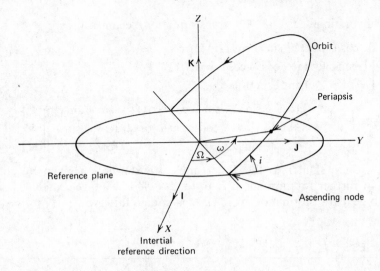

FIGURE 7.40 Unique specification of orbit orientation.

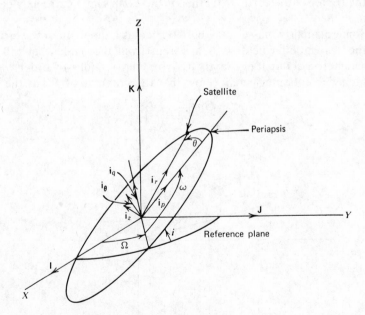

FIGURE 7.41 Unit vectors for the orbital frame.

series of rotations, Ω, i, ω. The transformation becomes

$$
\begin{bmatrix} \mathbf{i}_p \\ \mathbf{i}_q \\ \mathbf{i}_z \end{bmatrix} = \begin{bmatrix} (\cos\omega\cos\Omega - \sin\omega\cos i\sin\Omega) \\ (-\sin\omega\cos\Omega - \cos\omega\cos i\sin\Omega) \\ (\sin\Omega\sin i) \end{bmatrix}
$$

$$
\begin{bmatrix} (\cos\omega\sin\Omega + \sin\omega\cos i\cos\Omega) & (\sin\omega\sin i) \\ (-\sin\omega\sin\Omega + \cos\omega\cos i\cos\Omega) & (\cos\omega\sin i) \\ (-\sin i\cos\Omega) & (\cos i) \end{bmatrix} \begin{bmatrix} \mathbf{I} \\ \mathbf{J} \\ \mathbf{K} \end{bmatrix} \quad (7.135)
$$

which is quite similar to the Euler angle transformation given by equation (1.24). A further rotation about $\mathbf{i}_z$ through θ will locate the satellite position, and the next transformation to $\mathbf{i}_r$, $\mathbf{i}_\theta$, $\mathbf{i}_z$ is straightforward.

EXERCISES

7.1 Carry out the steps in demonstrating that the attraction of a homogeneous spherical shell on an outside point is simply $F = GM/D^2$, as illustrated in Figure 7.2.

7.2 Prove that the attraction of a homogeneous spherical shell on a point inside is always zero. This implies that a person would be weightless at the center of the earth. Why then is the earth's core extremely dense and molten?

7.3 Some planetary masses may be modeled as a dipole in order to describe the gravitational field. It is apparent from the figure that this field is symmetric about the Z-axis. If $r \gg d$, contributions of order $(d/r)^3$ and higher may be neglected. Show that the potential at P has the form

$$
U(r, \phi) = \frac{Gm}{r}\left[2 + \left(\frac{d}{r}\right)^2 (3\cos^2\phi - 1)\right]
$$

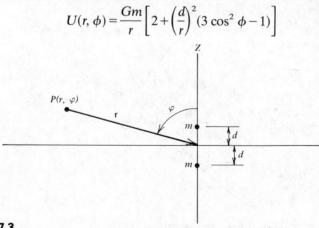

EXERCISE 7.3

7.4 Prove that the force of attraction $\mathbf{F}_i$ on the ith particle given by equation (7.5) is in fact conservative.

7.5 Use Rodriques' formula, equation (7.11), to derive expressions for $P_4(\cos \gamma)$ and $P_5(\cos \gamma)$.

7.6 Prove the identity given by equation (7.28).

7.7 Sketch a cross section of the moon's sphere of influence with respect to earth using expression (7.37).

7.8 Consider the restricted three-body problem discussed in Section 7.2.
   (a) Derive the correction to the earth's attraction used in obtaining set (7.45).
   (b) Derive set (7.45) noting that the origin of x, y is at the L_4 point which is both accelerating and rotating.

7.9 Expand the effective potential $U(x,y,z)$, given in the restricted three-body problem by equation (7.43), up to and including second order terms in the neighborhood of L_4. Then deduce the shape of nearby contours, $U(x,y,0) = c$. Use the nomenclature of Figure 7.12, but remember that equation (7.43) used the notation of Figure 7.10.

7.10 Apply the method of Section 7.3.2 to determine the true anomaly in a parabolic escape orbit from earth three hours after leaving a circular 200 km orbit. What are the position and velocity vectors at this point using the notation of Figure 7.14?

7.11 Kepler's equation, given as form (7.66) has been solved in many ways. One case of particular interest is that of low eccentricity orbits. Show that

$$\psi = M + \frac{e \sin M}{1 - e \cos M} - \frac{1}{2}\left(\frac{e \sin M}{1 - e \cos M}\right)^3$$

is a solution to this equation if fourth and higher powers of e are neglected.

7.12 Show that area CAQ in Figure 7.15 is just $a^2 H/2$.

7.13 Derive the expressions for position and velocity in hyperbolic paths, equations (7.71). Then derive the associated time equation, form (7.73). Start by assuming the conic equation and identity (7.68).

7.14 Consider an earth satellite observed to have position and velocity, as depicted in Figure 7.17,

$$\mathbf{r} = 10^4 \, \mathbf{I} \text{ km}$$
$$\mathbf{v} = 8.5 \, \mathbf{J} \text{ km/s}$$

at a reference point. Determine the position and velocity four hours later.

7.15 A comet is observed to have position and velocity

$$\mathbf{r}_0 = 1.4 \, \mathbf{I} \text{ A.U.}$$
$$\mathbf{v}_0 = -5.0 \, \mathbf{I} + 25.0 \, \mathbf{J} \text{ (km/s)}$$

in a heliocentric orbit. Assume **I** is directed toward the first point of Aries and **J** is normal to **I** and in the ecliptic. Determine its position and velocity six months later.

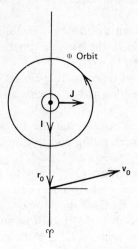

EXERCISE 7.15

7.16 In addition to the special case of minimum energy transfer between two specified points, it is interesting to consider an elliptic transfer which ends at the apoapsis. This corresponds to the vacant focus lying on r_2 in Figure 7.21, and is sometimes called a *tangential transfer*. Derive expressions for a, e, and p in terms of r_1, r_2, and ν.

7.17 Another elliptic transfer from P to Q in Figure 7.21 of interest is called the *symmetric transfer*. This occurs when FF^* is parallel to PQ. Derive corresponding expressions for a, e, and p in terms of r_1, r_2, and ν.

7.18 Derive expressions (7.95) and (7.96) for the semilatus rectum of the two parabolic transfers between two specified points.

7.19 Derive the hyperbolic transfer time, expression (7.113), by making proper substitutions into equations (7.108) and integrating.

7.20 Carry out the steps in the derivation of equation (7.116)

7.21 A trip from earth to a newly discovered planet can be approximated by the situation shown. Use

$$\mu_\odot = 3 \times 10^{-4} \, \text{AU}^3/\text{day}^2$$
$$r_1 = 1 \, \text{AU}$$
$$r_2 = 2 \, \text{AU}$$
$$\nu = 60°$$

(a) Compute the values of a, n, and t for minimum energy transfer.

(b) If the transfer time is specified as 500 days, determine the smallest appropriate value of a for the transfer and sketch the path.

(c) If a for the transfer path is given as 2.0 AU, calculate the minimum time to make this transfer.

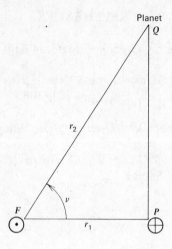

EXERCISE 7.21

7.22 The National Aeronautics and Space Administration is planning a Mariner Jupiter/Uranus flyby mission. The launch will take place on November 3, 1979 with Jupiter arrival at a flyby radius of 5.5 Jupiter radii on April 10, 1981. Using Jupiter gravity assist to accelerate and turn the trajectory, the spacecraft will proceed to encounter Uranus during July 1985.

(a) Determine the possible values of a for this transfer if $\nu = 147°$.

(b) Determine the minimum velocity increment required to leave low earth orbit (300 km altitude) and coast to Jupiter in the specified time.

(c) The heliocentric anomaly difference between Jupiter passage and Uranus arrival is 70°. Determine whether a Jupiter flyby at 5.5 radii will result in Uranus arrival in July 1985 without the use of onboard thrusters.

7.23 What is Julian Date corresponding to 6:00 p.m. on April 21, 1977?

7.24 The Skylab spacecraft passed over State College, Pa. ($\lambda_E = 282°$) on June 30, 1973 at 10:30 A.M. EST. What was the corresponding Julian Date and Greenwich mean time?

7.25 The flattening of Mars is about $\frac{1}{192}$ with an equatorial radius of 3410 km. Calculate the rectangular station coordinates of a Martian ground station at a geodetic latitude of 40° and elevation of 300 m. Express your answer in Mars radii (m.r.) to 6 decimal places.

REFERENCES

Battin, R. H., *Astronautical Guidance*, McGraw-Hill, 1964, Chapters 1, 2, 3.

Breakwell, J. V., R. W. Gillespie, and S. Ross, "Researches in Interplanetary Transfer", *ARS Journal*, Feb. 1961, pp. 201–208.

Escobal, P. R., *Methods of Orbit Determination*, Wiley, 1965, Chapters 1, 5, 6.

Szebehely, V., *Theory of Orbits, The Restricted Problem of Three Bodies*, Academic Press, 1967, Chapter 10.

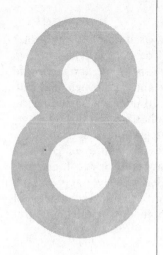

Orbital Perturbations

The effects of orbital perturbations have been mentioned casually several times in prior chapters and then put aside for later reference. This important topic is now treated to the depth appropriate for this text. Baker defines perturbative forces as *forces acting on an object other than those that cause it to move along some reference orbit.* First, methods of treating perturbative forces are reviewed, and then fundamental cases are considered. A perturbation, as opposed to a perturbative force, is a deviation from some normal or expected motion. Orbital perturbations arise from such sources as other attracting bodies, atmospheric pressure, asphericity of the central body, solar radiation, magnetic fields, and relativistic effects. Do not assume that perturbations are generally insignificant. Many interplanetary missions would miss their targets entirely if the perturbing effect of other attracting bodies were not taken into account. Ignoring the earth's oblateness effects on a typical satellite would result in complete failure to predict its position over a long period of time. Some perturbations are analytically describable with excellent accuracy, while others, such as atmospheric and solar pressure, are very difficult to anticipate over several orbital periods. The two fundamental categories of perturbation techniques are *special perturbations* and *general perturbations*. Special perturbations refers to techniques which deal with the direct numerical integration of the equations of motion, including all necessary perturbing accelerations. These are emphasized in this chapter. General perturbation techniques involve analytic integration of series expansions of the perturbing accelerations. Such methods are more difficult and lengthy for obvious reasons. However, they do permit better physical interpretation of effects. Therefore, generalizations are included whenever appropriate. Specific topics included under special perturbations are the techniques of Cowell and Encke, and the variation of parameters method. Numerical integration methods are briefly discussed.

8.1 COWELL'S METHOD

Early in this century, P. H. Cowell developed a simple and straightforward perturbation method to be used in determining the orbit of the eighth moon of Jupiter. Cowell and Crommelin also used it to predict the return of Halley's comet for a period up to 1910. This method has proven very useful with modern fast computers. The approach consists of writing the equations of motion, including all perturbations, and integrating them step-by-step numerically. Thus, for the perturbed two-body problem, equation (7.27) can be written as

$$\ddot{\mathbf{r}} + \frac{\mu}{r^3}\mathbf{r} = \mathbf{a}_p \tag{8.1}$$

where $\mathbf{a}_p$ is the perturbing acceleration. Numerical integration is carried out by reducing this to two first order equations,

$$\dot{\mathbf{r}} = \mathbf{v}, \qquad \dot{\mathbf{v}} = \mathbf{a}_p - \frac{\mu}{r^3}\mathbf{r}$$

each of which would further be treated in component form. Acceleration $\mathbf{a}_p$ could be the result of other attracting bodies such as the moon and sun in an earth orbital case. The main advantage of this method is its simplicity of formulation and implementation. Any number of perturbations can be handled at the same time. However, several disadvantages are also associated with this procedure. When motion is near a large attracting body, smaller integration steps must be taken to guarantee accuracy. This leads to long computer runs and large roundoff errors. As a basis for future comparison, Cowell's method is about 10 times slower than Encke's for lunar and interplanetary trajectories. For earth satellites the advantage of Encke's method over Cowell's is less, but it is still 2 or 3 to 1.

8.2 ENCKE'S METHOD

It is interesting to note that Encke developed his method in 1857, more than a half-century before the simpler one of Cowell. One might speculate that Cowell's method may have been thought of earlier but was not pursued because of a lack of computing capability. Encke's method is more sophisticated, but requires less computation. The primary distinction in the two approaches is that Encke integrated the difference between primary and perturbing accelerations, whereas Cowell integrated the sum of all accelerations. Thus, Encke's approach requires the use of a reference orbit along which the object would move in the absence of all perturbing accelerations. Presumably, this would be a conic section in an ideal Newtonian gravitational field (Keplerian orbit). Then all calculations are made relative to this reference orbit, called the *osculating orbit*. This name is appropriate, because *osculation* means *kissing*. The true and reference orbits do, in fact, make contact at one or more points. Thus, at any given time the osculating orbit becomes the true orbit if all perturbing accelerations are removed at that instant. Referring to Figure 8.1, at that instant the osculating and true orbits are in contact. Any osculating orbit is good until the actual orbit deviates too far from it. When this happens a process called *rectification* must be performed before further integration. This means that a new reference time (epoch) and starting point are chosen which correspond to the true orbit at that instant. A new osculating orbit is then chosen which is the Keplerian orbit corresponding to the position and velocity at the reference time and place.

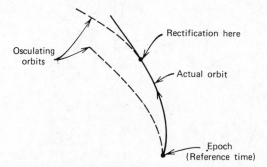

FIGURE 8.1 Osculating and actual orbits.

Consider now the implementation of Encke's method. The fundamental problem here is to determine an analytic expression for the difference between the true and reference orbits. Let $\mathbf{r}$ and $\boldsymbol{\rho}$ in Figure 8.2 be the radius vectors to the true orbit and osculating orbit, respectively, at a given time, $t - t_o$, since the preceding rectification. The equation of the true orbit is given by expression (8.1) and the osculating orbit is simply

$$\ddot{\boldsymbol{\rho}} + \frac{\mu}{\rho^3} \boldsymbol{\rho} = 0 \tag{8.2}$$

Note that

$$\mathbf{r}(t_0) = \boldsymbol{\rho}(t_0), \qquad \mathbf{v}(t_0) = \dot{\boldsymbol{\rho}}(t_0)$$

Deviation from the reference orbit is $\delta\mathbf{r}$, given by $\delta\mathbf{r} = \mathbf{r} - \boldsymbol{\rho}$ and

$$\delta\ddot{\mathbf{r}} = \ddot{\mathbf{r}} - \ddot{\boldsymbol{\rho}} \tag{8.3}$$

Combining forms (8.1) and (8.2) with (8.3) yields

$$\delta\ddot{\mathbf{r}} = \mathbf{a}_p + \left[\frac{\mu}{\rho^3}(\mathbf{r} - \delta\mathbf{r}) - \frac{\mu}{r^3}\mathbf{r} \right]$$

which is rewritten as

$$\delta\ddot{\mathbf{r}} = \mathbf{a}_p + \frac{\mu}{\rho^3}\left[\left(1 - \frac{\rho^3}{r^3} \right)\mathbf{r} - \delta\mathbf{r} \right] \tag{8.4}$$

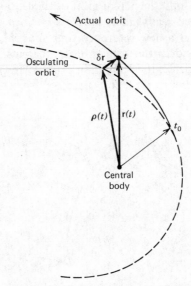

FIGURE 8.2 Measure of deviation from reference orbit.

This can now be integrated numerically from $t = t_0$ to obtain $\delta\mathbf{r}(t)$ with $\boldsymbol{\rho}$ as a known function of time. In principle, position and velocity can now be calculated as a function of time. However, the term $(1 - \rho^3/r^3)$ is the difference of two nearly equal quantities, especially near t_0. This would require excessive computer time and tend to defeat the advantage of more accuracy over Cowell's method. One method of handling such a situation is to define a small variable, ε as

$$2\varepsilon = 1 - \frac{r^2}{\rho^2} \tag{8.5}$$

which gives

$$\frac{\rho^3}{r^3} = (1 - 2\varepsilon)^{-3/2}$$

This permits equation (8.4) to be written as

$$\delta\ddot{\mathbf{r}} = \mathbf{a}_p + \frac{\mu}{\rho^3}\{[1 - (1 - 2\varepsilon)^{-3/2}]\mathbf{r} - \delta\mathbf{r}\} \tag{8.6}$$

It remains to express ε in some convenient numerical manner. This can be done through a binomial series expansion,

$$1 - (1 - 2\varepsilon)^{-3/2} = 3\varepsilon - \tfrac{15}{2}\varepsilon^2 + \tfrac{35}{2}\varepsilon^3 - \cdots \tag{8.7}$$

However, prior to the availability of high speed computers, this was a lengthy task. Thus, another approach was developed using a new function, defined as

$$f = \frac{1}{\varepsilon}[1 - (1 - 2\varepsilon)^{-3/2}] \tag{8.8}$$

This allows equation (8.6) to be written as

$$\delta\ddot{\mathbf{r}} = \mathbf{a}_p + \frac{\mu}{\rho^3}(f\varepsilon\mathbf{r} - \delta\mathbf{r}) \tag{8.9}$$

To implement this, values of f are obtained from expression (8.8). Of course, values of ε must first be obtained. Since

$$r^2 = x^2 + y^2 + z^2$$

in Cartesian coordinates, then

$$r^2 = (\rho_x + \delta x)^2 + (\rho_y + \delta y)^2 + (\rho_z + \delta z)^2$$

where ρ_x, ρ_y, and ρ_z are the components of $\boldsymbol{\rho}$. This leads to a new form of equation (8.5),

$$\varepsilon = -\frac{1}{\rho^2}[\delta x(\rho_x + \tfrac{1}{2}\delta x) + \delta y(\rho_y + \tfrac{1}{2}\delta y) + \delta z(\rho_z + \tfrac{1}{2}\delta z)] \tag{8.10}$$

The deviation from the reference orbit is small and should remain small between rectifications of osculating orbits. Thus, terms like δx^2, δy^2, δz^2 in result (8.10) can be neglected to give

$$\varepsilon \cong \frac{\rho_x \delta x + \rho_y \delta y + \rho_z \delta z}{\rho^2} \tag{8.11}$$

which is easily handled for known values of δx, δy, and δz.

Encke's formulation reduces the number of integration steps, because $\delta \mathbf{r}$ changes more slowly than $\mathbf{r}$, permitting larger step sizes with accuracy equivalent to using Cowell's method at smaller step sizes. Advantages of this method diminish if $\mathbf{a}_p$ becomes much larger than $\mu (f \varepsilon \mathbf{r} - \delta \mathbf{r})/\rho^3$ or $\delta r/\rho$ does not remain quite small. In the first instance the reference parameters or orbit need to be changed, since the perturbations are becoming primary. In the other case a new osculating orbit needs to be chosen. Rectification should be initiated when $\delta r/\rho$ is greater than or equal to some small constant depending on desired accuracy and usually of the order of 0.01.

A computational algorithm might have the following steps:

1. Start by defining an osculating orbit with $\delta \mathbf{r} = 0$. Use initial conditions $\mathbf{r}(t_0) = \boldsymbol{\rho}(t_0)$ and $\mathbf{v}(t_0) = \dot{\boldsymbol{\rho}}(t_0)$.
2. Calculate $\delta \mathbf{r}(t_0 + \Delta t)$ for the first integration step using an appropriate value of Δt. Note that $\varepsilon(t_0) = 0$.
3. Determine $\boldsymbol{\rho}(t_0 + \Delta t)$ for the osculating orbit, $\varepsilon(t_0 + \Delta t)$ from equation (8.10) or (8.11), and $f(t_0 + \Delta t)$ from expression (8.8).
4. Test $\delta r/\rho$ for size and rectify if a specified limit is reached.
5. Calculate $\mathbf{r} = \boldsymbol{\rho} + \delta \mathbf{r}$ and $\mathbf{v} = \dot{\boldsymbol{\rho}} + \delta \dot{\mathbf{r}}$.
6. Increment t by Δt and calculate a new $\delta \mathbf{r}$.
7. Repeat steps (3) through (6) until a satisfactory value of t is reached.

The perturbing accelerations are implicitly assumed to be expressible in analytic or empirical form. Tabular formats are also possible. Other considerations will arise in actual calculations. For example, a lunar flight will involve changing attraction centers as the moon is approached.

8.3 VARIATION OF PARAMETERS OR ELEMENTS

The methods of Cowell and Encke offer direct determination of position and velocity but do not permit easy interpretation of the associated changes in orbital elements. Furthermore these two approaches lend themselves only to numerical integration except in rare special cases. The *variation of parameters* method considers the influence of perturbing forces directly on the elements of a Keplerian orbit. This permits analytic descriptions of the rates of change of these elements. However, position and velocity of a point on the orbit is not directly extracted. Thus, this method may not be as attractive for actual orbital path prediction, but allows a great deal of physical insight into the nature of perturbation effects. The variation of parameters approach, as used here, is a special perturbation method, but the concept is actually the basis for general perturbations.

8.3.1 Geometrical Development

Consider first the in-plane perturbations as two components of impulses with tangential part $f_t\,dt$ and inward normal part $f_n\,dt$ as depicted in Figure 8.3. Note that f_t and f_n are the tangential and normal perturbing acceleration components (forces per unit mass) which are assumed to be small compared with that of the central attraction. Referring to Section 1.2.2 and remembering that only first order effects are retained here, only f_t can change the orbital energy and semimajor axis. This can be evaluated by considering energy equation (2.27) for a conic. Use this in form

$$\mathscr{E} = -\frac{\mu}{2a}$$

and take its differential to obtain the relation between small energy change and its effect on semimajor axis. This change is due to the tangential force acting

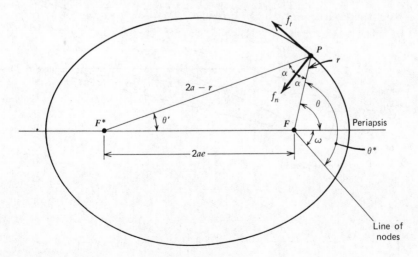

FIGURE 8.3 In-plane perturbing impulses.

over time dt,

$$d\mathscr{E} = \frac{\mu}{2a^2}\, da = f_t v\, dt \qquad (8.12)$$

Not only is a changed, but the vacant focus F^* is displaced away from its instantaneous position by an amount $2\, da$, because $FP + F^*P = 2a$ and FP does not change instantaneously by f_t. Thus

$$d(F^*P) = 2\, da \qquad (8.13)$$

The inward normal acceleration f_n leaves a unchanged, but the impulse $f_n\, dt$ changes the direction of the tangent by an amount $f_n\, dt/v$ and, hence, the direction from the instantaneous position to the vacant focus changes by an amount $2f_n\, dt/v$ counterclockwise. Referring to Figure 8.4, the rotation of F^*P is, thus,

$$d\theta' = -2\, d\alpha$$

The change in flight path angle $d\beta$ is just

$$\tan d\beta \cong d\beta \cong -\frac{v_n}{v} = -\frac{f_n\, dt}{v}$$

where f_n and v_n are positive inward and $d\beta$ is negative inward. Since $d\alpha = d\beta$

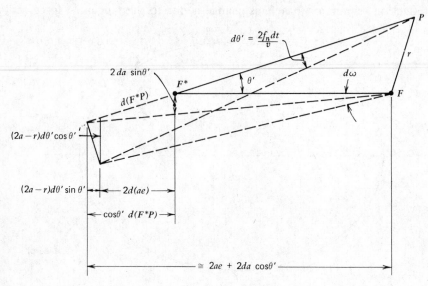

FIGURE 8.4 Effects of normal acceleration on major axis.

by geometry,

$$d\theta' = 2\frac{f_n \, dt}{v} \tag{8.14}$$

The other orbital parameters, e and ω can now be considered. Again referring to Figure 8.4, the change in distance between foci can be written to first order in small quantities as

$$2 \, d(ae) \cong d(F^*P) \cos \theta' - (2a - r) \, d\theta' \sin \theta'$$

or

$$2 \, d(ae) \cong 2 \, da \cos \theta' - (2a - r)\frac{2f_n \, dt}{v}\sin \theta' \tag{8.15}$$

Using the geometry of Figure 8.4,

$$d\omega \cong \frac{2 \, da \sin \theta'}{2ae} + \frac{(2a - r) \, d\theta' \cos \theta'}{2ae} \tag{8.16}$$

Expressions (8.15) and (8.16) can be written in more useful forms by noting that

$$(2a - r) \cos \theta' = 2ae + r \cos \theta$$

$$(2a - r) \sin \theta' = r \sin \theta$$

$$v^2 = \frac{\mu}{a}\frac{2a - r}{r}$$

$$d(ae) = e \, da + a \, de = da \cos \theta' - (2a - r)\frac{f_n \, dt}{v}\sin \theta'$$

The resulting expressions are

$$\dot{e} = \frac{2f_t}{v}(\cos \theta + e) - \frac{f_n}{v}\frac{r}{a}\sin \theta \tag{8.17}$$

$$e\dot{\omega} = \frac{2f_t}{v}\sin \theta + \frac{f_n}{v}\left(2e + \frac{r}{a}\cos \theta\right) \tag{8.18}$$

When determining perturbative accelerations it is more natural to use radial and transverse components than f_t and f_n. The relationship between these two sets of components is illustrated in Figure 8.5, and is expressed analytically as

$$f_r = f_t \sin \beta - f_n \cos \beta$$

$$f_\theta = f_t \cos \beta + f_n \sin \beta$$

Remembering that $v_\theta = v \cos \beta$ and $v_r = v \sin \beta$, equations (8.17) and (8.18)

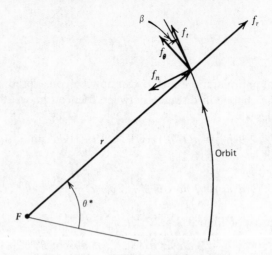

FIGURE 8.5 Nomenclature for radial and transverse perturbing accelerations.

may be combined to replace f_t and f_n. After some manipulation,

$$\dot{e} \cos \theta + e\dot{\omega} \sin \theta = \frac{2(1 + e \cos \theta)}{v_\theta} f_\theta$$

$$\dot{e} \sin \theta - e\dot{\omega} \cos \theta = (1 + e \cos \theta)\frac{f_r}{v_\theta} + e \sin \theta \frac{f_\theta}{v_\theta}$$

$\qquad(8.19)$

Set (8.19) may be solved directly for $\dot{e}$ and $\dot{\omega}$,

$$\dot{e} = \frac{r}{h}\{\sin \theta(1 + e \cos \theta)f_r + (e + 2\cos \theta + e\cos^2 \theta)f_\theta\} \qquad(8.20)$$

$$\dot{\omega} = \frac{r}{he}\{-\cos \theta(1 + e \cos \theta)f_r + \sin \theta(2 + e \cos \theta)f_\theta\} \qquad(8.21)$$

If f_r and f_θ can be expressed as analytic functions of time, these two equations can be integrated to yield e and ω over the period of interest.

Of course, the orbit normal component of perturbing force must be added to complete this discussion. This is the *out-of-plane* or *lateral* acceleration f_L, defined in Figure 8.6 as being positive to the left of the flight path. Impulse $f_L \, dt$ rotates the orbital plane through an angle of magnitude $f_L \, dt/v_\theta$ about the radius vector. Hence,

$$di = \frac{f_L \, dt}{v_\theta} \cos \theta^*$$

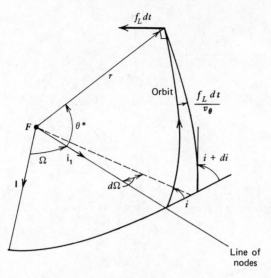

FIGURE 8.6 Out-of-plane perturbation effects.

$$\sin i \, d\Omega = \frac{f_L \, dt}{v_\theta} \sin \theta^*$$

and

$$d\theta^* = -\cos i \, d\Omega + \frac{h}{r^2} \, dt$$

Although it is true that f_L effects ω, this is not taken into account because the position of periapsis does not actually rotate in the orbital plane. Thus, it is sufficient to summarize with

$$\frac{di}{dt} = \overset{\circ}{i} = \frac{f_L}{v_\theta} \cos \theta^* \qquad (8.22)$$

$$\dot{\Omega} = \frac{f_L}{v_\theta} \frac{\sin \theta^*}{\sin i} \qquad (8.23)$$

$$\dot{\theta} = \frac{h}{r^2} - \dot{\Omega} \cos i \qquad (8.24)$$

Equations (8.12), (8.20)–(8.24) represent a complete set of element rates due to perturbing accelerations. It is typically necessary to transform this acceleration from an inertial frame to the orbit frame of r, θ^*, and L as depicted in Figure 8.7. Noting that

$$\sin L = \mathbf{i}_r \cdot \mathbf{K} = \sin i \sin \theta^*$$

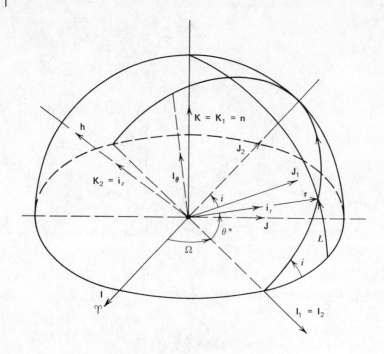

FIGURE 8.7 Transformation geometry to orbital frame.

the transformation consists of three rotations, Ω, i, and θ^* and is taken in the following sequence:

$$\begin{bmatrix} \mathbf{I} \\ \mathbf{J} \\ \mathbf{K} \end{bmatrix} = \begin{bmatrix} \cos \Omega & -\sin \Omega & 0 \\ \sin \Omega & \cos \Omega & 0 \\ 0 & 0 & 1 \end{bmatrix} \begin{bmatrix} \mathbf{I}_1 \\ \mathbf{J}_1 \\ \mathbf{K}_1 \end{bmatrix}$$

$$\begin{bmatrix} \mathbf{I}_1 \\ \mathbf{J}_1 \\ \mathbf{K}_1 \end{bmatrix} = \begin{bmatrix} 1 & 0 & 0 \\ 0 & \cos i & -\sin i \\ 0 & \sin i & \cos i \end{bmatrix} \begin{bmatrix} \mathbf{I}_2 \\ \mathbf{J}_2 \\ \mathbf{K}_2 \end{bmatrix}$$

and

$$\begin{bmatrix} \mathbf{I}_2 \\ \mathbf{J}_2 \\ \mathbf{K}_2 \end{bmatrix} = \begin{bmatrix} \cos \theta^* & -\sin \theta^* & 0 \\ \sin \theta^* & \cos \theta^* & 0 \\ 0 & 0 & 1 \end{bmatrix} \begin{bmatrix} \mathbf{i}_r \\ \mathbf{i}_\theta \\ \mathbf{i}_z \end{bmatrix}$$

Combining these leads to the orthogonal transformation

$$
\begin{bmatrix} \mathbf{I} \\ \mathbf{J} \\ \mathbf{K} \end{bmatrix} = \begin{bmatrix} (\cos \Omega \cos \theta^* - \sin \Omega \cos i \sin \theta^*) \\ (\sin \Omega \cos \theta^* + \cos \Omega \cos i \sin \theta^*) \\ (\sin i \sin \theta^*) \end{bmatrix}
$$

$$
\begin{matrix} (-\cos \Omega \sin \theta^* - \sin \Omega \cos i \cos \theta^*) & (\sin \Omega \sin i) \\ (-\sin \Omega \sin \theta^* + \cos \Omega \cos i \cos \theta^*) & (-\cos \Omega \sin i) \\ (\sin i \cos \theta^*) & (\cos i) \end{matrix} \begin{bmatrix} \mathbf{i}_r \\ \mathbf{i}_\theta \\ \mathbf{i}_z \end{bmatrix} \qquad (8.25)
$$

Inertial components may be expressed in the orbital frame by

$$
\begin{bmatrix} \mathbf{i}_r \\ \mathbf{i}_\theta \\ \mathbf{i}_z \end{bmatrix} = \boldsymbol{\beta}^T \begin{bmatrix} \mathbf{I} \\ \mathbf{J} \\ \mathbf{K} \end{bmatrix} \qquad (8.26)
$$

where $\boldsymbol{\beta}$ is the large matrix of transformation (8.25).

8.3.2 Earth Oblateness Effects

An important application of the variation of parameters method is the evaluation of earth oblateness effects on orbits of its satellites. The gravitational potential of earth is given by equation (7.16) and when limited to oblateness only, becomes

$$
U_\oplus(r,\phi) = \frac{\mu_\oplus}{r} \left[1 - \sum_{k=2}^{\infty} \left(\frac{R_\oplus}{r} \right)^k J_k P_k (\cos \phi) \right] \qquad (8.27)
$$

Only the J_2 term is of interest here. Thus, gravitational attraction becomes

$$
\nabla U_\oplus = -\frac{\mu_\oplus}{r^3} \mathbf{r} + \mathbf{f}
$$

where

$$
\mathbf{f} = -\frac{\mu_\oplus J_2 R_\oplus^2}{2} \left\{ \frac{6(\mathbf{r} \cdot \mathbf{n})\mathbf{n}}{r^5} + \left[\frac{3}{r^4} - \frac{15(\mathbf{r} \cdot \mathbf{n})^2}{r^6} \right] \frac{\mathbf{r}}{r} \right\} \qquad (8.28)
$$

where $\mathbf{n}$ is a unit vector normal to the equator, defined in Figure 8.7. Taking radial, transverse, and lateral components of $\mathbf{f}$ yields

$$
f_r = -\frac{3\mu_\oplus J_2 R_\oplus^2}{2r^4} (1 - 3 \sin^2 i \sin^2 \theta^*)
$$

$$
f_\theta = -\frac{3\mu_\oplus J_2 R_\oplus^2}{r^4} \sin^2 i \sin \theta^* \cos \theta^* \qquad (8.29)
$$

$$
f_L = -\frac{3\mu_\oplus J_2 R_\oplus^2}{r^4} \sin i \cos i \sin \theta^*
$$

where f_r is along $\mathbf{i}_r$, f_θ is along $\mathbf{i}_\theta$, and f_L is along $\mathbf{i}_z$. The precession of the node line is of critical importance in many orbital missions. Adapting equation (8.23) to the oblateness acceleration gives this precession rate, $\dot\Omega$ as

$$\frac{d\Omega}{dt} = \frac{f_L}{v_\theta}\frac{\sin\theta^*}{\sin i} = -\frac{3\mu_\oplus J_2 R_\oplus{}^2}{r^4 v_\theta}\sin^2\theta^* \cos i \tag{8.30}$$

However, it is more convenient to express node precession in terms of orbital position. Note that $v_\theta = h/r$ and expression (8.24) yields

$$(\dot\theta)^{-1} = \left(\frac{h}{r^2}\right)^{-1}\left[1 + \frac{\dot\Omega(\cos i)r^2}{h} + \cdots\right]$$
$$= \left(\frac{h}{r^2}\right)^{-1}[1 + O(J_2)]$$

This permits equation (8.30) to be written as

$$\frac{d\Omega}{d\theta} = \frac{d\Omega}{dt}(\dot\theta)^{-1} = -\tfrac{3}{2}J_2\left(\frac{R_\oplus}{p}\right)^2 \cos i(1 + e\cos\theta)(1 - \cos 2\theta^*) + O(J_2{}^2)$$

or

$$\frac{d\Omega}{d\theta} \cong -\tfrac{3}{2}J_2\left(\frac{R_\oplus}{p}\right)^2 \cos i[1 + e\cos(\theta^* - \omega)](1 - \cos 2\theta^*) \tag{8.31}$$

To obtain the variation of Ω over an entire orbit, assume variations of p, e, and ω over one circuit contribute only second order terms. Thus, an average value of $d\Omega/d\theta$ can be used to determine the change in Ω per revolution by holding p, e, and ω fixed during the orbit, that is

$$\left(\frac{d\Omega}{d\theta}\right)_{av} = \frac{1}{2\pi}\int_0^{2\pi}\left(\frac{d\Omega}{d\theta}\right)d\theta \tag{8.32}$$

and the rate of change per revolution is given by

$$\frac{d\Omega}{dN} = 2\pi\left(\frac{d\Omega}{d\theta}\right)_{av} \tag{8.33}$$

This form can be applied to any of the orbital elements for at least several revolutions, assuming $|\mathbf{f}|$ is small. A process of rectification analogous to that of Encke's method is possible in order to extend the usefulness of this approach. Combining equations (8.31) and (8.33) gives

$$\left(\frac{d\Omega}{d\theta}\right)_{av} = -\tfrac{3}{2}J_2\left(\frac{R_\oplus}{p}\right)^2 \cos i \tag{8.34}$$

which is accurate to first order in J_2. Similarly, applying this approach to the argument of perigee leads to

$$\frac{d\omega}{d\theta} = \tfrac{3}{2}J_2\left(\frac{R_\oplus}{p}\right)^2 \frac{1}{e}\{(\cos\theta + e(1+\cos 2\theta)$$

$$+ e^2[\tfrac{3}{4}\cos\theta + \tfrac{1}{4}\cos(3\theta^* - 2\omega)])[1 - \tfrac{3}{2}\sin^2 i\,(1 - \cos 2\theta^*)]$$

$$- \sin^2 i \sin 2\theta^*\,[2\sin\theta + \tfrac{3}{2}e\sin 2\theta + e^2(\tfrac{1}{4}\sin\theta + \tfrac{1}{4}\sin 3\theta)]\}$$

$$- \frac{d\Omega}{d\theta}\cos i \qquad\qquad\qquad (8.35)$$

Taking the average over one orbit provides

$$\left(\frac{d\omega}{d\theta}\right)_{av} = \tfrac{3}{2}J_2\left(\frac{R_\oplus}{p}\right)^2 (1 - \tfrac{3}{2}\sin^2 i) - \cos i\left(\frac{d\Omega}{d\theta}\right)_{av}$$

Combining this with result (8.34) gives a relatively simple form,

$$\left(\frac{d\omega}{d\theta}\right)_{av} = \tfrac{3}{2}J_2\left(\frac{R_\oplus}{p}\right)^2 (2 - \tfrac{5}{2}\sin^2 i) \qquad\qquad (8.36)$$

Note that $d\omega/d\theta$ is negative for inclinations between $63.4°$ and $116.6°$. Of course, if $i > 90°$ the orbit is retrograde. When i is near zero consider $(\Omega + \omega)$ rather than Ω, ω separately. In such cases

$$\lim_{i \to o}\left[\frac{d}{d\theta}(\Omega + \omega)\right]_{av} = \tfrac{3}{2}J_2\left(\frac{R_\oplus}{p}\right)^2 \qquad\qquad (8.37)$$

If this method is again applied to the remaining elements, the results would be

$$\left(\frac{di}{d\theta}\right)_{av} = \left(\frac{dh}{d\theta}\right)_{av} = \left(\frac{de}{d\theta}\right)_{av} = \left(\frac{da}{d\theta}\right)_{av} = 0 \qquad\qquad (8.38)$$

Therefore, only Ω and ω are effected to first order by J_2. Precession of the node line is sometimes used to maintain an orbit in sunlight year round. Such an orbit is referred to as *sun synchronous*.

8.3.3 Solar-Lunar Attraction

Another important problem which lends itself to treatment by the variation of parameters approach is evaluation of sun and moon attraction effects on a near-earth satellite. The *effective attraction* of the moon on a unit mass near earth is just the moon's attraction at the mass minus the moon's attraction at the earth's center. Referring to Figure 8.8 this is written as

$$\mathbf{f}_\mathbb{C}^{eff} = \frac{\mu_\mathbb{C}(\mathbf{r}_\mathbb{C} - \mathbf{r})}{|\mathbf{r}_\mathbb{C} - \mathbf{r}|^3} - \frac{\mu_\mathbb{C}\mathbf{r}_\mathbb{C}}{r_\mathbb{C}^3} \qquad\qquad (8.39)$$

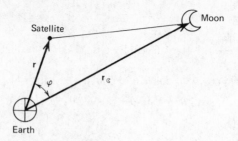

FIGURE 8.8 Effective attraction of the moon on an earth satellite.

Notice that this is consistent with the definition of effective attraction of the moon given in Section 7.2.2, except that distances were equal and normalized in the restricted three-body problem. Thus, the effective potential of the moon is

$$U_{\mathbb{C}}^{\text{eff}} = \frac{\mu_{\mathbb{C}}}{|\mathbf{r}_{\mathbb{C}} - \mathbf{r}|} - \frac{\mu_{\mathbb{C}}}{r_{\mathbb{C}}^3}(\mathbf{r}_{\mathbb{C}} \cdot \mathbf{r})$$

The second term compensates for the fact that the earth's center is accelerating around the barycenter. Now the perturbing acceleration due to lunar attraction is simply

$$\mathbf{f}_{\mathbb{C}}^{\text{eff}} = \frac{\partial}{\partial \mathbf{r}} U_{\mathbb{C}}^{\text{eff}}$$

Since $r \ll r_{\mathbb{C}}$ for near-earth satellites, expand the first part of $U_{\mathbb{C}}^{\text{eff}}$ as

$$\frac{1}{|\mathbf{r}_{\mathbb{C}} - \mathbf{r}|} = (r_{\mathbb{C}}^2 - 2\mathbf{r}_{\mathbb{C}} \cdot \mathbf{r} + r^2)^{-1/2}$$

which could be used as a generating function of a Legendre polynomial series. However, convergence is so fast that a simple expansion will suffice. Use

$$\frac{1}{|\mathbf{r}_{\mathbb{C}} - \mathbf{r}|} = \frac{1}{r_{\mathbb{C}}}\left\{1 + \left[\left(\frac{-2\mathbf{r}_{\mathbb{C}} \cdot \mathbf{r}}{r_{\mathbb{C}}^2}\right) + \left(\frac{r}{r_{\mathbb{C}}}\right)^2\right]\right\}^{-1/2}$$

$$= \frac{1}{r_{\mathbb{C}}}\left\{1 + \frac{\mathbf{r}_{\mathbb{C}} \cdot \mathbf{r}}{r_{\mathbb{C}}^2} + \left[\frac{3}{2}\left(\frac{\mathbf{r}_{\mathbb{C}} \cdot \mathbf{r}}{r_{\mathbb{C}}^2}\right)^2 - \frac{1}{2}\left(\frac{r}{r_{\mathbb{C}}}\right)^2\right] + \cdots\right\}$$

and the effective potential becomes

$$U_{\mathbb{C}}^{\text{eff}} = \frac{\mu_{\mathbb{C}}}{r_{\mathbb{C}}}\left\{1 + \left[\frac{3}{2}\left(\frac{\mathbf{r}_{\mathbb{C}} \cdot \mathbf{r}}{r_{\mathbb{C}}^2}\right)^2 - \frac{1}{2}\left(\frac{r}{r_{\mathbb{C}}}\right)^2\right] + O\left(\frac{r}{r_{\mathbb{C}}}\right)^3\right\} \qquad (8.40)$$

When the gradient operation is performed the constant term is dropped and

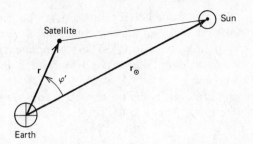

FIGURE 8.9 Effective attraction of the sun on an earth satellite.

the effective attraction becomes

$$\mathbf{f}_{\mathbb{C}}^{\text{eff}} = \frac{\mu_{\mathbb{C}}}{r_{\mathbb{C}}^{2}}\left\{3\left(\frac{\mathbf{r}_{\mathbb{C}}\cdot\mathbf{r}}{r_{\mathbb{C}}^{2}}\right)\frac{\mathbf{r}_{\mathbb{C}}}{r_{\mathbb{C}}} - \frac{\mathbf{r}}{r_{\mathbb{C}}} + O\left(\frac{r}{r_{\mathbb{C}}}\right)^{2}\right\} \tag{8.41}$$

Notice that the potential can be rewritten in geometric form

$$U_{\mathbb{C}}^{\text{eff}} \cong \frac{\mu_{\mathbb{C}} r^{2}}{2r_{\mathbb{C}}^{3}}(3\cos^{2}\phi - 1) \tag{8.42}$$

The effective potential of the sun is similarly derived. Referring to Figure 8.9,

$$U_{\odot}^{\text{eff}} \cong \frac{\mu_{\odot} r^{2}}{2r_{\odot}^{3}}(3\cos^{2}\phi' - 1) \tag{8.43}$$

In order to compute perturbation effects, vectors must be expressed in component form. Using Figure 8.10, the moon's position vector becomes

$$\frac{\mathbf{r}_{\mathbb{C}}}{r_{\mathbb{C}}} = \cos\delta_{\mathbb{C}}\,(\cos\alpha_{\mathbb{C}}\,\mathbf{I} + \sin\alpha_{\mathbb{C}}\,\mathbf{J}) + \sin\delta_{\mathbb{C}}\,\mathbf{K} \tag{8.44}$$

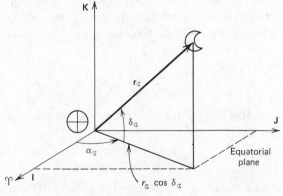

FIGURE 8.10 Components of the moon's position vector.

where $\alpha_{\mathbb{C}}$, $\delta_{\mathbb{C}}$ are right ascension and declination of the moon, respectively. Then $\mathbf{f}_{\mathbb{C}}^{\text{eff}}$ can be resolved in $\mathbf{i}_r$, $\mathbf{i}_\theta$, and $\mathbf{i}_z$ components using transformation (8.26). A similar treatment applies to $\mathbf{f}_{\odot}^{\text{eff}}$. These components are then used in the variation of parameter equations derived in Section 8.3.1 to determine effects on orbital elements.

An effect of particular interest is the precession of the equinoxes due to solar attraction on an oblate earth. The torque exerted by the sun on the earth is the vector

$$\mathbf{Q}_\odot = \mathbf{r}_\odot \times \left[-M_\odot \frac{\partial}{\partial \mathbf{r}_\odot} U_\oplus(r_\odot, \delta_\odot) \right]$$

Since only the first order oblateness is significant,

$$U_\oplus = \frac{\mu_\oplus}{r_\odot} \left\{ 1 - \frac{J_2}{4} \left(\frac{R_\oplus}{r_\odot} \right)^2 [1 - 3 \cos 2\delta_\odot] \right\}$$

where $\mathbf{r}_\odot$ is the position of the sun relative to earth, $\delta_\odot$ is the sun's declination relative to the equator, and $M_\odot$ is the sun's mass. Torque becomes

$$\mathbf{Q}_\odot = \frac{3\mu_\odot M_\oplus J_2 R_\oplus{}^2}{r_\odot{}^3} (-\sin i \cos i \sin \theta^* \mathbf{i}_\theta + \sin^2 i \sin \theta^* \cos \theta^* \mathbf{i}_z) \quad (8.45)$$

Transforming this into the $\mathbf{I}_1$, $\mathbf{J}_1$, $\mathbf{K}_1$ frame and noting that oblateness is related to a difference between axial and transverse inertias, that is,

$$M_\oplus J_2 R_\oplus{}^2 = I_Z - I_X$$

yields

$$\mathbf{Q}_\odot = \frac{3\mu_\odot (I_Z - I_X)}{r_\odot{}^3} \sin i (\cos i \sin^2 \theta^* \mathbf{I}_1 - \sin \theta^* \cos \theta^* \mathbf{J}_1) \quad (8.46)$$

where $i = 23.5°$, the *obliquity of the ecliptic*, and θ^* is the angular position of the sun past the vernal equinox. Vector $\mathbf{Q}_\odot$ must equal the rate of change of the earth's angular momentum, $I_Z \omega_\oplus \mathbf{n}$. Averaging over a one-year cycle of θ^* gives

$$\left(\frac{d\mathbf{n}}{dt} \right)_{av} = \frac{3}{2} \frac{\mu_\odot}{r_\odot{}^3} \left(\frac{I_Z - I_X}{I_Z \omega_\oplus} \right) \sin i \cos i \, \mathbf{I}_1 \quad (8.47)$$

with $\mathbf{I}_1$ along the intersection of the equatorial and ecliptic planes and defined by

$$\mathbf{I}_1 = -\frac{\mathbf{i}_z \times \mathbf{n}}{\sin i}$$

where $\mathbf{i}_z$ is the fixed direction of the pole of the ecliptic. Hence, the earth's axis precesses westward around $\mathbf{i}_z$, and the vernal equinox moves westward along the ecliptic at the rate

$$\frac{3}{2}\frac{\mu_\odot}{r_\odot{}^3}\left(\frac{I_Z - I_X}{I_Z}\right)\frac{\cos i}{\omega_\oplus}$$

Of course, the moon's effect must be superimposed, with inclination varying between 18.5° and 28.5° in an 18 year cycle. Thus, if $\mu_\odot/r_\odot^3$ is replaced by

$$\frac{\mu_\odot}{r_\odot{}^3} + \frac{\mu_{\mathbb{C}}}{r_{\mathbb{C}}{}^3}$$

and $i_{\mathbb{C}} = 23.5°$, then the observed precession rate of 1 revolution in 25,800 years leads to a value

$$\frac{I_Z - I_X}{I_Z} = 0.0032$$

Using $J_2 = 0.0010827$, leads to

$$I_Z = 0.34 M_\oplus R_\oplus{}^2$$

which is slightly less than that for a homogeneous sphere. Therefore, the earth must have a core which is denser than the outer part.

8.4 GENERAL PERTURBATIONS

This chapter concentrates on special perturbations and techniques for handling them. However, a treatment of orbital perturbations is not complete without at least a cursory discussion of general perturbations. Analytical integration of series expansions (or integrable expressions) of the perturbative accelerations is particularly useful in the prediction of orbits extending over long periods of time. This principle is, of course, useful in studying long-term environmental effects on the attitude of a passive body. Such analytical integrations are usually based on perturbative derivatives derived by the variation of parameters method. Thus, general perturbations permit solutions to orbit problems that are appropriate to various missions for given initial conditions. In addition, this approach permits better physical interpretation of perturbation effects and a clearer interpretation of the sources of perturbative forces. For example, an observed harmonic deviation or anomaly can be easily associated with its source. Discovery of the earth's *pear shape* by J. A. O'Keefe and A. Eckels was made possible through an analysis of long period general perturbation terms in orbital eccentricity. Jeffreys had used this procedure even earlier to determine the gravitational potential of Saturn from a general perturbation analysis of its six inner moons.

There are a large number of general perturbation schemes available in the literature. An attempt is made here to categorize them. The interested reader should survey other texts for further details.

1. Methods employing a reference orbit.
2. Methods using Keplerian elements. These typically involve true anomaly in the argument of periodic terms.
3. Methods based on the *Vinti potential*. These use spheroidal coordinates.
4. Methods using a short power series in eccentricity.
5. Methods employing an averaging process.
6. Methods using rectangular coordinates to handle the perturbations.

Of course, there are a number of combinations of general and special perturbations. Such methods are really special in nature, because they apply to specific situations.

8.5 NUMERICAL METHODS

Special perturbation methods must inevitably be supported by numerical integration techniques. Although the analytic formulation of a perturbation problem may be elegant, numerical results may be meaningless if the technique is not accurate enough. Many such cases occur because limitations of the numerical approach were not understood. Thus, a combination of experience and education in numerical analysis is essential. This section summarizes several of the available methods in an effort to assist in selection of integration techniques which are most appropriate for a given problem.

The selection process should lead to the most efficient method. A screening procedure should ask several questions: (a) Will a wide range in independent variables be required? This is associated with a large number of integration steps. (b) Will the problem be sensitive to small errors? (c) Will the dependent variables change rapidly? This is associated with a constant step-size. Primary factors to be considered for any given method are speed, accuracy, and storage requirements. However, all desirable qualities usually cannot be obtained in one method for any problem at hand. A compromise is necessary to make a practical selection. Some of the qualities to seek are: (a) large step-sizes to minimize computing time, (b) variable step-size provisions, (c) stability to avoid exponential growth of errors, (d) insensitivity to roundoff errors, and (e) controllability of truncation errors. These are certainly important factors to consider when evaluating an integration method.

Prior to selecting a method of numerical integration it is advisable to anticipate the kinds of errors involved. There are two primary sources of error that will always be present to some degree. These consist of roundoff and truncation errors. *Roundoff error* results from the number of digits carried in the computer. For example, if two nine-digit numbers are to be added in a machine which carries only eight digits, then the result is an eight-digit number. Consider the two numbers

$$72.5635686 + 53.5413727 = 126.1049413$$

which would appear in the computer as

$$72.563569 + 54.541373 = 126.104942$$

Thus, both numbers would be rounded off before addition, which in fact compounds the resulting roundoff error. The repeated process of roundoff in an integration process can result in significant errors. Brouwer and Clemence give a formula for the probable error after n steps of a double integration as $0.1124n^{3/2}$ in units of the last decimal, which means that in an extended number of examples about half the errors should be larger than this, and half smaller. This is equivalent to $\log_{10}(0.1124n^{3/2})$ decimal digits for n integration steps. For example, if a precision of six significant figures is required after 200 integration steps, then $\log_{10}[0.1124(200)^{3/2}]$ is about 2.5. This means that nine figures should be carried in the numerical integration process in order to be confident that accuracy to six figures is attained. One way to inhibit large roundoff error is to use double precision arithmetic.

Truncation error can be thought of as the result of an inexact solution to the differential equation. In other words, numerical integration is an exact solution of some difference equation which does not exactly represent the true differential equation. Thus, truncation error is the unused part of the series expression employed in the integration. Larger step sizes lead to greater truncation errors. On the other hand, smaller step sizes lead to large roundoff errors. Hence, errors are unavoidable in numerical integration and it is important to search for a way to minimize the sum of roundoff and truncation errors. Remember that roundoff error is a function of the computer used, while truncation error is a function of the integration method used.

In view of unavoidable truncation errors in numerical integration, it is advisable to survey all available methods before committing a great deal of effort to a solution. These schemes can be divided into single-step and multi-step categories. *Multi-step* methods usually require a single-step method to get them started at the beginning and after each step size change. *Single-step* methods include Runge-Kutta, Gill, Euler-Cauchy and Bowie. Multi-step methods include Milne, Adams-Moulton, Gauss-Jackson, Obrechkoff, and Adams-Bashforth. These multi-step techniques are sometimes called *predictor-corrector* methods. Most of these are designed to handle first-order differential equations but some will solve second-order systems directly. A few methods are quite popular and will be mentioned here. Runge-Kutta methods use a fundamental approach which approximates a Taylor series extrapolation of a function by several evaluations of the first derivatives at points within the interval of extrapolation. A variation on this is the Runge-Kutta-Gill approach; developed to minimize storage registers and roundoff errors. In general, Runge-Kutta methods are stable and do not require a starting procedure. They are simple, have a relatively small truncation error, and the step size is easy to change. Unfortunately, there is no simple way to determine truncation error, which makes it difficult to select an appropriate step size. This shortcoming led to development of multi-step or predictor-corrector methods. Such techniques take advantage of the history of the function being integrated. They are generally faster than single-step methods but at a cost of greater complexity and a requirement for a starting procedure. Some multi-step methods calculate a predicted value for the dependent variable x_{n+1} and then substitute it into the differential equation to get $\dot{x}_{n+1}$. This is used to calculate a corrected value of x_{n+1}, thus the name *predictor-corrector*.

In order to use this type of scheme, it is apparent that the predicted value is computed first. Then the derivative of this value permits determination of the corrected value using a corrector formula. It is further possible to repeat steps until there is no significant change in x_{n+1} for successive iterations. Of course, the step size can be changed if the truncation error is larger than desired. A Runge-Kutta method is used quite often to start this procedure. The Gauss-Jackson method is one of the most popular for trajectory problems of the Cowell and Encke type. It is designed to handle second-order systems of equations, and it is faster than integrating two first-order equations. The associated predictor is generally more accurate than the predictor and corrector of other methods, even though it also includes a corrector of its own. Furthermore, it exhibits particularly good control of accumulated roundoff errors. Of course, it has the expected disadvantage of increased complexity.

EXERCISES

8.1 If gravitational force were proportional to $1/r^3$, then equation (8.4) would be

$$\delta\ddot{\mathbf{r}} = \mathbf{a}_p + \frac{\mu}{\rho^4}\left[\left(1 - \frac{\rho^4}{r^4}\right)\mathbf{r} - \delta\mathbf{r}\right]$$

What would be the correct form of equation (8.8) for f?

8.2 Show the steps in the derivation of expressions (8.17) and (8.18) from (8.15) and (8.16).

8.3 Verify set (8.19) starting with equations (8.17) and (8.18).

8.4 Verify the expression for perturbing acceleration due to oblateness, equation (8.28). Use the gravitational potential given by form (8.27).

8.5 Show that the three components of perturbing acceleration due to oblateness, f_r, f_θ, and f_L are given correctly by set (8.29).

8.6 A satellite initially in circular orbit is subject to the constant (in magnitude and inertial direction) perturbation $\mathbf{f}$ due to solar radiation pressure. Assume the sun is in the orbit plane and ignore the earth's shadow. Describe the net motion per circuit, if any, of the vacant focus. How does this change the orbit?

8.7 One inclination for stationary perigee is $116.6°$. If $r_p = 6700$ km, select values of a and e consistent with a *sun synchronous* orbit. Assume first order oblateness only.

8.8 Demonstrate that first order oblateness (J_2 only) has no average effects on inclination, momentum, eccentricity, and orbital energy. That is, show that

$$\left(\frac{di}{d\theta}\right)_{av} = \left(\frac{dh}{d\theta}\right)_{av} = \left(\frac{de}{d\theta}\right)_{av} = \left(\frac{da}{d\theta}\right)_{av} = 0$$

8.9 Assume that general relativity theory yields

$$\left(\frac{d\rho}{d\theta}\right)^2 + \rho^2 = \frac{2}{h^2}(\mathscr{E} + \mu_\odot \rho) + \frac{2\mu_\odot \rho^3}{c^2}$$

where $\rho = 1/r$ and $c =$ speed of light. Show that the perihelion of a planet advances at a rate of

$$\frac{6\pi\mu_\odot}{c^2 a(1-e^2)} \quad \text{(rad/revolution)}$$

8.10 Prove that the moon's effective attraction is correctly defined by expression (8.39). Start by writing the equation of motion with respect to the barycenter. Then transfer to the earth-centered system.

8.11 Derive an expression for di/dt due to solar-lunar attraction. Integrate for one orbit to obtain $(di/d\theta)_{av}$.

8.12 Integrate torque expression (8.46) over one circuit of the earth about the sun and show that result (8.47) is correct.

8.13 Assume the moon's orbit is circular with inclination $i = 5.2°$ relative to the ecliptic.

(a) Ignore earth oblateness and show that the sun causes the longitude of the ascending node to precess as

$$\left(\frac{d\Omega}{d\theta}\right)_{av} = -\alpha \frac{\mu_\odot}{\mu_\oplus}\left(\frac{r_{\mathbb{C}}}{r_\odot}\right)^3 \cos i \sin^2(\Omega - \lambda_\odot)$$

and determine the value of α.

(b) Average this result over the yearly cycle of $\lambda_\odot$ (treating Ω on the right hand side as constant) to determine the period (in years) of westward movement along the ecliptic of Ω.

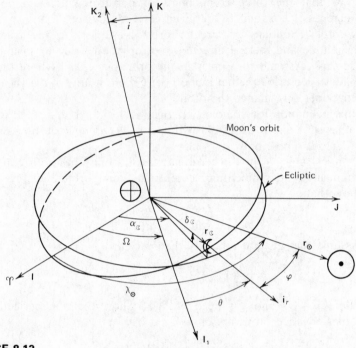

EXERCISE 8.13

REFERENCES

Baker, R. M. L., Jr., *Astrodynamics, Applications and Advanced Topics*, Academic Press, 1967, Chapters 2, 3, 4.

Bate, R. R., D. D. Mueller, and J. E. White, *Fundamentals of Astrodynamics*, Dover, 1971, Chapter 9.

Brouwer, D., and G. M. Clemence, *Methods of Celestial Mechanics*, Academic Press, 1961, Chapters 4, 5, 13.

Hamming, R. W., *Numerical Methods for Scientists and Engineers*, McGraw-Hill, 1962, Chapters 2, 11, 12, 15, 16.

Kelly, L. G., *Handbook of Numerical Methods and Applications*, Addison-Wesley, 1967, Chapter 14.

Special
Problems

Material on attitude dynamics, control, and astrodynamics presented to this point has been concerned with establishing fundamentals and illustrative examples to explain applications. This chapter is concerned with extending these principles to more sophisticated applications requiring somewhat unusual methods of attack, while remaining within the scope of this text. Three topics are included, two of which deal with unusual attitude maneuvers, i.e., not typically executed during the normal operations of a satellite. The final topic deals with a special attitude-sensing problem which arises during certain orbit correction maneuvers. All three cases have at least one thing in common; they address satellite control problems of practical concern.

9.1 ATTITUDE ACQUISITION MANEUVER OF A BIAS MOMENTUM SATELLITE

9.1.1 Sequence of Events

Many geostationary communications satellites and other high altitude spacecraft employ body-stabilized configurations with bias momentum devices and incorporate apogee kick motors. Specialized acquisition maneuvers have been developed in order to establish on-orbit attitude with a minimum of propellant and sensor penalties. During the apogee transfer phase spin stabilization is used to maintain stable orientation prior to apogee burn, to minimize kick motor thrust misalignment effects, to ensure solar power, and to minimize thermal gradients. As soon as orbit injection is complete, it is desirable to despin the vehicle, reorient it, and spin up the momentum wheel. Since this wheel is usually mounted along an axis transverse to the apogee motor axis, this sequence may be complicated. An attitude acquisition sequence which uses only the wheel spin-up motor to transfer momentum simultaneously from the initial spin axis to the pitch-axis-mounted wheel is now considered. The basic acquisition process consists of the three steps illustrated in Figure 9.1. At an appropriate apogee, with the kick

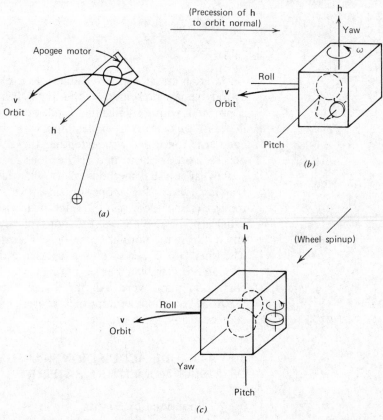

FIGURE 9.1 Transverse wheel acquisition sequence. (a) Pre-apogee kick. (b) Pre-acquisition. (c) Post-acquisition.

motor axis properly aligned in the desired thrust direction, circularization and inclination change take place. This situation is depicted in Figure 4.16. Spacecraft momentum is then decreased to minimize propellant expenditure for precession of the momentum vector. A value of momentum is selected which is close to that of the final bias momentum of the wheel. Some excess is usually included to permit a slewing rate about pitch after wheel spin-up. This will permit earth acquisition and minimize thermal gradients before final deployment of arrays, antennae, etc. After this adjustment of momentum magnitude, a precession is required to orient **h** along the orbit normal. The satellite is still spinning about its yaw axis with the wheel aligned along the pitch axis. Then the wheel motor may be turned on and attitude acquisition begins. Spacecraft momentum is absorbed by the wheel as the pitch axis aligns itself with the orbit normal. As the wheel approaches its final speed nutation begins, and dampers are used passively to decrease this motion in a reasonable time. No attitude sensors are required during this reorientation maneuver. Only the acceleration of the wheel with respect to the spacecraft must be limited. Otherwise, the maneuver is done in an *open loop* manner.

9.1.2 Equations of Motion and Stability

Equations describing the motion of a spacecraft with internal momentum appear in Section 5.2. Fortunately, the configuration of interest here corresponds to that of Figure 5.14 with two exceptions. The rotor is much smaller for this application, and the axes should be relabeled to avoid confusion. Assume that the roll, pitch, and yaw axes are nominally the principal set, and replace x, y, and z in Figure 5.14 with y, r, and p, respectively. This new nomenclature is presented in Figure 9.2. A linear damper is included to permit nutational damping and for stability considerations. No damping is expected from the small wheel. Therefore, equations (5.55) to (5.59) are rewritten here in the new nomenclature,

$$I_y \dot{\omega}_y - \omega_r \omega_p (I_r - I_p) + I_w \Omega \omega_r + m(1-\mu)\dot{\omega}_y z^2 - m(1-\mu)\omega_r \omega_p z^2$$
$$+ 2m(1-\mu)\omega_y \dot{z}z - mb\dot{\omega}_p z - mb\omega_y \omega_r z = 0 \quad (9.1)$$

$$I_r \dot{\omega}_r - \omega_p \omega_y (I_p - I_y) - I_w \Omega \omega_y + m(1-\mu)\dot{\omega}_r z^2 + m(1-\mu)\omega_y \omega_p z^2$$
$$+ 2m(1-\mu)\omega_r z\dot{z} - mb\ddot{z} + mb\omega_y^2 z - mb\omega_p^2 z = 0 \quad (9.2)$$

$$I_p \dot{\omega}_p - \omega_y \omega_r (I_y - I_r) + I_w \dot{\Omega} + mb\omega_r \omega_p z - 2mb\omega_y \dot{z} - mb\dot{\omega}_y z = 0 \quad (9.3)$$

$$I_w (\dot{\omega}_p + \dot{\Omega}) = T \quad (9.4)$$

$$m(1-\mu)\ddot{z} + c\dot{z} + kz - m(1-\mu)(\omega_y^2 + \omega_r^2)z + mb\omega_y \omega_p - mb\dot{\omega}_r = 0 \quad (9.5)$$

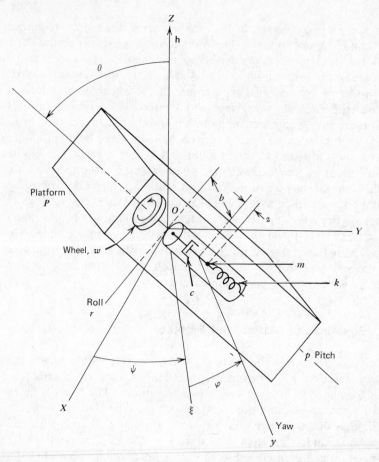

FIGURE 9.2 Nomenclature and configuration considered for attitude acquisition maneuver.

where I_w is the wheel moment of inertia about its axis and

$$\mu = \frac{m}{m + M_p + M_w}$$

Equations (9.1) to (9.5) constitute a complete description of attitude motion for a body containing a symmetric wheel and damper as specified above.

The special solution of primary interest corresponds to the one given by set (5.60) for nominal orientation during operational life of the satellite. Rewritten in the more convenient notation, this becomes

$$\omega_p = \omega_p = \text{constant (usually, orbital rate)}$$
$$\Omega = \Omega_R = \text{constant}$$
$$\omega_r = \omega_y = z = 0$$
$$T = 0$$

An identical stability argument to that of Section 5.2.3 can be made for this case. Therefore, conditions (5.69) apply here as

$$h - I_r\omega_p > 0, \; h - I_y\omega_P > 0 \tag{9.6}$$

where h is assumed positive by sign convention selection. If a situation arises in which $\omega_P > 0$, an unstable condition can exist. Although the desired state is one in which the wheel possesses almost all momentum, it is possible that the orientation maneuver could generate a relatively large value of ω_P. If the wheel cannot absorb the original momentum of the vehicle, then this instability is highly likely. The qualitative conditions for a successful acquisition maneuver and performance equations are the primary objectives here.

9.1.3 Simulations

Digital computer simulations offer a method of testing several combinations of parameter values in order to gain insight in selecting an analytical approach. A large number of cases were tested to study *maneuver success,* that is, proper reorientation without an instability. All cases assumed initial spin about the positive yaw axis. A constant net torque, T was applied to the momentum wheel, which was initially at rest relative to the vehicle body. The wheel was allowed to acquire its final momentum with respect to the vehicle, h_w. Moments of inertia, motor torque, and initial yaw rate were varied. Equations (9.1) to (9.5) were solved simultaneously and integration was carried out until h_w was reached.

The maneuver phase of primary interest is the initiation of momentum transfer. Restriction of consideration to this aspect permitted some simplification of the equations, because damping was proven to be a minor factor. This was demonstrated by comparing two digital computer programs, one with damping and one without. No appreciable differences could be observed before h_w was reached. An example case is illustrated in Figure 9.3. Angular rates about the body axes and nutation angle are plotted. The wheel speed profile is included as an indicator of maneuver progress relative to wheel momentum. Initial body momentum is $h = 41 \text{ N·m·s}$, requiring that $\omega_y(0) = 0.581$ rad/s. Several observations may be noted. The case shown corresponds to a vehicle with inertia ratios similar to those expected of future communications satellites.

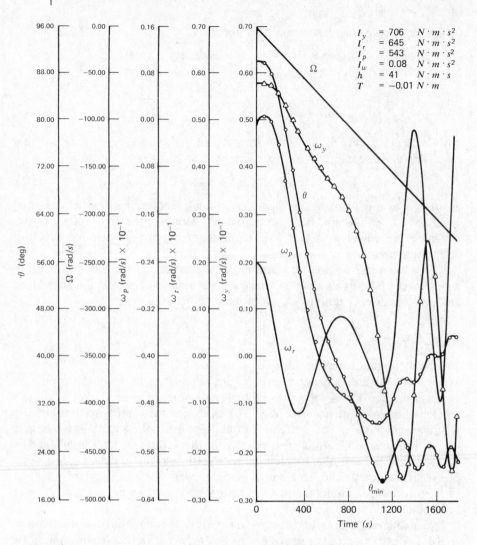

FIGURE 9.3 Example simulation of the acquisition maneuver.

Pitch rate starts upward in a positive manner as a direct response to wheel torque turn-on, but quickly reverses its sign. Nutation angle decreases monotonically until θ_{min} is reached. At this point the period of decreasing yaw rate and negatively increasing pitch rate ends. Beyond this point motion is essentially spin about the pitch axis with nutation and precession. Notice also that as the yaw rate first passes zero, θ_{min} is reached.

9.1.4 Physical Arguments

Dynamics of initial wheel start up will now be considered by two heuristic arguments. First, an interpretation of the equations of motion as the wheel speeds up is employed to derive a basic condition required of the inertia ratios. Second, a direct interpretation of internal torques and momentum balance which uses a dumbell inertia model is discussed. This technique permits derivation of a very basic and useful performance equation for the wheel spin-up portion of the acquisition maneuver.

Since damping can be ignored during wheel startup, equation (9.3) for pitch axis motion becomes

$$I_p \dot{\omega}_p + I_w \dot{\Omega} - \omega_y \omega_r (I_y - I_r) = 0$$

Let $I_p^P = I_p - I_w$ and use equation (9.4) to get

$$I_p^P \dot{\omega}_p = (I_y - I_r)\omega_y \omega_r + |T| \tag{9.7}$$

where T is always taken as negative in this analysis. Initial wheel startup conditions are developed by noting that the first term on the right-hand side is the *inertial torque*. It must have a negative value large enough to exceed the motor torque for a period of time near the beginning of the momentum wheel spin-up to prevent platform spin-up in the opposite direction. Assume that the vehicle is initially spinning about the yaw axis with $\omega_y > 0$. The roll rate is zero when the motor torque is first applied. Therefore, for at least a short period of time, the motor torque will dominate and ω_p will be positive. Once the wheel begins gathering momentum, ω_r becomes negative, because pitch motion induces a negative component of $\boldsymbol{\omega}$ along the roll axis. The sign of the inertial torque term thus depends on the sign of $(I_y - I_r)$. If the initial spin axis has greater inertia than the roll axis, inertial torque will have the required negative sign. This appears to be a necessary condition for proper initiation of the maneuver. In addition to being a negative quantity, the inertial torque must have a sufficiently large magnitude to exceed the motor torque. It might also be expected that parameter variations which tend to increase the magnitude of the inertial torque relative to the motor torque will result in the buildup of a greater net angular momentum along the pitch axis and therefore a smaller value of θ_{min}. Computer simulations have confirmed that decreases in $|T|$ and increases in $(I_y - I_r)$ and $\omega_y(0)$ do not produce smaller final nutations.

This interpretation can be visualized through the use of dumbbell models, originally used to gain insight into processes occurring during *flat-spin recovery* of a dual-spin satellite. The approach depicts platform and rotor asymmetries as pairs of dumbbells oriented to represent mass asymmetries (differences in transverse inertias). Figure 9.4 shows a basic dumbbell representation. If this were a dual-spin satellite both rotor and platform could be modeled in this

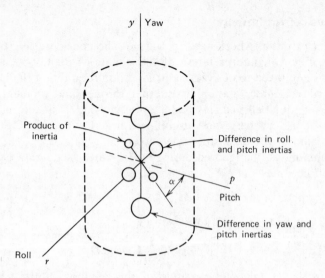

FIGURE 9.4 Dumbbell representation of inertia distribution.

way. It is also possible to represent dynamic inbalances (products of inertia) by using additional pairs of dumbbells. Such a case is included in Figure 9.4. The small pair is in the pitch-yaw plane at an angle α from the p-axis. This represents I_{yp}, leaving I_r as a principal axis.

Dynamic torques produced by centrifugal forces when the motor is engaged generally result in an upper bound on motor torque to achieve despin. This upper limit is associated with platform spin-up. To transfer momentum from a transverse axis to the wheel axis, the platform must resist being spun up while the rotor or wheel increases its momentum. Thus, an upper bound on motor torque is just the dynamic torque on the platform due to initial spin rate and inertia asymmetries of the platform itself. Since any bias momentum wheel to be used will be axially symmetric, there will be no dynamic torque to restrain the wheel from spinning up under any motor torque. Therefore, there is no lower bound on this applied torque. It is important to note that the upper bound does not necessarily determine the success of the reorientation, but only the nutation angle from which active or passive damping must bring the vehicle to the final orientation. Results indicate that this nutation angle generally increases with increased motor torque.

In order to describe the process of momentum transfer and the nature of restraining torques, consider a balanced asymmetrical platform depicted in Figure 9.5. This configuration corresponds to an inertia distribution in which $I_y > I_r > I_p$. Initially the spacecraft is spinning about the yaw axis and the wheel is stationary with respect to the body. When torque is applied to the wheel its

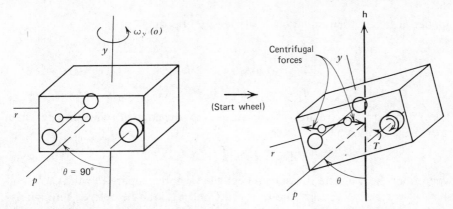

FIGURE 9.5 Wheel startup situation.

speed steadily increases. This torque is experienced by the platform in an opposite direction and a pitching motion begins. However, this motion is restrained by a restoring torque due to centrifugal forces acting on the small dumbbell pair. These forces are normal to the instantaneous angular velocity vector of the platform, which is initially close to **h**. Thus, if the motor torque is small, pitch motion is essentially stopped at an angle which produces an equal and opposite centrifugal torque. Meanwhile, the wheel continues to collect momentum. In order to obey the law of conservation of momentum, the wheel axis must tend to line up with the original yaw axis direction and the body rates must be decreasing. Thus, a rolling motion also takes place. The large dumbbells are deflected, but the yaw angular rate produces a centrifugal torque to restrain these. However, gyrocompassing torque from the wheel, given by equation (5.40), counters this restraining torque, and roll motion continues as the wheel spins up. As the body rates decrease, centrifugal torques diminish until they cannot match the motor torque. At this point nutation reaches a minimum and thereafter oscillates between two limits until damping dissipates this motion. Thus, pitch continuously adjusts until the orientation associated with maximum centrifugal torque is reached, at which time this torque just equals that of the wheel spin-up motor. The best situation is one in which initial yaw rate is high and the platform is highly asymmetrical about the pitch axis. Both of these factors contribute to increased dynamic torque, thus permitting higher motor torque.

9.1.5 Development of Performance Equations

Now that the physical basis for the transition maneuver has been established, performance expressions may be developed. The equations of motion are rewritten without damping. If the wheel is considered to be a producer of applied torques, then equations (9.1) to (9.3) become

$$I_y\dot{\omega}_y - \omega_r\omega_p(I_r - I_p) = -I_w\Omega\omega_r$$

$$I_r\dot{\omega}_r - \omega_p\omega_y(I_p - I_y) = I_w\Omega\omega_y$$

$$I_p\dot{\omega}_p - \omega_y\omega_r(I_y - I_r) = -I_w\dot{\Omega}$$

The pitch equation relates motor torque to dynamic torque directly when rewritten in form (9.7),

$$I_p\dot{\omega}_p - \omega_y\omega_r(I_y - I_r) = -T \tag{9.8}$$

where the wheel inertia is assumed negligible when compared to I_p. A relationship between inertia ratios, motor torque, and θ_{min} can be derived through use of expression (9.8). First, rewrite it in terms of Euler angles and rates using the transformation given by equation (1.27),

$$-T = I_p(\ddot{\phi} + \ddot{\psi}\cos\theta - \dot{\psi}\dot{\theta}\sin\theta)$$

$$-\tfrac{1}{2}(I_y - I_r)[\dot{\psi}^2\sin^2\theta\sin 2\phi + 2\dot{\psi}\dot{\theta}\sin\theta\cos 2\phi - \dot{\theta}^2\sin 2\phi] \tag{9.9}$$

Consider the physical situation as θ decreases for a given motor torque. The following events occur simultaneously: (1) θ approaches a minimum value (i.e., $\dot{\theta} \to 0$); (2) $\dot{\psi}$ approaches a stationary value to maintain overall momentum; and (3) $\dot{\phi}$ is essentially zero until $\dot{\theta}$ comes to zero due to torque balances. Thus, if θ just stops at the maximum specified value, beyond which damping can take over, then $T = T_{max}$. Equation (9.9) gives T_{max} by setting $\dot{\theta} = \ddot{\psi} = \ddot{\phi} = 0$,

$$T_{max} \cong \tfrac{1}{2}\dot{\psi}^2[(I_y - I_r)\sin^2\theta\sin 2\phi] \tag{9.10}$$

The appropriate value of ϕ at θ_{min} is the one which maximizes the right side of this equation. This angle changes slowly from 90° as θ decreases and continues to approach either 45° or 135° as θ stops. After $\ddot{\theta}$ reaches zero, $\dot{\varphi}$ changes significantly. Thus, replace $\sin 2\phi$ by 1. Now only $\dot{\psi}$ must be related to the physical situation at θ_{min}. Remembering that the value of θ is generally determined by

$$\sin^2\theta = \frac{h_y^2 + h_r^2}{h^2} \tag{9.11}$$

where $h_y = I_y\omega_y$ and $h_r = I_r\omega_r$, the situation at θ_{min} leads to

$$\dot{\psi} = \frac{\sqrt{2}h}{\sqrt{I_y^2 + I_r^2}} \tag{9.12}$$

Torque, momentum, and nutation may now be related through a single expression. Combine equations (9.10) and (9.12) to obtain the *transverse wheel performance equation,*

$$T^* = \sin^2 \theta_{\min} \tag{9.13}$$

where

$$T^* = \frac{T_{\max}(I_y^2 + I_r^2)}{h^2(I_y - I_r)} \tag{9.14}$$

This result has proven consistent with simulations. Thus, $\theta_{\min}$ is simply related to a nondimensionalized torque. Higher values of torque are permitted for higher values of $\dot{h}$ if $\theta_{\min}$ is specified. Notice that if $I_r > I_y$, definition (9.14) does not make sense. Thus, suspicion is aroused as to the validity of this result and the success of a maneuver if $I_r > I_y$. Remember that initial startup conditions required that $I_y > I_r$ to prevent the body from spinning up in the opposite direction. Simulations confirm that the only acceptable inertia ratios for success are those for which $I_y > I_r$. Furthermore, the best situations correspond to $I_y > I_p > I_r$ and $I_y > I_r > I_p$.

With the benefit of insight gained from the dumbbell model concept, results may be physically explained. The case in which the pitch axis is the intermediate principal axis with yaw as major axis would appear to give the lowest values of $\theta_{\min}$ for the same motor torque and values of inertias. This was confirmed by simulation. A worst case might be anticipated by considering other dumbbell arrangements. Refer to Figure 9.6 (case 6) in which yaw is the minor principal axis and roll the major axis. Here the restraining pitch torque is not available, but there is a pitching torque to help speed up pitch rate in the wrong direction. The pitch rate immediately becomes positive and remains positive with increasing magnitude. Nevertheless, simulations indicate a low value of $\theta_{\min}$ is reached, but the spacecraft has in fact turned itself upside down such that all wheel momentum is balanced by excess body momentum. The final pitch rate is sufficient to contain initial momentum plus wheel momentum. To summarize all the possible combinations, cases 1, 2, and 3 do maneuver to an upright position, but case 3 results in very large values of $\theta_{\min}$. Cases 4, 5, and 6 turn upside down.

To illustrate the stability problem of an upside down maneuver, apply criteria (9.6) to the final state stability of case 4, corresponding to $I_p > I_r > I_y$. Assume the wheel momentum magnitude is just equal to the initial body momentum about yaw. After wheel spin-up the distribution of momentum is

$$h_{\text{platform}} = 2h_{\text{wheel}} = 2h$$

Thus, net momentum h is

$$h = \frac{I_p \omega_p(f)}{2}$$

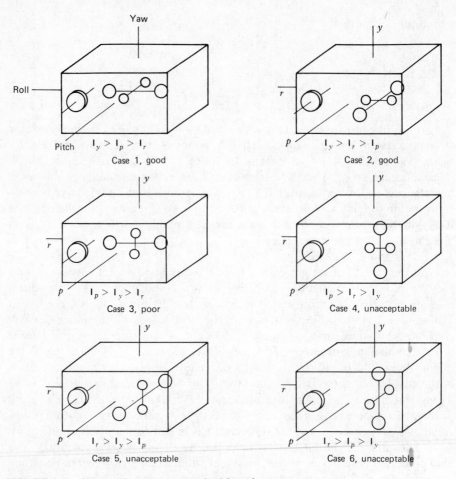

FIGURE 9.6 All possible asymmetrical inertia cases.

where $\omega_p(f)$ is the final platform rate. The stability conditions become

$$\left(\frac{I_p}{2} - I_r\right)\omega_p(f) > 0 \qquad \text{and} \qquad \left(\frac{I_p}{2} - I_y\right)\omega_p(f) > 0$$

or

$$I_p > 2I_r \qquad \text{and} \qquad I_p > 2I_y$$

This implies that stability about the pitch axis is only possible for this inertia distribution when the pitch inertia is at least twice that of roll and yaw, a physical impossibility. Wheel cross-coupling torques cause this instability which, in fact, would not exist without the wheel, because pitch is the major axis here.

In summary, analyses, physical arguments, and simulations indicate that the concept of simultaneous despin, reorientation, and momentum wheel spin-up appears practical for cases in which the initial spin axis is the major principal axis of inertia. Figure 9.6 may serve as a selection chart. Only cases 1 and 2 are acceptable. Fortunately, to maintain attitude stability during the transfer orbit, spin about the axis of maximum inertia is highly desirable. Spin about the minor axis is possible with active damping, but this would preclude the acquisition scheme proposed here. The RCA domestic communications satellite, launched in December 1975, was the first to employ this maneuver. It has an inertia ratio corresponding to case 1.

9.2 AUTOMATIC DETUMBLING OF A SPACE STATION

In the operation of manned space vehicles there is always a finite probability that an accident will occur which results in uncontrolled tumbling of a spacecraft. Such uncontrolled motion creates a hazardous environment to the crew, which would experience oscillating accelerations. The structural integrity of the disabled vehicle may be jeopardized by prolonged tumbling, presenting additional danger. A rescue vehicle probably would not arrive for at least 24 hours because of fueling, launching, and rendezvous operations. Attempts at hard docking by a manned rescue craft with the disabled vehicle would not be advisable, because the rescue crew would then be exposed to the same hazardous environment as the tumbling ship. Furthermore, the rescue craft may be large and difficult to maneuver, thus requiring excessive propellant usage. Detumbling could be accomplished by external means, for example, a remotely controlled robot or by impinging fluid jets on the disabled spacecraft. Of course, it would be desirable to develop an internal autonomous control system to either completely detumble the ship or lessen the tumbling motions until the rescue craft arrived. Such a device would become active upon loss of control. Many schemes could be successful, but the one to be considered here is a movable mass control device which reduces tumbling to simple spin. Then crew evacuation would be simplified significantly, and final despin by the rescue craft is easier. The system presented moves a control mass according to a selected control law, in the acceleration environment created by tumbling motion. By moving the mass properly, the rotational kinetic energy of the system may be increased or decreased creating simple spin states about the minimum or maximum inertia axis, respectively. The control system described here is designed for the latter case because this leaves the craft in a stable spin state.

9.2.1 Equations of a Tumbling Vehicle with Moving Mass

Complete equations of motion for a rigid spacecraft with attached control mass can be developed using the general formulation given by expression (5.47). The vehicle of interest here consists of a rigid main body and attached control mass, and is shown in generalized form in Figure 9.7. An appropriate reference point is the main body center of mass. Applying equation (5.47) yields a set of three coupled nonlinear differential equations for vehicle dynamics in terms of angular rates, and the movable mass position, velocity, and acceleration. All of these quantities are measured with respect to the x, y, z body-fixed principal axes. The equations for torque-free motion become

$$[I_x + \mu(y^2 + z^2)]\dot{\omega}_x + [I_z - I_y + \mu(y^2 - z^2)]\omega_y\omega_z + \mu[-xy\dot{\omega}_y - xz\dot{\omega}_z$$
$$+ (2y\dot{y} + 2z\dot{z})\omega_x + yz(\omega_z^2 - \omega_y^2) - 2\dot{x}y\omega_y - 2\dot{x}z\omega_z - xz\omega_x\omega_y$$
$$+ xy\omega_x\omega_z + y\ddot{z} - z\ddot{y}] = 0 \quad (9.15a)$$

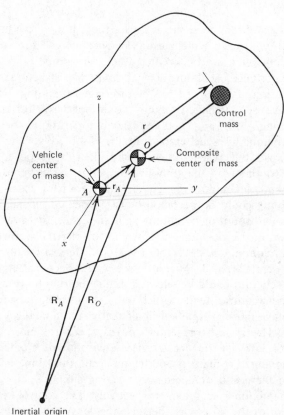

FIGURE 9.7 Main body and attached mass system geometry.

$$[I_y + \mu(z^2 + x^2)]\dot{\omega}_y + [I_x - I_z + \mu(z^2 - x^2)]\omega_z\omega_x$$
$$+ \mu[-yz\dot{\omega}_z - yx\dot{\omega}_x + (2z\dot{z} + 2x\dot{x})\omega_y + zx(\omega_x^2 + \omega_z^2) - 2\dot{y}z\omega_z - 2\dot{y}x\omega_x - yx\omega_y\omega_z$$
$$+ yz\omega_y\omega_x + z\ddot{x} - x\ddot{z}] = 0 \quad (9.15b)$$

$$[I_z + \mu(x^2 + y^2)]\dot{\omega}_z + [I_y - I_x + \mu(x^2 - y^2)]\omega_x\omega_y$$
$$+ \mu[-zx\dot{\omega}_x - zy\dot{\omega}_y + (2x\dot{x} + 2y\dot{y})\omega_z + xy(\omega_y^2 - \omega_x^2) - 2\dot{z}x\omega_x - 2\dot{z}y\omega_y - zy\omega_z\omega_x$$
$$+ zx\omega_z\omega_y + x\ddot{y} - y\ddot{x}] = 0 \quad (9.15c)$$

where ω_x, ω_y, ω_z are the angular velocity components of the main body, I_x, I_y, I_z are the principal moments of inertia of the main body, x, y, z are components of $\mathbf{r}$ in the body-fixed reference frame, and μ is the reduced mass $mM/(m+M)$ with m, M the control and main body masses, respectively. This set of equations is valid irrespective of the physical mechanism driving the control mass.

In view of the fact that equation (5.47) might have been written as

$$\dot{\mathbf{h}}_B + \mu\mathbf{r} \times \ddot{\mathbf{r}} = 0$$

where $\mathbf{h}_B$ is the main body angular momentum about its center of mass, the force acting on the control mass is defined as $\mathbf{f} = \mu\ddot{\mathbf{r}}$. This can be expressed in component form as

$$f_x = \mu[\ddot{x} - 2\dot{y}\omega_z + 2\dot{z}\omega_y - y\dot{\omega}_z + z\dot{\omega}_y + y\omega_x\omega_y + z\omega_x\omega_z - x(\omega_y^2 + \omega_z^2)] \quad (9.16a)$$

$$f_y = \mu[\ddot{y} - 2\dot{z}\omega_x + 2\dot{x}\omega_z - z\dot{\omega}_x + x\dot{\omega}_z + z\omega_y\omega_z + x\omega_y\omega_x - y(\omega_z^2 + \omega_x^2)] \quad (9.16b)$$

$$f_z = \mu[\ddot{z} - 2\dot{x}\omega_y + 2\dot{y}\omega_x - x\dot{\omega}_y + y\dot{\omega}_x + x\omega_z\omega_x + y\omega_z\omega_y - z(\omega_x^2 + \omega_y^2)] \quad (9.16c)$$

Rotational kinetic energy of the system about the composite center of mass may be generally expressed as

$$T_{\text{rot}} = \tfrac{1}{2}[\boldsymbol{\omega} \cdot \mathbf{I} \cdot \boldsymbol{\omega} + M\dot{\mathbf{r}}_c \cdot \dot{\mathbf{r}}_c + m\dot{\mathbf{r}}_m \cdot \dot{\mathbf{r}}_m] \quad (9.17)$$

By the definition of center of mass

$$\dot{\mathbf{r}}_c = m\dot{\mathbf{r}}/(m+M), \qquad \dot{\mathbf{r}}_m = M\dot{\mathbf{r}}/(m+M)$$

so that equation (9.17) becomes

$$T_{\text{rot}} = \tfrac{1}{2}[\boldsymbol{\omega} \cdot \mathbf{I} \cdot \boldsymbol{\omega} + \mu\dot{\mathbf{r}} \cdot \dot{\mathbf{r}}]$$

The rate of energy change is obtained by differentiation

$$\dot{T}_{\text{rot}} = \boldsymbol{\omega} \cdot \mathbf{I} \cdot \dot{\boldsymbol{\omega}} + \dot{\mathbf{r}} \cdot \mathbf{f} \quad (9.18)$$

Notice that

$$\mathbf{I} \cdot \dot{\boldsymbol{\omega}} + \boldsymbol{\omega} \times \mathbf{I} \cdot \boldsymbol{\omega} = -\mathbf{r} \times \mathbf{f}$$

which leads to

$$\boldsymbol{\omega} \cdot \mathbf{I} \cdot \dot{\boldsymbol{\omega}} = -\boldsymbol{\omega} \cdot \boldsymbol{\omega} \times \mathbf{I} \cdot \boldsymbol{\omega} - \boldsymbol{\omega} \cdot \mathbf{r} \times \mathbf{f}$$

The first term on the right-hand side is zero, giving

$$\dot{T}_{\text{rot}} = -\boldsymbol{\omega} \cdot \mathbf{r} \times \mathbf{f} + \dot{\mathbf{r}} \cdot \mathbf{f}$$

Remembering that

$$\dot{\mathbf{r}} = [\dot{\mathbf{r}}]_b + \boldsymbol{\omega} \times \mathbf{r}$$

permits

$$\dot{T}_{\text{rot}} = [\dot{\mathbf{r}}]_b \cdot \mathbf{f} \tag{9.19}$$

Therefore, the rate of change of kinetic energy is found to be independent of the vehicle inertia properties and dependent only on the relative velocity of the control mass and the force applied to the mass. Motion of the control mass is specified by a control law, which is selected using this relation.

9.2.2 Control Law Selection

Equations (9.15) determine the attitude response of the spacecraft for specified motion of the control mass. The control law relates motion of this mass to measurable vehicle parameters so that kinetic energy decreases. A satisfactory control law should not be unnecessarily complicated, nor have excessive power or sensor requirements. It should, however, require determination only of measurable vehicle parameters, produce stable responses, and result in a final state of simple spin about the axis of maximum inertia. The space station is generally assumed to have three distinct moments of inertia, I_x, I_y, and I_z, defined so that $I_z > I_y > I_x$. Since initial tumble rates may be equally large about all three axes, the equations of motion cannot be simplified by linearization. Fortunately, a limited number of simple cases may be identified which permit development of a suitable control law.

For assistance in formulating this control law the *Liapunov stability theory* appears to be appropriate. It states that a system of differential equations plus the mass motion represented by the control law will be completely stable and approach its minimum state if there exists a scalar function $V(u)$ such that

$$V(u) > 0 \qquad \text{for all } u \neq 0 \tag{9.20a}$$

$$\dot{V}(u) \leq 0 \qquad \text{for all } u \tag{9.20b}$$

$$V(u) \rightarrow \infty \qquad \text{as } \|u\| \rightarrow \infty \tag{9.20c}$$

where u is the state vector and $\|u\|$ is the *Euclidean norm* (length) of the state vector. It turns out that a convenient scalar to use here as the Liapunov function is just the system kinetic energy. Due to the nature of this function, conditions (9.20a) and (9.20c) are automatically satisfied. Thus, if a control law is chosen such that condition (9.20b) is satisfied, the system will be completely stable and approach its minimum state. To implement this, the limits of control mass movement must be specified. Assume a case in which this mass is restricted to move along a track parallel to the z axis, and offset from this axis as shown in Figure 9.8. The rate of change of rotational kinetic energy given by equation (9.19) becomes simply

$$\dot{T}_{rot} = \dot{z} f_z \qquad (9.21)$$

If, for example, the force applied to the control mass is selected as

$$f_z = -\mu c \dot{z} \qquad (9.22)$$

then the energy rate becomes

$$\dot{T}_{rot} = -\mu c \dot{z}^2$$

which satisfies condition (9.20b). Equating expressions (9.16c) and (9.22) yields the mass equation of motion

$$\ddot{z} + c\dot{z} - (\omega_x{}^2 + \omega_y{}^2)z = a\dot{\omega}_y - b\dot{\omega}_x - a\omega_z\omega_x - b\omega_y\omega_z \qquad (9.23)$$

This form suggests a second order system driven by motion of the tumbling vehicle. However, vehicle dynamics which produce this forcing function are determined by differential equations which are also functions of control mass motion. Since $m \ll M$, the effect of mass motion over one cycle is small, and the vehicle is experiencing nearly uncontrolled tumbling. This observation permits the forcing function to be thought of as dominated by vehicle motions over any

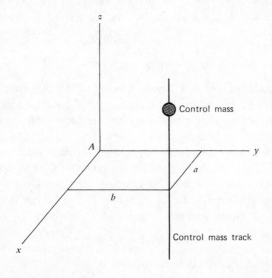

FIGURE 9.8 Control mass path.

given cycle. The resulting form of mass equation of motion may then be used to determine effects of control law on mass dynamics.

The control law given by equation (9.22) satisfies the required stability conditions and decreases rotational kinetic energy of the system. The forcing function vanishes when a final spin about the z axis is established. However, due to the negative, decaying coefficient of the z term in equation (9.23), the mass would not necessarily return to its initial zero position upon reaching a simple spin state. A *reset* to zero is necessary to restore full control capability of the system should another tumbling situation arise. This can be achieved by modifying the force applied to the control mass,

$$f_z = -\mu c_1 \dot{z} - \mu (c_2 + \omega_x^2 + \omega_y^2) z \qquad (9.24)$$

The equation of mass motion becomes

$$\ddot{z} + c_1 \dot{z} + c_2 z = a\dot{\omega}_y - b\dot{\omega}_x - a\omega_z \omega_x - b\omega_y \omega_z \qquad (9.25)$$

This is a conventional second order differential equation with damping and return to zero position. The associated rate of energy dissipation is given by

$$\dot{T}_{\text{rot}} = -\mu c_1 \dot{z}^2 - \mu z \dot{z}(c_2 + \omega_x^2 + \omega_y^2) \qquad (9.26)$$

This formulation departs from the Liapunov method since $\dot{T}_{\text{rot}}$ is not necessarily negative semidefinite. The first term decreases kinetic energy of the system monotonically, while the second term is oscillatory in nature. If the control system constants c_1 and c_2 are chosen properly the secular negative semidefinite term will dominate over each complete mass cycle. If every mass cycle has a net negative value for $\dot{T}_{\text{rot}}$, the system will approach its minimum energy state and simple spin.

9.2.3 Parameter Sizing Arguments

For the selected control law, expression (9.24), the rate of change of rotational kinetic energy is given by equation (9.26), as a function of two control system parameters, c_1 and c_2. Values for these two parameters should be chosen such that dissipation rate is high and simple spin is approached quickly. Of course, mass amplitude and power constraints must be satisfied. Guidelines for selection of these parameters are now developed, starting with some qualitative observations concerning effects of c_1 and c_2 on control system performance. The right-hand side of equation (9.25) is considered the forcing function of mass dynamics so that

$$\ddot{z} + c_1 \dot{z} + c_2 z = F$$

while the force on m due to vehicle motion is

$$F = a\dot{\omega}_y - b\dot{\omega}_x - a\omega_z \omega_x - b\omega_y \omega_z \qquad (9.27)$$

An analogy can be established between the mass motion and a simple spring-mass-damper system in which c_1 corresponds to a damping constant and c_2 corresponds to a spring constant. From expression (9.26) it would appear that the value of c_1 should be large. However, the c_1 term is also a function of mass relative velocity, which is influenced by c_1. In the spring-mass-damper system, a large value of c_1 corresponds to a strong damper, which would limit the speed of the mass. Thus, an increase of c_1 could result in a net decrease in the magnitude of secular dissipation. A similar tradeoff results for parameter c_2. The second term in equation (9.26) is oscillatory in nature, but it insures the return of the control mass to its zero position once a simple spin state is reached. To limit energy addition during parts of the mass cycle it would appear that c_2 should be small. However, the spring-mass-damper analogy indicates a small value of c_2 corresponds to a weak spring, allowing a relatively large amplitude of mass oscillation. In summary, it would appear that an optimum set of control system parameters may exist and should be pursued.

A completely analytical solution is not possible due to the nature of the equations involved. On the other hand a numerical solution is possible, but this provides no insight concerning effects of c_1 and c_2 selections. Therefore, a *hybrid* approach is taken in which both analytical and numerical aspects are employed. This is based on the assumption that the net change in rotational kinetic energy over one mass cycle is small. For a selected tumbling state the free dynamics may be solved numerically, and these results may then be used to calculate the forcing function profile from equation (9.27). Although F changes as tumbling diminishes, the initial tumble state provides a design point for sizing control system parameters. The nature of F is expected to be oscillatory, permitting a Fourier series to be fitted to a numerically generated forcing function profile for the free motion case. This series fit provides information concerning the relationship between control parameters and system performance. The Fourier series used is

$$F = A_0 + \sum_{n=1}^{\infty} \left(A_n \cos \frac{2n\pi t}{\tau} + B_n \sin \frac{2n\pi t}{\tau} \right)$$

where τ is the period of the function F. Since the forcing function is expected to contain no secular terms, let $A_0 = 0$. Also, F is a smooth function, permitting the series to be truncated with sufficient accuracy

$$F = \sum_{n=1}^{p} \left(A_n \cos \frac{2n\pi t}{\tau} + B_n \sin \frac{2n\pi t}{\tau} \right)$$

Coefficients A_n and B_n are obtained numerically from the known solution of F. This expansion is rewritten for convenience as

$$F = \sum_{n=1}^{p} D_n \sin \left[s_n t + \tan^{-1} \left(\frac{A_n}{B_n} \right) \right] \tag{9.28}$$

where $D_n = [A_n^2 + B_n^2]^{1/2}$ and $s_n = 2n\pi/\tau$. Equation (9.25) becomes

$$\ddot{z} + c_1\dot{z} + c_2 z = \sum_{n=1}^{p} D_n \sin\left[s_n t + \tan^{-1}\left(\frac{A_n}{B_n}\right)\right] \qquad (9.29)$$

with particular solution

$$z_p = \sum_{n=1}^{p} E_n \sin \phi_n \qquad (9.30)$$

where

$$E_n = \frac{D_n}{[(c_2 - s_n^2)^2 + (c_1 s_n)^2]^{1/2}}$$

$$\phi_n = s_n t + \tan^{-1}\left(\frac{A_n}{B_n}\right) - \tan^{-1}\left(\frac{c_1 s_n}{c_2 - s_n^2}\right)$$

Control mass speed is obtained by differentiating expression (9.30),

$$\dot{z}_p = \sum_{n=1}^{p} E_n s_n \cos \phi_n \qquad (9.31)$$

These solutions for position and velocity histories are used to determine the rate of change of rotational kinetic energy. Equation (9.26) takes the form

$$\dot{T}_{\text{rot}} = -\mu \sum_{j=1}^{p} \sum_{k=1}^{p} [E_j E_k s_k \{c_1 s_j \cos \phi_j \cos \phi_k + (c_2 + \omega_x^2 + \omega_y^2) \sin \phi_j \cos \phi_k\}]$$

The secular part of this equation corresponds to $j = k$ terms. Thus,

$$\dot{T}_{\text{sec}} = -\frac{\mu}{2} \sum_{n=1}^{p} c_1 E_n^2 s_n^2$$

or

$$\dot{T}_{\text{sec}} = -\frac{\mu}{2} \sum_{n=1}^{p} \frac{c_1(A_n^2 + B_n^2)s_n^2}{[(c_2 - s_n^2)^2 + (c_1 s_n)^2]} \qquad (9.32)$$

Several observations can now be made regarding the effect of control system variables on the secular part of energy dissipation. An increase in the control mass (and in μ) will linearly affect $\dot{T}_{\text{sec}}$. Increasing coefficients A_n and B_n will increase $\dot{T}_{\text{sec}}$ quadratically. This is accomplished by increasing the amplitude of the forcing function, which corresponds to increasing the mass offset distances a and b. Therefore, the control mass track should be placed at the maximum allowable distance from the center of mass of the vehicle. The effect of c_1 and c_2 is more difficult to interpret, but it is evident that c_2 should be of the order of s_n^2. With this assumption equations (9.32) and (9.30) become, respectively

$$\dot{T}_{\text{sec}} \cong -\frac{\mu}{2c_1} \sum_{n=1}^{p} (A_n^2 + B_n^2) \tag{9.33}$$

$$z_p \cong \frac{1}{c_1} \sum_{n=1}^{p} \left(\frac{A_n^2 + B_n^2}{s_n} \right) \sin \phi_n \tag{9.34}$$

The first indicates that c_1 should have a small value. However, the second equation implies that decreasing c_1 results in increasing the maximum amplitude of the mass motion. These observations agree with the spring-mass-damper analogy, and it is concluded that c_1 should be the smallest value which limits control mass amplitude to its maximum allowable value. Equations (9.30) and (9.32) may be used to generate a nomograph for selection of control system parameters c_1 and c_2. Once the control mass, track position, and an estimate of initial tumble state are obtained, $\dot{T}_{\text{sec}}$ may be calculated for various values of c_1 and c_2. Using equation (9.30) the corresponding maximum mass amplitudes may be determined. Each selected maximum mass amplitude has an associated optimum set of c_1 and c_2 values which correspond to a maximum energy dissipation rate. Thus, once the maximum allowable mass amplitude has been determined, appropriate values of c_1 and c_2 may be selected from the nomograph associated with the specified situation.

For the control law expressed as equation (9.24) it is evident that several quantities must be sensed by the control logic circuits. These include f_z, z, $\dot{z}$, and $(\omega_x^2 + \omega_y^2)$. Measurement of f_z may be accomplished with a linear accelerometer mounted on the control mass. If the accelerometer is aligned with the z-component of mass acceleration, then the output is proportional to $\ddot{z}$. Mass position and speed are easily sensed using any of a number of simple devices. The combination $(\omega_x^2 + \omega_y^2)$ may be determined with a linear accelerometer mounted on the z-axis, because acceleration of a fixed point P located by a vector $\mathbf{d}$ from the vehicle center of mass is

$$\mathbf{a}_P = \boldsymbol{\omega} \times (\boldsymbol{\omega} \times \mathbf{d}) + \dot{\boldsymbol{\omega}} \times \mathbf{d}$$

Here $\mathbf{d} = d\mathbf{k}$ and the $\mathbf{k}$-component of $\mathbf{a}_p$ is

$$(a_p)_z = -d(\omega_x^2 + \omega_y^2) \tag{9.35}$$

Therefore, sensor requirements for implementation of the control law appear to be modest.

9.2.4 Example Application

To demonstrate the feasibility of a movable mass system, using the control law of equation (9.24), a modular space station, shown in Figure 9.9, was

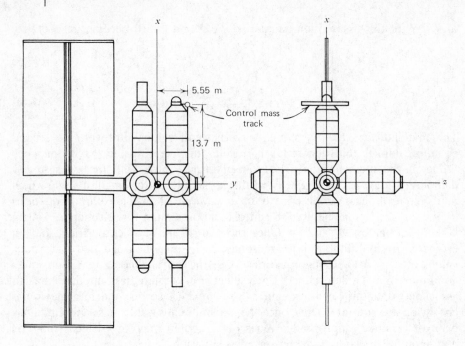

FIGURE 9.9 Modular space station configuration.

chosen as an example vehicle. The specified geometric axes are assumed to be principal ones and relevant properties of this spacecraft are given as

$$I_x = 5.15 \times 10^6 \text{ N·m·s}^2$$
$$I_y = 6.28 \times 10^6 \text{ N·m·s}^2$$
$$I_z = 6.74 \times 10^6 \text{ N·m·s}^2$$
$$M = 9.98 \times 10^4 \text{ kg}$$

This configuration was selected since it is a relatively large vehicle and does not have an artificial-g mode. A passive damping system such as a viscous ring or pendulum damper would not be practical tor this application. An estimated worst tumble state for this vehicle occurs as the result of a collision with a space shuttle orbiter. Initial rates would then be $\omega_x(0) = -2.86 \times 10^{-4}$ rad/s, $\omega_y(0) = -0.199$ rad/s, and $\omega_z(0) = 0.103$ rad/s. The control mass track is placed at the farthest allowable point from the vehicle center of mass and oriented parallel to the maximum inertia axis. For this spacecraft the mass track offset distances are taken as $a = 13.7$ m and $b = 5.55$ m. A control mass of 99.8 kg which corresponds to 0.1% of the vehicle mass was used to generate the nomograph of Figure 9.10. Different values of control mass will shift the curves

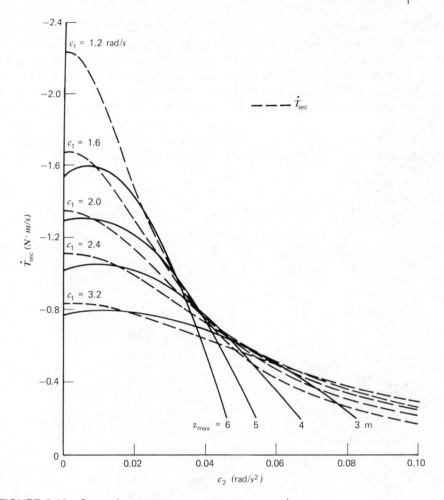

FIGURE 9.10 Control system parameter nomograph.

up or down but will not affect their shape. The dotted lines give the secular rate of change of rotational kinetic energy as a function of control system parameters c_1 and c_2. Smaller values of c_1 give larger values of $\dot{T}_{sec}$. Also shown by solid lines are curves which represent given maximum mass amplitudes. For each selected amplitude there appears to be an optimum set of c_1 and c_2 which results in near-maximum values of $\dot{T}_{sec}$. Although the nomograph shows relatively large energy dissipation rates for $c_2 = 0$, assumptions under which this was generated must be kept in mind. It was assumed that the forcing function would be purely oscillatory so that it could be approximated by a Fourier series. For the actual case, however, the system is damped, and a value of $c_2 = 0$ corresponds to having no spring in the spring-mass-damper analogy. This implies that the force is not oscillatory and the control mass would migrate away from the initial position at $z = 0$. Thus, a nonzero value of c_2 is required to insure return of the mass to $z = 0$. Once the control system parameters have been selected, mass and vehicle equations of motion may be solved numerically.

Before presenting results, the physical significance of the control law given by equation (9.24) is discussed briefly. Figures 9.11a and 9.11b show typical

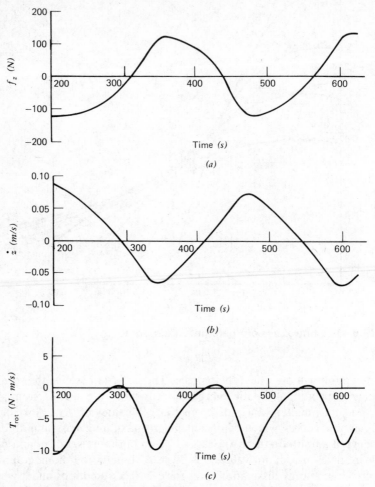

FIGURE 9.11 Typical cycles of control force, speed, and energy dissipation rate. (a) Control force. (b) Mass velocity. (c) Energy dissipation rate.

cycles of control force acting on control mass f_z and velocity relative to the body fixed axes $\dot{z}$, respectively. A comparison of these curves implies that the control law causes the force to be generally opposite to the relative velocity. Figure 9.11c presents the dissipation rate of the system, which is the product of force and relative velocity. It is apparent that $\dot{T}_{rot}$ is generally negative, corresponding to energy dissipation. The positive portions of $\dot{T}_{rot}$ correspond to energy addition. These intervals are due to the second term of equation (9.26) which was selected to ensure the control mass oscillates about its zero position and returns there after simple spin is established. This situation is further clarified by Figure 9.12 which shows typical cycles of control mass position over the same time period. Superimposed on each mass cycle are the directions of control mass speed and the force. It is evident that the force is generally a retarding one. In the first half cycle the mass is moving *up* and the applied control force is directed *down* which produces energy dissipation. Eventually, this force will be in the same direction, resulting in energy addition. However, this lasts for only a short time before force and velocity are again directed oppositely.

If a maximum mass amplitude of 3 m is selected, the appropriate values of c_1 and c_2 are selected from Figure 9.10 as $c_1 = 3.2$ rad/s and $c_2 = 0.02$ rad/s^2. Although this nomograph is based on a 0.1% control mass, it can be used for masses which represent up to a few percent of the vehicle mass. Simulations were run using a 1% control mass of 998 kg, with resultant angular velocity

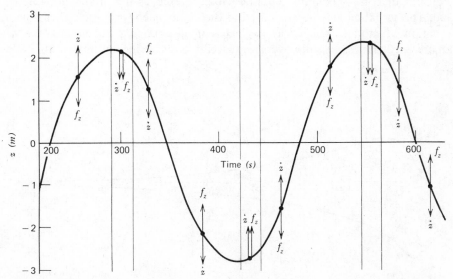

FIGURE 9.12 Typical cycle of control mass position.

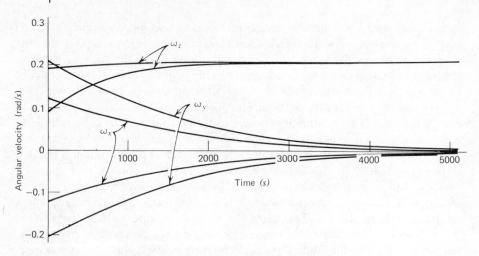

FIGURE 9.13 Envelopes of angular velocity oscillations.

histories shown in Figure 9.13. Since decay of oscillation amplitude is of primary importance, only envelopes formed by these oscillations are presented. The control system effectively collapses ω_x and ω_y envelopes to zero, as the mean value of ω_z increases to its steady spin value, consistent with constant total angular momentum. Figure 9.14 illustrates the displacement envelope and indicates that the maximum amplitude slightly exceeds the predicted value of 3 m. This variation may be attributed to the homogeneous solution of the mass equation of motion which was neglected in determining the nomograph of Figure 9.10. However, this overshoot occurs only during the first mass cycle

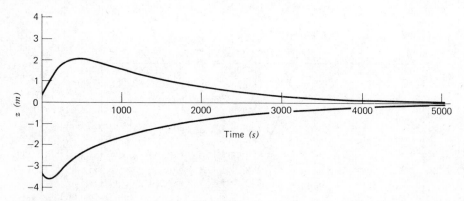

FIGURE 9.14 Envelope of control mass oscillations.

and is small. Asymmetry of the envelope is also associated with the homogeneous solution, since oscillations become symmetric as transients decay.

Figures 9.13 and 9.14 indicate that for the stated initial conditions, a maximum mass amplitude of approximately 3 m, and a control mass of 998 kg, this movable mass control system is capable of converting tumbling motion into simple spin within 2 hr. It is enlightening to investigate the effect of various parameters on control system performance via a time constant, τ_c, defined as the time required for the control system to collapse the ω_x or ω_y envelope to $1/e$ times its initial value. Sensitivity of system performance on mass size is illustrated in Figure 9.15. Above about 0.2% an increase in control mass has a marked effect on τ_c. The shape of the curve is as expected since an extremely small mass produces very little dissipation while an extremely large mass produces a large dissipation rate. However, peak power and force required also increase with increased control mass size. Thus, the control mass should be selected as large as possible, consistent with imposed mass, power, and force limitations. The total energy required to operate the control system does not vary appreciably with control mass weight. Finally, to demonstrate the effect of mass displacement amplitude on control system performance, two other sets of c_1, c_2 values were selected from Figure 9.10. The results are shown in Figure 9.16 in the form of time constant versus maximum mass amplitude for a control mass of 499 kg. Table 9.1 presents the values of c_1 and c_2 used along with resulting maximum amplitudes during first and second mass cycles and predicted values given by Figure 9.10.

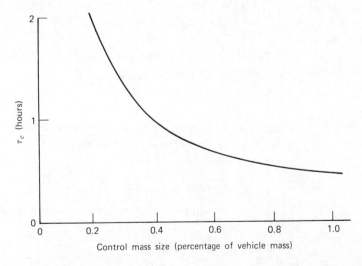

FIGURE 9.15 Variation of time constant with magnitude of control mass.

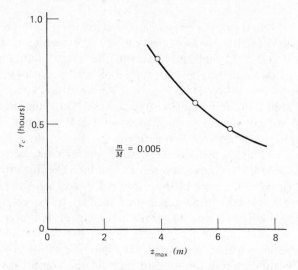

FIGURE 9.16 Variation of time constant with maximum mass amplitude.

In summary, for a large space station that is tumbling as a result of a collision with a shuttle orbiter, a movable mass system is capable of detumbling the vehicle within a period of two hours for the assumptions used. This can be accomplished with 1% of the vehicle mass and a maximum displacement amplitude of approximately 3 m. The several conclusions regarding design of such control systems must include the following: (1) The mass track should be placed as far as possible from the vehicle center of mass and oriented parallel to the maximum inertia axis; (2) The control mass size should be as large as is consistent with peak force and power limitations; (3) Better performance is available through larger mass amplitudes.

TABLE 9.1 Predicted and Computed Maximum Mass Amplitudes for Modular Space Station Simulation

c_1	c_2	Predicted z_{max}	Computed z_{max} (m)	
(rad/s)	(rad/s²)	(m)	First Cycle	Second Cycle
3.2	0.020	3.0	3.7	2.8
2.4	0.014	4.0	5.1	3.7
1.9	0.012	5.0	6.3	4.6

9.3 YAW SENSING STRATEGY FOR INCLINATION CONTROL

Body-stabilized, synchronous communications satellites present unique problems in many areas. Of particular interest here is the interaction of orbital inclination control impulses with attitude control performance. High pointing accuracy must be maintained throughout such maneuvers. However, these spacecraft typically employ bias momentum devices and hydrazine propulsion systems. Anticipated thrust misalignment torques due to inclination correction thrusters exceed solar pressure torques, which are the basis for sizing bias momentum devices. This will result in excessive yawing unless direct yaw sensing is provided. Several techniques for measuring yaw error include the use of rate gyros, direct yaw sensing by star tracking, and radio frequency methods employing associated ground equipment. It is possible to avoid the need for direct yaw measurement by using smaller and more frequent thrust impulses for inclination correction or by increasing momentum wheel stiffness through increased angular momentum. Sun sensors can provide yaw information when the satellite is near the dawn and evening terminators. However, any technique which provides continuous yaw sensing is costly in terms of complexity, weight, and operations. Alternate methods which avoid direct yaw sensing are also costly and tend to counteract the advantages of using bias momentum wheels and hydrazine thrusters. One example is the use of bias momentum values which are larger than needed for normal operations between inclination corrections. Since the basic attitude control system does not nominally require continuous yaw sensing, it seems advantageous to consider a yaw sensing technique that requires a minimum of complexity, weight, and added operations during inclination corrections. Thus, if yaw measurement is required only for short intervals of time, a simple device which can operate during those times is quite desirable. This motivation has led to a technique for using sun sensors so that yaw requirements are satisfied. In other words, the part-time capability of sun sensors can be matched to inclination correction operations by considering orbital geometry and mission requirements.

9.3.1 Orbit Normal Drift and Solar Geometry

For synchronous orbits with inclinations less than about 5°, the orbit normal vector is a more convenient representation than inclination and node position. This vector has components of inclination i and azimuth position, and is usually represented by its projection on the equatorial plane. Figure 9.17 illustrates zones of desirable normal vector position immediately after an inclination correction. Perturbing forces generally tend to precess this vector toward the first point of Aries, Υ. The azimuthal position of the orbit normal lags the ascending node position by 90°. It is possible to maximize the time interval

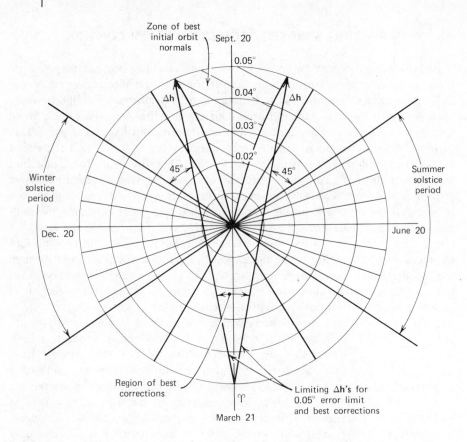

FIGURE 9.17 Qualitative plot of orbit normal for 0.05° limit and ±45° field of view.

between successive inclination corrections for a given tolerance. In addition to illustrating the sun's position at different times of the year, this figure indicates that maximizing this interval generally requires positioning the orbit normal away from ♈. Referring to Section 3.3, the ideal firing points on the orbits are at the nodes. Therefore, *solstice periods* are associated with very poor geometry for yaw sensing via the sun sensors. The lengths of these periods, which are centered around June 20 and December 20, Correspond to the sensor field of view, which is ±45° in Figure 9.17. Note that these periods can be narrowed somewhat without incurring a propellant penalty by firing slightly off node. The angular momentum change vector, **Δh**, represents Δv requirements per unit mass. The impulse per unit mass is merely the quantity of Δv imparted. Since this quantity is applied as a torque to the orbit plane, the associated value of Δv

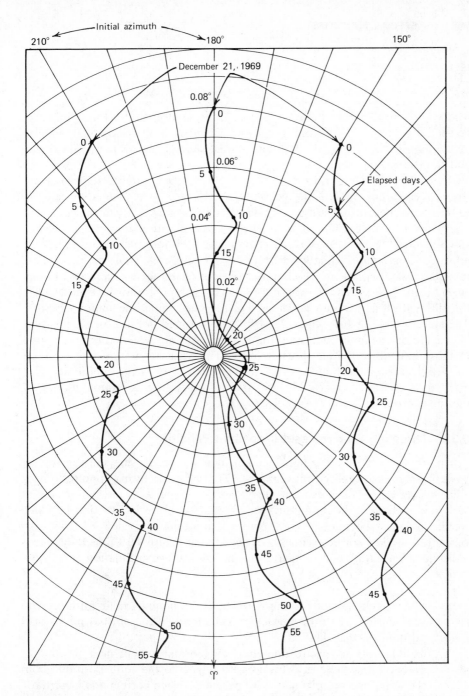

FIGURE 9.18 Examples of orbit normal precession for an inclination limit of 0.08°.

is

$$\Delta v = v\left(\frac{\Delta h}{h}\right) = v\sqrt{i_b^2 + i_a^2 - 2i_b i_a \cos \Delta\Omega} \qquad (9.36)$$

where i_b, i_a are the inclination values just before and just after thrust impulse, respectively, v is the orbital velocity (3070 m/s), h is the orbital angular momentum (v^2/ω_o), and $\Delta\Omega$ is the azimuthal shift of the orbit normal vector and the shift in the node line.

Notice that the thruster firing position is always ±90° azimuthally away from the corresponding $\Delta\mathbf{h}$. Qualitative limits on $\Delta\mathbf{h}$ for a 0.05° inclination holding value and for best propellant use are depicted in Figure 9.17. Normals to all $\Delta\mathbf{h}$ vectors which correspond to ideal propellant utilization fall in the solstice zones, indicating that sun sensors cannot give yaw information during these intervals if optimum inclination corrections are required. Thus, the problem is to determine alternate procedures for use during these two periods of each year. Figure 9.18 shows the orbit normal precession process over several weeks for different initial azimuth positions, but the same initial inclination of 0.08°. The starting date of December 21, 1969 was selected because the lunar-solar effects are maximum at that time. Initial conditions are assumed to correspond to those immediately after thruster firing. Thus, the maximum time between corrections for a 0.08° limit is about 48 days under worst-case conditions. This interval is generally longer and varies due to the cyclic precessional motion of the lunar orbit, with a period of 18.3 years.

9.3.2 Thrusting Strategy and Propellant Penalties

The objective of minimizing propellant penalty while using sun sensors for yaw information is realized through a series of off-node thrust impulses during each solstice period. The exact strategy and time depend on the sensor field of view and inclination holding limit. for example, consider a situation in which inclination must be held to 0.08° with a sensor field of view of ±45°. An optimized summer solstice sequence is depicted in Figure 9.19. The following is a list of events for 1974, which is a year of average inclination precession activity:

(a) May 7 is the last day to apply thrust at the nominal node position with yaw sensing. The orbit normal must be brought to point 1 from its drift position on that day.

(b) On May 30, to ensure that the orbit normal is kept near the line of precession, symmetrical firings with regard to the sun's position about the node line are planned. The period between corrections is 41 days. A firing on May 30 which brings the orbit normal to Point 2 guarantees the sun will pass the solstice point midway in this period.

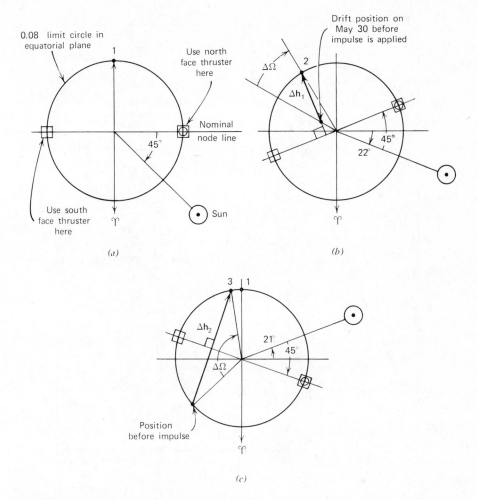

FIGURE 9.19 Basic thrusting strategy for a sensor field of view of ±45°. (a) May 7. (b) May 30. (c) July 12.

(c) On July 12, the inclination limit is reached and another impulse is applied. The sun is now leaving the solstice region and the orbit normal can be brought to Point 3, which is very close to ideal Point 1.

Figure 9.20 shows the path of the orbit normal over this period. If the availability of yaw sensing were independent of sun position, then the reference propellant situation would be established. Computer simulations were used to determine propellant usage for a limited number of cases with and without yaw

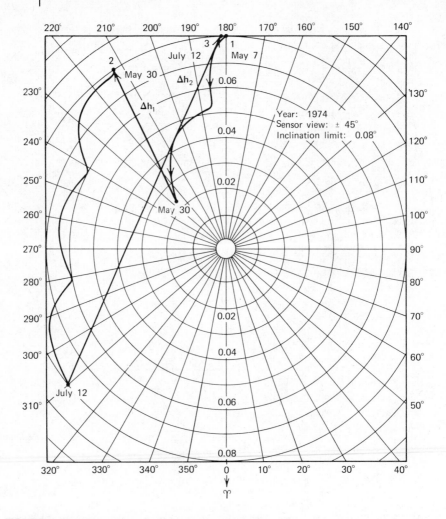

FIGURE 9.20 Example of orbit normal history.

constraints. The sequence described for 1974 indicated that the difference in velocity increments was only 0.30 m/s for the entire summer solstice interval. For an 810 kg satellite with a thruster of 200 second specific impulse, the propellant penalty was 0.124 kg; for all of 1974, it would be 0.248 kg. A similar sequence of thrusts was applied to years of high and low inclination precession, 1969 and 1978, respectively. An average of these data resulted in a 7-year propellant penalty estimate of 1.61 kg (0.1 percent of the usual propellant load for this size satellite.) A similar strategy was used for selected cases with fields of view up to ±60° and inclination limits of 0.06° to 0.1°. Because of the 1° per day motion of the sun, the thrusting strategy becomes more complicated when the maximum number of days between corrections is less than approximately 90 minus the number of degrees in the sensor field of view. For example, in the preceding case, the number of degrees is 45. Thus, the associated strategy is relatively simple because the maximum time between corrections is 47 days.

The maximum number of days between corrections is a function of perturbation magnitude and inclination holding limit. Solar-lunar attraction precesses the orbit normal at an average rate of about 0.003° per day. Thus, a sensor field of view of ±45° implies that thrusting strategy increases in complexity for inclination holding limits below about 0.07°. For example, consider a holding limit of 0.06° with this sensor field of view during 1969, a year of maximum orbit normal precession. Figure 9.21 illustrates the precession and correction history. It is important to notice that some of the applied impulses are used to move the orbit normal only a partial distance away from the limit circle. These maneuvers are essential to permit more time for the sun to leave the solstice region and thus return the strategy to normal. In the case shown in Figure 9.21

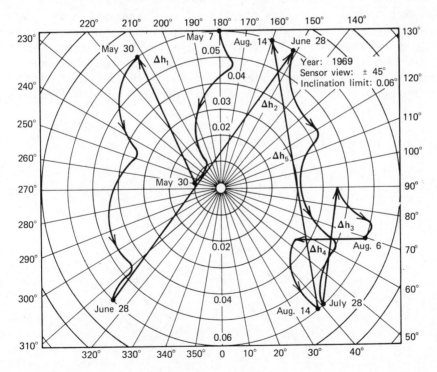

FIGURE 9.21 Example of complicated strategy.

the basic strategy can be used after August 14. Associated penalties are significantly higher for such cases. Thus, complicated strategies should be avoided through increased inclination holding limits and/or increased sensor fields of view.

Propellant penalties for optimum situations can be determined by using the simulation procedure outlined in the preceding section. This procedure should be used for the years of expected orbital life and for various fields of view and inclination holding limits; however each data point requires a substantial amount of time. To obtain conservative penalty figures, a simplified method was used to determine near-optimum values. The perturbation is modeled by a steady drift of 0.003° per day toward ♈. Figure 9.22 illustrates the simplified procedure for a sensor field of view of ±45° and inclination limit of 0.08°. On May 7 a maneuver brings the orbit normal back to Point 1. Drift is permitted until May 30, when another impulse is applied and Point 3 is reached. Drift is again permitted (in this case, for 45 days) until Point 4 is reached on July '14. An impulse then brings the orbit normal back to the centerline at Point 5. Further maneuvers do not result in propellant penalties until late fall. The difference in velocity increments is 0.15 m/s as opposed to 0.082 m/s for the optimum case in 1974. Thus, the simplified technique yields conservative estimates of propellant penalties. Results of calculations for inclination limits of 0.06° to 0.1° and a sensor field of view of ±45° are shown in Figure 9.23. An 810 kg satellite and thrusters with 200 second specific impulses have been assumed. The result of similar calculations for an inclination limit of 0.08° and varying fields of view, between ±45° and ±60° are plotted in Figure 9.24.

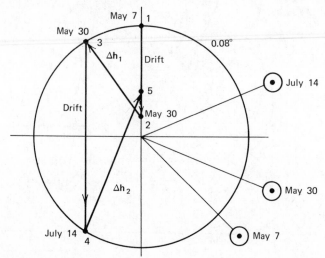

FIGURE 9.22 Simplified strategy for propellant penalty calculations.

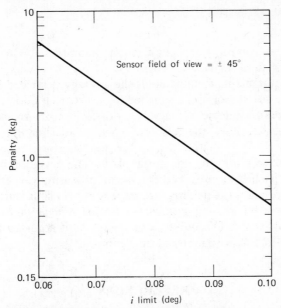

FIGURE 9.23 Effect of inclination on propellant penalty.

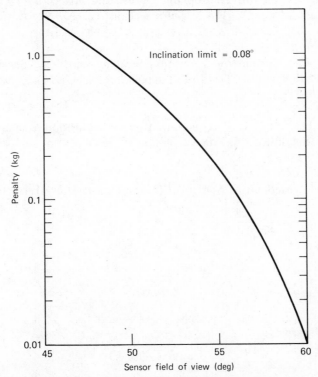

FIGURE 9.24 Effect of sensor field of view on propellant penalty.

These two figures confirm that penalties are minimized for larger inclination limits and sensor fields of view.

Hardware requirements to implement the strategy described here include a minimum of one sun sensor and associated logic circuits if both north and south facing thrusters are available. Otherwise, two sun sensors are necessary.

During solstice periods, the absolute value of declination of the sun is between 20° and 23.5°. Thus, the technique described here employs the projection of the sun onto the equatorial plane. The deviation in the expected declination has itself been proposed for use in measuring yaw error. However, sensitivity is poor and high pointing accuracy is not possible during the solstice periods. During spring and fall months, sun sensor application to yaw measurement is straightforward. Of course, sensor alignment accuracy and sensitivity must be compatible with year-round operations.

REFERENCES

Edwards, T. L., and M. H. Kaplan, "Automatic Spacecraft Detumbling by Internal Mass Motion," *AIAA Journal*, Vol. 12, *No. 4*, April 1974, pp. 496–502.

Kaplan, M. H., "Inclination Correction Strategy with Yaw Sensing via Sun Angle Measurements," *COMSAT Technical Review*, Vol. 5, *No. 1*, Spring 1975, pp. 15–27.

Kaplan, M. H., and T. C. Patterson, "Attitude Acquisition Maneuver for Bias Momentum Satellites," *COMSAT Technical Review*, Vol. 6, *No. 1*, Spring 1976, pp. 1–23.

Meirovitch, L., *Methods of Analytical Dynamics*, McGraw-Hill, 1970, Chapter 6.

ANSWERS TO SELECTED EXERCISES

CHAPTER 1

1.5	0.0053, −1.798, 10.038
1.6	2.406 m, −2.766 m, 0.750 m
1.7(a)	1.0 rad/s, 0, 181.73 rad/s
1.12	$\tan^{-1}(\Omega^2 R_E \sin\lambda \cos\lambda/g)$
1.13	650.5 m, west of muzzle

CHAPTER 2

2.2(d)	$(\pi/2 - e)/(2\pi)$
2.11(a)	0.479
(b)	106.3°
2.12	$\tau = 91.98$ min, $e = 0.037$
2.13(a)	3×10^4 **K** km^2/s
(b)	-13.7 km^2/s^2
(c)	2258 km
(d)	0.92
(e)	14,547 km
2.19(a)	30 N·m·s^2
(b)	3000 N·m
2.20(c)	13.82, 36.18, 40.0 (N·m·s^2)
2.22(a)	25, 25, 50
(b)	$\cos(1, x) = 1.0$, $\cos(1, y) = 0$, $\cos(1, z) = 0$
	$\cos(2, x) = 0$, $\cos(2, y) = 0.8$, $\cos(2, z) = 0.6$
	$\cos(3, x) = 0$, $\cos(3, y) = 0.6$, $\cos(3, z) = 0.8$

CHAPTER 3

3.1(b)	7000 km
3.3(b)	$6\pi H\sqrt{a/\mu}$
3.4(a)	$\theta_o = 28°$, $e = 0.56$
(b)	$a = 14,771$ km, $r_p = 6513$ km
(c)	0.089
3.5(a)	$\theta_0 = 110°$, $e = 0.34$
(b)	$\beta = 30°$, $v = 7.95$ km/s
3.7	1.49 km/s
3.8(a)	26,610 km
(b)	$e = 0.583$, $\Delta v = 3.32$ km/s
3.10	0.66 km/s

3.11(a) 2.0
   (b) 120°
   (c) −99,650 km
   (d) 93,272 km
   (e) 3.46 km/s
3.15 $\Delta v = 6.3$ km/s, $\theta_{\infty/\oplus} = 116.1°$
3.17(a) $e = 1.64$, $\delta = 75.0°$, $\theta_{\infty/\delta} = 127.5°$
3.19(a) 3.14 km/s
   (b) 2997.4 km
   (c) 10 days
3.22(a) $\Delta = 4700$ km, $e = 1.36$
   (b) 3.14 km/s
   (c) 0.84 km/s
3.28(a) $\dot{x}_0 = 0.0814$ km/s, $\dot{y}_0 = -0.123$ km/s

CHAPTER 4

4.2 4.51%
4.3(a) 98
   (b) 2.6 min
4.4(a) 800 s
   (b) 20 rad/s
4.9(a) $\psi = 53.8°$, $\sigma_1 = 50.7°$
   (b) 49.5°

CHAPTER 5

5.6(a) Set $\alpha = 0$
   (b) 32.4°
5.7(a) 6.43°/day
5.9 $k > 20$
5.10(a) 0.93 rad/s
   (b) 0.23 s
   (c) 48.8 rad/s^2
5.11 $l = 1.47 \times 10^{13}$ km
   $t = 3.7 \times 10^8$ days
5.16 $I_1 = I_{yaw}$, $I_2 = I_{roll}$, $I_3 = I_{pitch}$

CHAPTER 6

6.2(a) $G(s)/[1 + H_1(s)G(s)]$
6.7 $0 < K < 86$
6.8 One pole at $s = 1.35$
6.15 $|\Delta T_z| = 1.27 \times 10^{-5}$ N·m

CHAPTER 7

7.10 $\theta = 139°$

7.14 $\mathbf{r} = -4.23 \times 10^4 \, \mathbf{I} + 3.11 \times 10^4 \, \mathbf{J}$ (km)

          $\mathbf{v} = -2.78 \, \mathbf{I} + 0.034 \, \mathbf{J}$ (km/s)

7.21(a) $a = 1.183$ A.U.

          $n = 0.0135$ rad/day

          $t = 218.35$ days

7.24 $J.D. = 2441863.937500$

CHAPTER 8

8.7 $e = 0.37, \; a = 10,700$ km

8.13(a) $\alpha = 3/2$

  (b)   18.3 yrs

Index